Urban Planning Pathways to Carbon Neutrality

Urban Planning Pathways to Carbon Neutrality

Editor

Yan Li

MDPI

Basel • Beijing • Wuhan • Barcelona • Belgrade • Novi Sad • Cluj • Manchester

Editor
Yan Li
Zhejiang University
Hangzhou
China

Editorial Office
MDPI AG
Grosspeteranlage 5
4052 Basel, Switzerland

This is a reprint of articles from the Special Issue published online in the open access journal *Land* (ISSN 2073-445X) (available at: https://www.mdpi.com/journal/land/special_issues/urban_carbon_neutrality).

For citation purposes, cite each article independently as indicated on the article page online and as indicated below:

Lastname, A.A.; Lastname, B.B. Article Title. *Journal Name* **Year**, *Volume Number*, Page Range.

ISBN 978-3-7258-2455-7 (Hbk)
ISBN 978-3-7258-2456-4 (PDF)
doi.org/10.3390/books978-3-7258-2456-4

Contents

land

MDPI

Article

A Big Data-Based Commuting Carbon Emissions Accounting Method—A Case of Hangzhou

Song Li, Fei Xue, Chuyu Xia *, Jian Zhang, Ao Bian, Yuexi Lang and Jun Zhou

Faculty of Architecture, Civil and Transportation Engineering, Beijing University of Technology, Beijing 100124, China; lisong@emails.bjut.edu.cn (S.L.); xuefei@bjut.edu.cn (F.X.); zhjian@bjut.edu.cn (J.Z.); amhood1998@emails.bjut.edu.cn (A.B.); lyx@emails.bjut.edu.cn (Y.L.); zhouj@emails.bjut.edu.cn (J.Z.)

* Correspondence: chuyu.xia@bjut.edu.cn

Abstract: Commuting carbon emissions are an essential component of urban carbon emissions, and determining how to reduce them is an area of great debate among researchers. The current research lacks a tool and instrument that can extensively account for residents' commuting. Traditional methods are mainly based on questionnaire surveys, which have low accuracy at spatial and temporal aspects. High accuracy carbon emission accounting methods can effectively assist urban planning and achieve precise urban emissions reductions. This study applies a taxi commuting carbon emissions accounting method divided into two main steps. Firstly, the carbon emissions of taxi trajectories are calculated using taxi trajectory data and a carbon emission calculation method developed based on VSP. Secondly, the taxi trajectory and POI data are used to filter the commuter trajectory with the help of a two-step moving search method. In this way, the taxi commuting carbon emissions were obtained. Then, the spatial distribution characteristics of residential taxi commuting carbon emissions are analysed by spatial autocorrelation tools, which could facilitate low carbon zoning management. A typical working day in Hangzhou was selected as the research object of this study. The results show that (1) morning peak commuting carbon emissions in the main urban area of Hangzhou reached 2065.14 kg per hour, accounting for 13.73% of all taxi travel carbon emissions; and evening peak commuting carbon emissions reached 732.2 kg per hour, accounting for 4% of all taxi travel carbon emissions; (2) At the grid level, the spatial distribution of commuting carbon emissions in Hangzhou shows a single central peak that decays in all directions; and (3) The results at the resident community scale show that urban public transport facilities influence resident community commuting carbon emissions. In areas such as at the urban-rural border, resident community commuting carbon emissions show high levels of aggregation, and in the main urban area, resident community commuting carbon emissions show low levels of aggregation. This study not only provides a new method of commuting investigation but also offers constructive suggestions for future carbon emission reduction under Hangzhou's urban planning.

Keywords: commuting carbon emissions; taxi GPS data; VSP model; spatial-temporal emissions distribution; Hangzhou

Citation: Li, S.; Xue, F.; Xia, C.; Zhang, J.; Bian, A.; Lang, Y.; Zhou, J. A Big Data-Based Commuting Carbon Emissions Accounting Method—A Case of Hangzhou. *Land* **2022**, *11*, 900. https://doi.org/10.3390/land11060900

Academic Editor: Muhammad Shafique

Received: 21 May 2022
Accepted: 6 June 2022
Published: 13 June 2022

1. Introduction

The rapid growth of cities over the past 20 years has resulted in very significant carbon emissions [1–5]. Among various sectors, the transport sector accounts for 9% of total carbon emissions and has become a major contributor to China's environmental pollution [6–13]. Although commuting carbon emissions are not the largest share of transport carbon emissions, they have increased rapidly over the last 20 years. Furthermore, carbon emissions from transport and pm10 (10% of the total) are very harmful to humans [1]. At the same time, urban structures and carbon emissions of taxi commuting travel are closely related in terms of transportation carbon emissions, so it has become a major focus of research to study the characteristics of carbon emissions from taxi commuting travel and to optimize urban structures to reduce carbon emissions.

Past studies on transport carbon emissions and residential commuting have relied heavily on research data [14]. These studies have suffered from poor data precision and are unrepresentative. In past estimates of transport carbon emissions, researchers have often estimated carbon emissions from large-scale data on overall fuel use and local research data; for example, vehicle and road statistics from relevant statistical yearbooks have been used to calculate vehicle transport carbon emissions [15], and some researchers have collected data from video road surveys lasting two weeks, as well as from parking data [16]. The data from these studies are not only unrepresentative on a large scale due to the small amount of data considered, but also have low precision due to the difficulty of obtaining technical information on vehicles and driving parameters as well as exhibiting low timeliness due to the time and effort required to collect data [17]. In studies on commuting, residential travel questionnaires and surveys have been widely used to investigate residential transport behaviour, with questionnaire data including commuting information, starting points, etc. but the data are very coarse, e.g., starting point information is only accurate to the street level, representing the smallest administrative unit in China [6]. For example, researchers at Peking University used data from the 2017 Resident Travel Survey and collected over 500 questionnaires from different groups. Differences in commuting distances, times, and modes of transport between different income classes and different distributions of people were studied [6]. Wang et al. investigated the characteristics of urban commuting and the factors that influence them through a resident questionnaire [18]. Moreover, ref. [19] used the data from the National Household Travel Survey in 2017 to find the relationship between commuters' backgrounds and the external built environment through structural modelling. These studies suffer from a lack of representativeness on a large scale due to the use of small samples, and they also suffer from time-consuming and costly data collection methods, inaccurate data, and poor timeliness.

With the prevalence of information technology, an increasing amount of big data is being used by researchers, among which taxi big data are of great interest [20,21]. First, taxi data contain high-precision trajectory points of passengers' entire journeys, such as data on instantaneous speed, mileage, direction, and other important information, allowing for accurate carbon emission estimation [21]. Second, taxis are closely related to residents' lives. Numerous studies have found a strong daily routine regardless of the location and frequency of boarding and arrival [22–27]. Third, taxi data are controlled by transport sector companies, which are private, and the possession of such data by the transport sector has a direct effect on reducing emissions. Some studies have relied on taxi GPS data to estimate the spatial and temporal distribution of taxi carbon emissions and their characteristics in Shanghai [8]. Finally, taxi data also have the potential to be used to make trip-type judgements, as taxi estimation data include pick-up/drop-off point data. For example, types of taxi trips have been determined by the distributions and types of taxi drop-off points and surrounding POIs [27,28]. Due to the many advantages of taxi GPS data for carbon emission studies, we propose using these data for subsequent studies.

Current research supports the use of taxi GPS data to analyse urban travel patterns and cab travelling's carbon emissions. However, there are limitations to the current related research. This study uses big data to mine the characteristics of carbon emissions from the taxi commuting travel of Hangzhou residents. A framework for estimating taxi commuting carbon emissions using multiple big data sources, such as POI and taxi GPS data with G2SFCA and VSP models is first proposed. In this study, the GPS trajectory data of 1752 taxis in the city of Hangzhou on a typical weekday in 2017 were extracted for trajectory classification and trajectory carbon emission estimation, and commuting carbon emissions were aggregated to resident communities. Finally, the ArcGIS 10.6 platform and Geoda 1.18 software were used to analyse geographical and spatiotemporal characteristics and, finally, the characteristics of Hangzhou taxi commuting carbon emissions were derived.

The remainder of the article is arranged as follows. Section 2 shows the big data and methods that this research used; data used include taxi trajectory and POI data and methods used include the VSP model-based motor vehicle carbon emission accounting method, the estimation method of taxi trajectory classification based on G2SFCA, Geoda1.18-based spatial autocorrelation analysis, and ArcGIS 10.6-based carbon emission model mapping. Sections 3–5 present the study's experimental results and related discussion and conclusions.

2. Materials and Methods

2.1. Framework

The research process of the article is shown in Figure 1, which is divided into three main parts, the calculation of taxi carbon emissions, the identification of taxi commuting trajectories, and the spatial analysis of taxi commuting carbon emissions. The first is to calculate the carbon emissions of each trajectory based on the taxi trajectory data using a VSP-based carbon emission model. This method will be described in Section 2.4.1, and the results will be presented in Section 3.1.1. The G2SFCA (two-step floating catchment area method) will then be then used to identify commuter taxi tracks by using the pick-up and drop-off points of the taxi tracks and the POI data. This part of the methodology will be described in Section 2.4.2, and the results will be presented in Section 3.1.2. Finally, we aggregate taxi commuting carbon emissions to residential communities. We use a spatial autocorrelation tool to analyze the spatial characteristics of commuter carbon emissions. The methodology used in this section will be described in detail in Section 2.4.3. The results will be presented in Section 3.2.

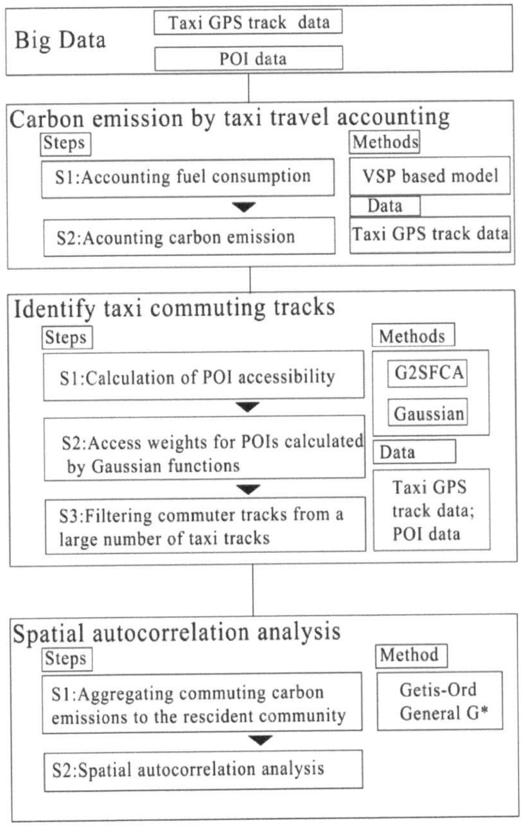

Figure 1. Research Process.

2.2. Study Area

The chosen study area is Hangzhou, Zhejiang Province, China, which, as the capital of Zhejiang Province, is one of the most prosperous cities in China, with an area of 16,596 km and a population of 7,539,000 (2017). We selected the main urban area of Hangzhou as the study area in this research, The residents are mainly concentrated in the city. Hangzhou is one of the first low-carbon pilot cities identified by the National Development and Reform Commission, so studying commuting carbon emissions in this region is essential. The location and urban structure of Hangzhou are shown in Figure 2.

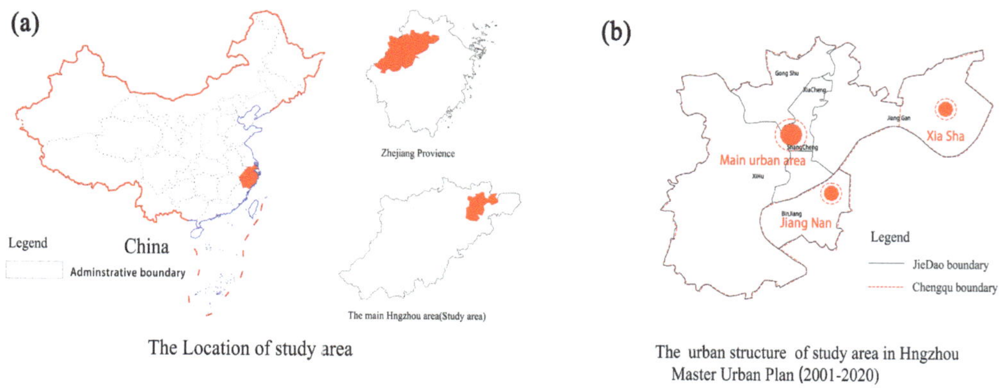

Figure 2. Location of the study area and Hangzhou Master Urban Plan of 2001–2020. ((**a**) shows the location of the study area in China, Zhejiang and Hangzhou respectively; (**b**) shows the structure of the region, the study area consists of three main centres, which are described in detail below).

According to the spatial structure of the study area under the Hangzhou Urban Master Plan, the spatial structure of Hangzhou includes one central area and three secondary areas of which the study area mainly contains the "one main area" in the spatial structure plan i.e., the central city and the northern areas of Xiasha Cheng and Jiangnan Cheng, which are approximately 15 km away from the main urban area.

The central city includes the West Lake District, Gongshu District, and Shangcheng District. The central city mainly contains the functions such as residential and administration offices; commerce and finance; science and technology; education, culture, and urban high-tech industries, forming the core of Hangzhou. Xiasha Cheng is a new city dominated by the Hangzhou Economic and Technological Development Zone and college and university campuses. Jiangnan Cheng is a modern city focused on science and technology and a distant commercial centre with high-tech industrial parks as its major backbone, with urban industry, teaching, and research and development occurring in coordination with each other. The study area includes the northern area of Jiangnan Cheng; the Binjiang District, which functions mainly as a high-tech industrial park hosting high-tech firms such as such as Alibaba; and a small number of residential communities.

2.3. Data Collection

2.3.1. Taxi GPS Data

The taxi GPS data for this study was acquired from the Comprehensive Information Centre of the Hangzhou Municipal Bureau of Transportation. The GPS trajectory data of taxi trips occurring from 6:30 a.m.–12:00 a.m. on a typical working day in 2017 were selected. The taxi trajectory GPS data were transmitted from each vehicle's GPS to the server, including real-time location (latitude and longitude), instantaneous speed, vehicle status, license plate number, vehicle status information, time, and other taxi data. The time interval between each upload of GPS information by a taxi ranged between 30 and 40 s.

This study used data for 6:30 a.m.–12:00 a.m., as shown in Table 1, amounting to 1,111,291 entries.

Table 1. Original taxi trajectory information (*** is a hidden taxi license plate).

Point Number	Licence Plate	Type	Longitude	Latitude	Speed (km/h)	Direction	State	ROAD_ ID	GPS_TIME	UPDATE_ TIME
1370	AUT ***	1	120.195694	30.35249	55	246	1	−1	05-6-2017 13:36:42	05-6-2017 13:36:47
1887	AUT ***	1	120.18566	30.306992	52	268	1	7550	05-6-2017 10:22:00	05-6-2017 10:22:02
1903	AUT ***	1	120.20334	30.308031	0	276	1	7479	05-6-2017 10:17:00	05-6-2017 10:17:03
2095	AUT ***	1	120.23065	30.36269	53	290	1	795	05-6-2017 15:02:22	05-6-2017 15:02:24

To study residents' commuting trips, taxi travel data for the hours of 6:30–10:00 and 16:30–20:00 were screened, and intermittent trajectory information was processed into coherent taxi trajectory information using the TRACK tool and PYTHON in the GIS platform. The trajectories that were not operational and lasted less than 2.5 min were then excluded. A total of 25,089 trajectories (6:30–10:00 and 16:30–20:00) were obtained in this study, and the time information of the trajectory points was used to recognize the pick-up and drop-off points, as shown in Table 2.

Table 2. Taxi trajectory data.

Trajectory ID	Start Time	End Time	Pick up Point Number	Drop off Point Number	Carbon Emissions (kg)	Distance (km)
34	05-6-2017 17:49:26	05-6-2017 17:56:07	34,677	63,090	0.551884	1.553232
46	05-6-2017 16:29:48	05-6-2017 16:34:08	7324	83,889	0.412152	1.748736
57	05-6-2017 17:30:25	05-6-2017 17:47:08	101,532	78,115	1.365418	3.655797

2.3.2. POI Data

The POI data that we used comes from the Gao-De platform, crawled through the API interface of the Gaode developer platform, and include the POIs for the whole area of Hangzhou, covering a total of 20 major categories (first-level classification). We use POIs with the two-step floating catchment method to classify taxi trajectories (as explicitly described in the Section 2). Therefore, the initial screening and classification of Hangzhou POIs according to their functions and opening times [27] and other specifications are detailed in Section 2.4.2 below.

2.4. Method

2.4.1. Taxi Carbon Emission Modelling

Among the models for estimating transport carbon emissions, carbon emission models based on the VSP model mainly applied to estimate the relationship between vehicle operating conditions and fuel consumption, are becoming increasingly popular among researchers. On the one hand, the VSP model reflects the relationships between specific power and acceleration, speed, and gradients. On the other hand, the VSP model also strongly correlates with fuel exhaustion, so this study develops a fuel consumption model for taxis based on the VSP model.

According to [29], the VSP model for light vehicles is simplified and expressed in the following equation.

$$VSP = v \cdot [1.1a + 0.98a \cdot grade(\%) + 0.132] + 0.000302 \cdot v^3 \tag{1}$$

where v is the speed of the motor vehicle in m/s. a is the transient acceleration of the motor vehicle in m/s^2. *grade* is the gradient (dimensionless parameter), which should be taken as 0.0 in the application.

We used an improved VSP-based model to develop a vehicle carbon emission estimation model [28]. There are two main steps. First, the fuel consumption of vehicle driving is derived from speed and mileage information from GPS data. Since VSP strongly correlates with fuel consumption, it is difficult to acquire the vehicle operating conditions required for VSP in this study. Therefore the fuel consumption rates in this study use the results of Zhao t et al. as shown in Table 3 [30]. Second, the corresponding carbon emissions are calculated from fuel consumption.

Table 3. Time constraints of POI points.

Types	Time Periods	Pick Up/Drop off Points	POI Types	POI Names
From home to work	6:30 a.m.–10:00 a.m.	Pick-up points	Residence	Commercial Residence
		Drop-off points	Work	Catering Services
				Road appurtenances
				Public Facilities
				Enterprises
				Shopping facilities
				Financial and Insurance Services
				Scientific, educational and cultural services
				Motorbike services
				Car Services
				Car repairs
				Automotive sales
				Live Services
				Sports and leisure services
				Government Agencies and Societies
				Accommodation services
			Medical	Health Care Services
			Transport	Transport Facility Services
From work to home	4:30 p.m.–8:00 p.m.	Pick-up points	Work	Catering Services
				Road appurtenances
				Public Facilities
				Enterprises
				Shopping facilities
				Financial and Insurance Services
				Scientific, educational and cultural services
				Motorbike services
				Car Service
				Car repairs
				Automotive sales
				Life Services
				Sports and leisure services
				Government Agencies and Societies
				Accommodation services
			Medical	Health Care Services
			Transport	Transport Facility Services
		Drop-off points	Restaurant	Catering Services
			Shopping	Shopping facilities
			Transport	Accommodation services
			Education	Scientific, educational and cultural services
			Residence	Commercial Residence
			Medical	Health Care Services
			Entertainment	Scenic areas
				Sports and leisure services
				Life Services

First step: Calculating fuel consumption for taxi journeys [31].

$$Q = ER_0 \cdot NER_I \cdot TI \tag{2}$$

where Q is the actual fuel exhaustion of the taxi in grams. ER_0 is the fuel exhaustion rate when VSP equals zero using parameter 0.274 g/s. NER_i is the standard fuel exhaustion rate in Table 4 when the speed is interval i and reflects dimensionless data. T_I is the total length of time, in seconds, that the taxi actually travels at an average speed in interval i.

Table 4. Standard fuel exhaustion rate.

i	Average Speed Interval i (km/h)	NER	i	Average Speed Interval i (km/h)	NER
1	0–2	1.085	20	38–40	2.395
2	2–4	1.258	21	40–42	2.441
3	4–6	1.311	22	42–44	2.47
4	6–8	1.477	23	44–46	2.538
5	8–10	1.573	24	46–48	2.566
6	10–12	1.645	25	48–50	2.581
7	12–14	1.729	26	50–52	2.595
8	14–16	1.807	27	52–54	2.68
9	16–18	1.841	28	54–56	2.716
10	18–20	1.923	29	56–58	2.756
11	20–22	1.997	30	58–60	2.81
12	22–24	2.045	31	60–62	2.864
13	24–26	2.092	32	62–66	2.956
14	26–28	2.163	33	66–70	3.05
15	28–30	2.187	34	70–80	3.289
16	30–32	2.251	35	80–	3.551
17	32–34	2.329			
18	34–36	2.338			
19	36–38	2.391			

Second step: Calculation of taxi carbon emissions based on fuel consumption.

$$E = Q \cdot EF_K \tag{3}$$

E denotes the carbon emissions generated per trajectory, and EF_k is the emission factor for emissions, which is 2.98 kg/L for CO_2 and a conversion factor of 0.27 between CO_2 and carbon, according to Xia et al. in their 2017 study [28,31].

2.4.2. Identification of Travel Origins and Destinations

As the time characteristics of commuting behaviour are more pronounced, this study defines two modes of commuting behaviour by referring to Zhao et al.'s time constraint for POIs [27]. They include travel from home to work, which occurs between 6:30–10:00, and travel from work to home, which occurs between 16:30–20:00. To filter out trajectories generated by commuting among all taxi trajectories, this study uses the time and space-based and POI point-of-interest travel inference method proposed by [27]. The method used in this study calculates the probability of visits for each POI within walking distance (500 m) of the drop-off point through the G2SFCA method with a Gaussian decay function. Some modifications are made to Zhao et al.'s method, which is also divided into three steps: identifying candidate POIs, calculating POI accessibility, and calculating POI visitation weights.

In identifying candidate POIs, mainly considering the more mixed distribution of POI points in urban space, this study introduces temporal constraints to filter POIs [27,28]. The rationale is that different functions provide specific services at different times. For example, in the case of restaurants and shopping centres, although the two are very similar in terms of space taken, restaurants only offer food and beverage services at midday.

Shopping centres are open for longer hours (10:00–21:00). Therefore, at specific times of the day, e.g., in the morning, restaurant POIs are set as noncandidate POIs, while shopping POIs are identified as candidate POIs. The classification by Zhao et al. is limited to determining drop-off points. In this study, destination POIs are extended to departure point and destination POIs with separate time constraints based on commutes. See Table 3 for details.

For the measurement of the access weight of a POI, there are three steps: first, for each pick-up or drop-off points j, search for POI points i within a certain radius and calculate supply/demand ratio R_j. Second, for each POI point i, search for pick-up or drop-off point j within walking radius and sum supply/demand ratios R_j to obtain the accessibility A_i of the POI. Finally, access weight W_{ij} of drop-off/boarding point j for the POI points within the search radius is calculated with a Gaussian function, where the enormous access weight is the type of destination/starting point. The following four main steps identify the trip type.

First step: Calculate POI point supply/demand ratio R_j within search threshold d_0 for single drop-off/pick-up point j.

$$R_j = \frac{1}{\sum\limits_{i \in \{d_{ij} \leq d_0\}} G(d_{ij}, d_0) \times 1} \quad (4)$$

Second step: Calculate supply/demand ratio A_i for each POI facility point i.

$$A_i = \sum_{j \in \{d_{ij} \leq d_0\}} G(d_{ij}, d_0) \times R_j \quad (5)$$

where R_j is the supply/demand ratio of the boarding/disembarking points for the POI points in the range. Equation (6) is a commonly used Gaussian distance decay function and is widely used with the 2SFCA method [32–34].

$$G(d_{ij}, d_0) = \begin{cases} \frac{e^{-\frac{1}{2} \times \left(\frac{d_{ij}}{d_0}\right)^2} - e^{-\frac{1}{2}}}{1 - e^{-\frac{1}{2}}}, & d_{ij} \leq d_0 \\ 0, & d_{ij} > d_0 \end{cases} \quad (6)$$

Third step: Calculate the visit weight of each pick-up/drop-off point j for facility point i within search radius d_0 and determine types of start- and endpoints.

$$W_{ij} = G(d_{ij}, d_0) \times A_i \quad (7)$$

For POI point i within the search range of pick-up or drop-off points j, the attenuation of the reachability Ai of the POI on d_{ij} is calculated. The value determined after the attenuation calculation is noted as access weigh t W_{ij} for visit i of j (Equation (7)). The attenuation is done using the Gaussian attenuation function mentioned above, which has a normal distribution of the computed results and is widely used in various calculations regarding distance attenuation [32–34].

For the POI points within the search range of pick-up/drop-off points, the POI point in which j has the largest visit weight to i is selected as the type of pick-up/drop-off point.

Final step: The type of track is determined by the type of POI to which the start- and endpoints belong: From home to work and from work to home.

2.4.3. Spatial Autocorrelation Analysis of Carbon Emissions from Residential Communities

This study uses the ArcGIS 10.6 platform to visualize the carbon emissions of taxi commuting, aggregating the carbon emissions of each small segment of the trajectory to the 1 km × 1 km grid to which it belongs.

Subsequently, the spatial distribution characteristics of carbon emission by taxi commuting travel in residential communities were explored. The carbon emissions of the total taxi trajectory were aggregated to the residential community. The spatial autocorrelation software GeoDa1.18 was used to conduct a Getis-Ord Gi analysis of residential taxi commuting carbon emissions to obtain the spatial aggregation characteristics of residential commuting carbon emissions.

The spatial autocorrelation mainly involves the Getis-Ord General Gi* assessment with the following correlation Equation (8):

$$Gi^* = \frac{\sum_{j=1}^{N} W_{ij} \cdot X_j - \overline{X} \sum_{j=1}^{N} W_{ij}}{\sqrt{\frac{\sum_{j=1}^{N} X_j^2}{N} - (\overline{X})^2} \cdot \sqrt{\frac{N \sum_{j=1}^{N} W_{ij}^2 - (\sum_{j=1}^{N} W_{ij})}{N-1}}}, \forall j \neq i \quad (8)$$

In Equation (8), N is the number of residential points. X_i and X_j are the carbon emissions of residential points i and j, respectively. $\overline{X}$ is the average of carbon emissions. W_{ij} is the spatial connection matrix between residential points i and j.

3. Results

3.1. Accounting for Residential Commuting Carbon Emissions of Taxis

3.1.1. Carbon Emissions of Taxi Travel

In this section, we apply the taxi carbon emission model mentioned in Section 2.4.1 to estimate the total carbon emissions by a cab's trajectory on a typically weekday in 2017, The results are shown in Figure 3 below.

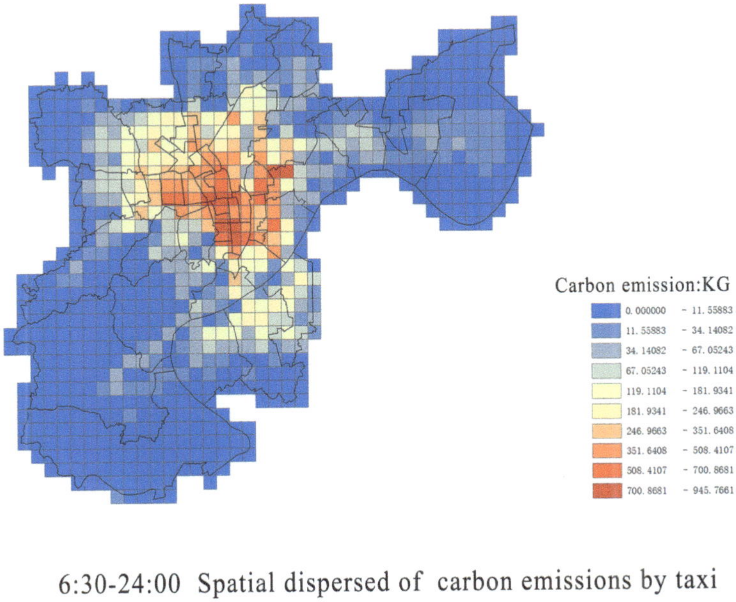

Figure 3. Spatial dispersion of all taxi carbon emissions.

The carbon emissions of taxis for the whole day from 6:30 a.m.–12:00 a.m. are 52,934 kg. From Figure 3, it can be observed that the carbon emissions of taxis are mainly concentrated near the main urban area centred on West Lake, and the total carbon emission trend belongs to the high centre and gradually spreads to the periphery. The subsequent higher carbon emissions are found in the Binjiang district on the north side of Jiangnan Cheng, while carbon emissions in Xiasha Cheng are lower.

In terms the spatial and temporal distributions of taxi emissions in the three cities the main urban area and Jiangnan Cheng have higher carbon emissions. In contrast Xiasha Cheng, a 'secondary city', has lower carbon emissions than Jiangnan Cheng, also a 'secondary city'. In terms of the functions of the main urban area, Xiasha Cheng and Jiangnan Cheng, and the spatial distance between them, the main urban area is positioned for technology production, business, finance, and living and residential functions. Xiasha Cheng is positioned as a high-tech industrial park and high education park.

3.1.2. Estimated Types of Residential Trips

In this section, the two-step mobile search method is applied to classify the taxi trip trajectories of a typical weekday in 2017, and the results are shown in Table 5.

Table 5. Results of Taxi Commute Carbon Emissions.

Duration	Type	Quantity	Ration (Count)	Carbon Emissions (kg)
6:30–10:00	Commuting (From home to work)	8738	78.60%	7228
	Others	2320	20.89%	1929
	Total	11,103	100%	9157
16:30–20:00	Commuting (From work to home)	2610	17.74%	2563
	Others	12,096	82.25%	11,694
	Total	14,706	100%	14,257

Table 5 shows that taxi commuting accounts for approximately 78% of all taxi travel from 6:30 a.m.–10:00 a.m. From 4:0 p.m.–8:00 p.m., taxi commute trips account for only 17.74% of all taxi travel. The number of trajectories from 6:30 a.m.–10:00 a.m. is similar to the number of trajectories collected for 4:30 p.m.–8:00 p.m. Thus, most taxi commuting is done in the morning from 6:30 a.m.–10:00 a.m., i.e., from home to work, while from 4:30 p.m.–8:00 p.m., when travel from work to home occurs, commuting is no longer the primary purpose of taxi travel.

3.1.3. Estimated Residential Commuting Carbon Emissions of Taxi Travel

From the results in Sections 3.1.1 and 3.1.2, we obtain residents' taxi commute carbon emissions. The spatial autocorrelation of the residents' taxi commuting carbon emission are shown in Figure 4 below.

For commuting purposes, total carbon emissions accounted for 18.6% of total carbon emissions (without screening short trajectories) at 9157 kg. The total commuting carbon emissions in the morning peak were 7226 kg, averaging 2064.57 kg per hour; the total carbon emissions for commuting in the evening peak were 2562 kg, averaging 732 kg per hour. As shown in Figure 4e,f, during the morning peak period, the number of taxis for commuting purposes accounted for 78% of the morning peak taxi trajectory, when most taxi trips were of the commuting type. As a result, as shown in Figure 4a,c, the spatial dispersal of taxi's commuting carbon emission from 6:30 a.m.–10:00 a.m. is mainly concentrated in the main urban area and the north side of Jiangnan Cheng, where the peak carbon emissions reach an average of over 100 kg in the grids. As shown in Figure 4b,d, from 4:30 p.m.–8:00 p.m., the taxi trajectory for commuting purposes accounted for 17.74% of the total taxi trajectory in the evening peak. The main purpose of taxi trips is not commuting, and the spatial distribution of carbon emissions in this period is still concentrated in the main urban area, with sporadic distribution in XiaSha Cheng.

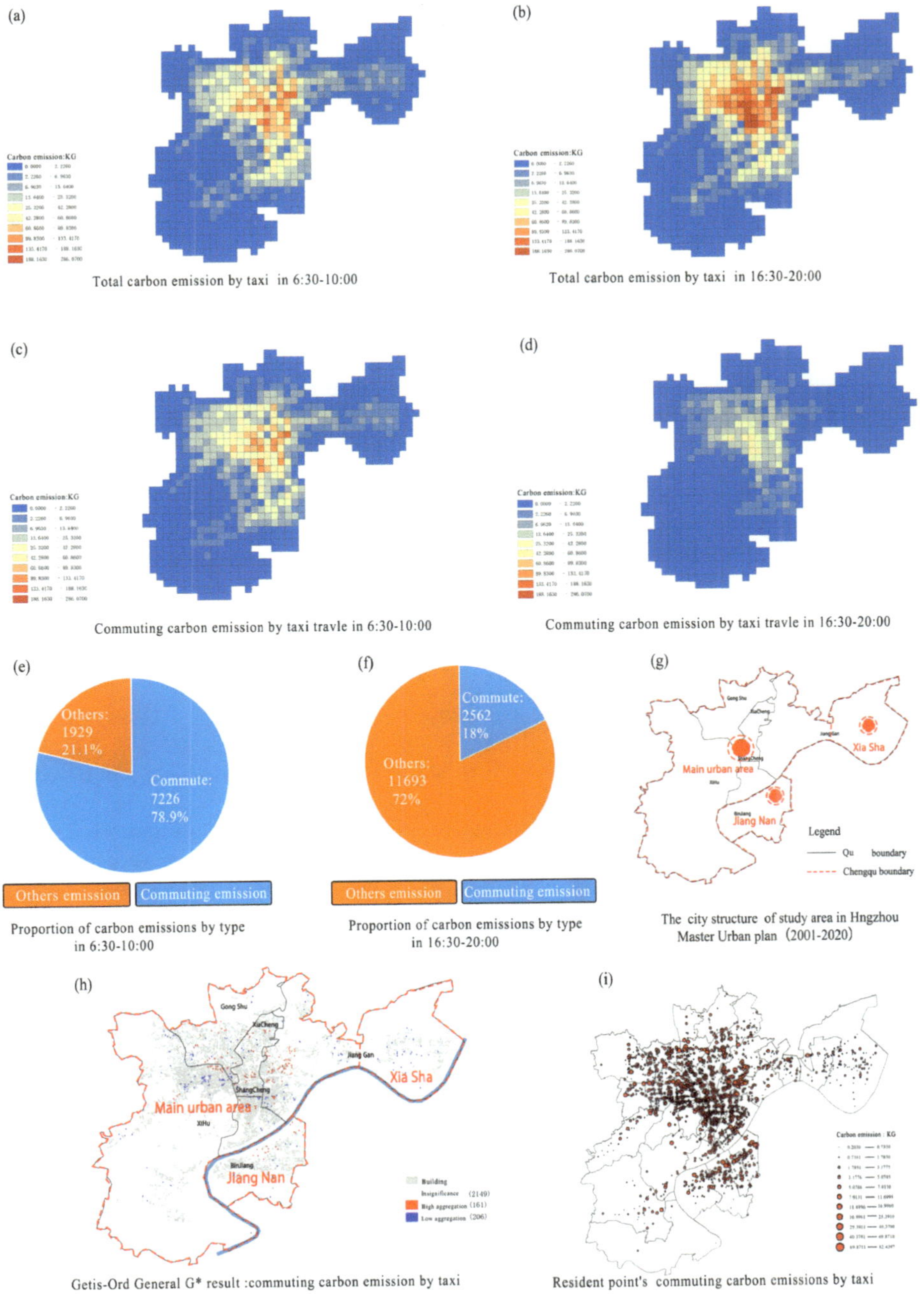

Figure 4. Spatial distribution of carbon emission by different type of taxis travel on a typical work day and its Getis—Ord General G* result. ((**a**–**d**) shows the spatial distribution of carbon emission in different types, (**e**,**f**) shows the proportion of carbon emission by type. (**g**) is the city structure oof study area. (**h**,**i**) shows the resdient communities carbon emission and it's result of Spatial autocorrelation).

3.2. Spatial Autocorrelation Analysis of Carbon Emissions from Residential Point Taxi Commuting

In this section, taxi commuting carbon emissions are aggregated to residential points. A spatial autocorrelation analysis explores the relationship between the spatial aggregation and spatial distribution of taxi commuting carbon emissions at residential communities. The spatial autocorrelation results are shown in Figure 4h,i.

In Figure 4, (i) is the sum of commuting carbon emissions by taxi travel from 6:30 a.m.–10:00 a.m. and 4:30 p.m.–8:00 p.m. on a typical weekday and is the result of a hierarchical symbolic visualization derived from ArcGIS, where the more carbon emissions there are at a round point, the larger its rounded area is; (h) shows the result of the Getis-Ord General Gi* of point (i).

Two results from this subsection are worth noting. First, the results of the distribution of the high and low clustering of carbon emissions from residential commuting (Figure 4h), the low and low clustering of carbon emissions is mainly concentrated within the main urban area, with a small number of clusters located in areas far from the main urban area.

Second, the majority of residential communities with high and high carbon emissions are concentrated within the urban-rural interface of the main urban area. For example, in the western side of the Jianggan District, in the urban-rural area near agricultural land as shown in Figure 4h, buildings are sparse. Some residential points in Xiasha Cheng, approximately 15 km away from the main urban area, show a clear low-low agglomeration pattern. On the other hand, Jiangnan Cheng, a subcity, is only separated from the main urban area by the geographical factor of the Qiantang River, but its residential community distribution shows distinctly high clustering.

4. Discussion

4.1. Discussion of the Results of the Proposed Carbon Emissions Accounting for Taxi Trips Based on the VSP Model

4.1.1. Comparison of Traditional Carbon Emission and VSP Model-Based Carbon Emission Estimations

The method proposed offers the advantages of high research accuracy, large-scale analysis, broad representation, and real-time performance relative to the use of traditional research data such as yearbook and fuel consumption data for carbon emission estimation.

The data used for carbon emission estimation are the big data of taxi trajectories from taxi companies. Based on the VSP model, carbon emissions are calculated from acceleration, speed, mileage, time, and vehicle driving condition data from GPS information to calculate fuel consumption data for each small section of a taxi trajectory. Due to the high refresh rate (20–40 s) and accuracy of GPS positioning, the data volume is large and representative of a broad range.

Using traditional methods of calculating motor vehicle carbon emissions, Ref. [35] accounted for transportation carbon emissions in Florida in 2011. Using data on population, road miles, numbers of cars and trucks, and fuel economy for 67 counties, the VMT model was applied to estimate carbon emissions in Florida [35]. However, this method's lowest resolution of transport carbon emission results is only of the county scale. Data are also available on an annual basis and are time-lagged. Compared to the VSP model-based method proposed in this study, this method can accurately account for carbon emissions to a hundred-meter level and is more time sensitive than the previous method. The same problem affects other studies, such as those based on vehicle and road statistics from relevant statistical yearbooks [15], video road survey data, and parking data [16]. Therefore, the present method has the advantages of achieving high levels of research accuracy, a large scale, wide representation, and real-time performance.

4.1.2. Comparison of Taxi and Traditional Commuting Investigations

This study uses a two-step floating catchment method and taxi track data to screen a large number of taxi commuters. Compared with traditional commuting research, this method has the advantages of using a wide range of samples, obtaining process data, using accurate data, and obtaining data in a timely manner. In particular, by combining the identification of commuting trajectories with the accounting method of commuting carbon emissions by taxi travel, commuting carbon emissions can be accurately obtained.

There are few studies on commuting and commuting carbon emissions. Previous studies mainly use residents' travel questionnaires and other tools to obtain information on residents' travel destinations, starting places, and transportation modes. However, the accuracy of starting places and destinations is usually poor and usually accurate to the street level [6], and critical data such as traffic trajectories are lost. Wang et al. studied the characteristics of resident commuting research in China using questionnaires in 2018 [18]. Lou et al. studied the commuting patterns of different income groups, and their new town research examines the commuting characteristics of people of different income groups and from different types of new towns [6]. These studies accurately portray the effects of different commuting groups' social and economic characteristics on commuting [19]. However, they do not accurately account for the carbon emissions of commuting, and they are inadequate in terms of sample size, the precision of commuting and carbon emissions of the commuting process.

This study uses a two-step floating search method to calculate types of taxi trajectories by using the POIs of the pick-up/drop-off points of taxi trajectories and surrounding areas. It is possible to filter out a large volume of commuting samples from a large number of taxi tracks. On the one hand, this makes up for the lack of information about the commuting process in questionnaire research, and on the other hand, it avoids the problems of small sample sizes and the loss of commuting information in traditional research to obtain commuting samples. Combined with the VSP model based on the taxi carbon emission accounting method mentioned in the previous section, the approach can accurately obtain the carbon emissions of residents' taxi commuting.

4.2. The Impact of Urban Spatial Structure and the Construction of Urban Public Transport Facilities on Carbon Emissions by Taxis

This study finds that the low clustering of commuting carbon emissions in the main urban area and high clustering in the main urban area periphery may be due to a lack of public transport in the urban periphery (Figure 5). The results also show that urban structures may impact commuting carbon emissions. Future urban planning measures should increase the construction of public transport facilities at the urban-rural interface and, at the same time, focus on balancing jobs and housing to promote a rational mix of functions.

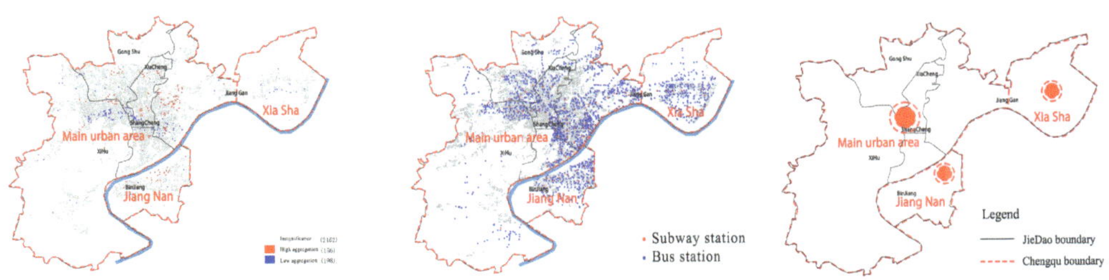

Getis-Ord General G* result : 6:30 - 11:00 — Bus station and Subway station in Hangzhou — The urban structure of study area in Hngzhou Master Urban plan (2001-2020)

Figure 5. Comparison of the spatial distributions of residential points' commuting carbon emissions and bus stops.

According to (i,h) in Figure 4, the high clustering of commuting carbon emissions from residential communities mainly occurs in the urban-rural border areas in the main urban area, such as on the western side of the main urban area and in the Jianggan District, which are urban-rural border areas with minor public infrastructure construction located farther from the main urban area. The main function is residential communities, and low-low aggregation of carbon emissions from commuting to residential locations mainly occurs in the core areas of the central city, such as West Lake and the south side of the Shangcheng District. These areas are mainly business cores with many residential locations. This may be caused by the insufficient number of public transport facilities in suburban and urban-rural areas, with the shortfall being filled by other forms of transport. The phenomenon is relevant to studies of areas such as Chengdu and Seoul. After Li et al. identified GWR models for online taxi trips and critical factors in Chengdu, they found that online taxi suburban public transport complements other services [36]. Nam et al. applied taxi GPS data signals in Seoul, South Korea to analyse taxi passengers' origins and spatiotemporal variation. Their results indicate that traditional taxis form a complementary relationship with underground trips in places where subways are less common [37]. Other studies on taxis conclude that net cars and taxis complement bus metro lines. When there is a shortage of bus metros, the ridership density of net cars and taxis is higher [38–40]. It may be that public transport systems do not have an advantage in terms of long-distance commuting with numerous studies showing that residents prefer to take taxis over buses for long distance commuting [41] and that high-income residents of new city areas prefer other types of travel such as car and taxi transport [6,42], as this not only eliminates the need for intermediate transfers but also makes travel more comfortable overall. When we compare Figure 5, the aggregated relationship between rail transit distribution and the commuting pollutions of carbon by residential taxi transports, we also find that residential commuting carbon emissions by taxi travel are lower in areas with a higher density of public transport facilities. This phenomenon is consistent with previous studies. Therefore, higher carbon emissions from taxi commuting in urban fringe settlements shown in Section 3.2 are, on the one hand, due to the inadequate construction of public infrastructure in the urban fringe and, on the other hand, likely due to the overall public transport system experience not being as optimised as that of taxis. Therefore, in the future, in Hangzhou's master urban planning, more transport facilities should be built within the urban-rural boundaries of the central city area, including on the northern side of Shangcheng District and the west side of the Jianggan District. At the same time, the operating standards of public transport operations should also be raised and their capacity should be increased.

The urban's spatial structure may also influence commuting carbon emissions of residents' taxi travel. Our experiments show that carbon emissions from urban settlements in Xiasha Cheng, which is located far from the main urban area, show a low-low aggregation pattern of carbon emissions. In contrast, on the north side of Jiangnan Cheng closer to the central city area, there is an apparent high aggregation of carbon emissions from residential locations. The study area consists mainly of the central and two secondary cities, Xiasha Cheng and Jiangnan Cheng. The main function of Xiasha Cheng is to serve as an integrated new city with an economic and technological development zone and a high education park as its mainstay. Under the visionary plan, Jiangnan Cheng functions as a high-tech industrial and residential area and business core. A common feature of these three areas is a difference in land use.

Previous studies have shown that land-use patterns and land-bearing functions significantly affect carbon emissions via taxi travel. For example, Ref. [43] investigated the relationship between land properties, land use practices and positive and negative carbon transformations. Ref. [44] determined the relationship between land use types and "source–sink" patterns from taxi GPS data for Shanghai, and their results show a strong correlation between these variables. Ref. [45] measured land use types with 95% accuracy using the pick-up and drop-off point data of 4000 taxi trajectories for a year. Ref. [46], in a study of urban clusters using taxi GPS data, showed that different urban functions might lead to travel clusters of different sizes. Refs. [47,48]'s research shows that cities with a better mix of urban structures and road structures have lower carbon emissions when they expand, whereas cities with a single core have higher carbon emissions when they expand. These results suggest that urban functions and function mixes may influence the residential travel carbon emissions of taxis and hence commuting carbon emissions. However, some studies have shown that excessive travel distance may lead to a preference for using taxis to travel with a threshold value of 15 km [49], which is similar to the distance from the central city area to Xiasha Cheng. From the results of previous experiments and studies, we speculate that this phenomenon may be due, on the one hand, to the fact that Xiasha Cheng is located far from the main urban area and is a fully functional integrated new city with better jobs, housing, and functions. On the other hand, this may be the case because the north side of Jiangnan Cheng is more homogeneous and closer to the main urban area, resulting in an imbalance of jobs and functions at the regional level. Therefore, in Hangzhou's future urban master plan, attention should be given to strengthening the balance of jobs and housing and to the coordination of urban functions in the Jiangnan Cheng area.

5. Conclusions

China's rapid urbanization has led to severe air pollution, and the transport sector has always accounted for a high proportion of CO_2 emissions. While representing a small proportion of transport carbon emissions, commuting carbon emissions have been growing at a relatively fast rate. At the same time, with the development of internet technology and big data technology, it is possible to conduct large-scale commuting surveys. As a new version of the urban master plan for Hangzhou is currently being prepared, the main purpose of this research is to provide a new way of studying commuting. The second purpose is to reveal the spatial and temporal characteristics of carbon emissions from commuting in Hangzhou and provide constructive suggestions for Hangzhou urban master planning to reduce commuting carbon emissions.

First, this study improves a method for measuring taxi commuting carbon emissions based on the VSP models with the G2SFCA method. We then use spatial autocorrelation analysis and Getis-Ord general Gi* regression analysis to explore the spatial and temporal characteristics of carbon emissions of residents' transport from taxi trips and the factors that may affect residents' taxi commuting carbon emissions. The study results indicate that carbon emissions in the central area of Hangzhou show a typical monocentric peak with a gradual decay towards the urban-rural border. The carbon emissions of taxis from 6:30–24:00 are 52,634 kg. Areas with high carbon emissions all day are mainly concentrated in the main urban area, with a small proportion concentrated on the northern side of Jiangnan Cheng and Xiasha Cheng. In the morning and evening peak hours, i.e., 6:30 a.m.–10:00 a.m. and 4:30 p.m.–8:00 p.m., taxi commuting carbon emissions reach 7228 kg and 2563 kg, respectively. Most people use taxis to commute from home to work.

We found that urban public transport facilities influences the commuting carbon emissions of residents. From the perspective of urban infrastructure development, residential locations with high-high aggregation of carbon emissions from taxi trips were evident near areas with poor public transport, especially in terms of metro density, such as in the urban-rural junction of the central city area. In contrast, areas with better public transport, such as the main urban area, appear as distinct settlements where carbon emissions from taxi travel are low-low aggregated. In terms of policy recommendations for urban planning at the level of urban presidential planning, we should strengthen the balance of jobs and housing and the construction of supporting jobs on the north side of Jiangnan Cheng. To strengthen the scale of population and urban spatial structure, the coordination of urban function distribution arrangement should be pursued to alleviate regional traffic pressure.

The coordinated arrangement of population size, urban spatial structure, and urban function distribution should be improved to ease traffic pressure in the region.

The public transit system in the city should be improved, and the density of public transit facilities in the city's fringe areas should be increased, such as the density of bus stops and railway stations. Public transport operation services must be enhanced and their capacity must be increased. A shift in urban transport from car-based transport to public transport will also be necessary to ease the pressure of carbon emissions from urban commuting.

Author Contributions: Conceptualization, C.X.; methodology, C.X. and S.L.; software, S.L. and J.Z. (Jun Zhou); formal analysis, S.L. and A.B.; resources, C.X. and J.Z. (Jun Zhou); data curation, S.L. writing—original draft preparation, S.L. and Y.L.; writing—review and editing, C.X., F.X., S.L. and J.Z. (Jian Zhang); visualization, S.L. and A.B.; supervision, C.X., F.X. and J.Z. (Jian Zhang); project administration, C.X. and F.X.; funding acquisition, C.X. All authors have read and agreed to the published version of the manuscript.

Funding: This work was supported by the National Natural Science Foundation of China (Grant No. 72004014).

Institutional Review Board Statement: Not applicable.

Informed Consent Statement: Not applicable.

Data Availability Statement: (1) The taxi GPS data for this study was acquired from Comprehensive Information Centre of the Hangzhou Municipal Bureau of Transportation. (2) The POI data that we used comes from the Gao-De platform, crawled through the API interface of the Gaode developer platform: https://lbs.amap.com/ [50] accessed on 19 May 2022.

Conflicts of Interest: The authors declare that they have no conflict of interest.

References

1. Pawłowski, L.; Litwińczuk, Z.; Zhou, G. (Eds.) *The Role of Agriculture in Climate Change Mitigation*, 1st ed.; CRC Press: Boca Raton, FL, USA, 2020; ISBN 978-1-00-300273-4.
2. Dong, Z.; Xia, C.; Fang, K.; Zhang, W. Effect of the Carbon Emissions Trading Policy on the Co-Benefits of Carbon Emissions Reduction and Air Pollution Control. *Energy Policy* **2022**, *165*, 112998. [CrossRef]
3. Wang, X.; Chen, S. Urban-Rural Carbon Footprint Disparity across China from Essential Household Expenditure: Survey-Based Analysis, 2010–2014. *J. Environ. Manag.* **2020**, *267*, 110570. [CrossRef] [PubMed]
4. Liu, Z.; Deng, Z.; Davis, S.J.; Giron, C.; Ciais, P. Monitoring Global Carbon Emissions in 2021. *Nat. Rev. Earth Environ.* **2022**, *3*, 217–219. [CrossRef] [PubMed]
5. He, Y.; Xia, C.; Shao, Z.; Zhao, J. The Spatiotemporal Evolution and Prediction of Carbon Storage: A Case Study of Urban Agglomeration in China's Beijing-Tianjin-Hebei Region. *Land* **2022**, *11*, 858. [CrossRef]
6. Li, S.; Zhao, P. Examining Commuting Disparities across Different Types of New Towns and Different Income Groups: Evidence from Beijing, China. *Habitat Int.* **2022**, *124*, 102558. [CrossRef]
7. Li, T.; Wu, J.; Dang, A.; Liao, L.; Xu, M. Emission Pattern Mining Based on Taxi Trajectory Data in Beijing. *J. Clean. Prod.* **2019**, *206*, 688–700. [CrossRef]
8. Luo, X.; Dong, L.; Dou, Y.; Zhang, N.; Ren, J.; Li, Y.; Sun, L.; Yao, S. Analysis on Spatial-Temporal Features of Taxis' Emissions from Big Data Informed Travel Patterns: A Case of Shanghai, China. *J. Clean. Prod.* **2017**, *142*, 926–935. [CrossRef]
9. Xia, C.; Li, Y.; Xu, T.; Chen, Q.; Ye, Y.; Shi, Z.; Liu, J.; Ding, Q.; Li, X. Analyzing Spatial Patterns of Urban Carbon Metabolism and Its Response to T Change of Urban Size: A Case of the Yangtze River Delta, China. *Ecol. Indic.* **2019**, *11*, 184–200. [CrossRef]
10. Xia, C.; Li, Y.; Ye, Y.; Shi, Z.; Liu, J. Decomposed Driving Factors of Carbon Emissions and Scenario Analyses of Low-Carbon Transformation in 2020 and 2030 for Zhejiang Province. *Energies* **2017**, *10*, 1747. [CrossRef]
11. Fang, K.; Li, C.; Tang, Y.; He, J.; Song, J. China's Pathways to Peak Carbon Emissions: New Insights from Various Industrial Sectors. *Appl. Energy* **2022**, *306*, 118039. [CrossRef]
12. Fang, K.; Zhang, Q.; Song, J.; Yu, C.; Zhang, H.; Liu, H. How Can National ETS Affect Carbon Emissions and Abatement Costs? Evidence from the Dual Goals Proposed by China's NDCs. *Resour. Conserv. Recycl.* **2021**, *171*, 105638. [CrossRef]
13. Song, J.; Qi, Z.; Gai, Y.; Chen, S. Shared Network and Supply Chain Features for Synergetic Control of Carbon and Air Pollutant Emissions. *Sci. Total Environ.* **2022**, *827*, 154391. [CrossRef]
14. Liu, Z.; Sun, T.; Yu, Y.; Ke, P.; Deng, Z.; Lu, C.; Huo, D.; Ding, X. Real-Time Carbon Emission Accounting Technology toward Carbon Neutrality. *Engineering* **2022**, S2095809922001400. [CrossRef]
15. Cai, H.; Xie, S. Estimation of Vehicular Emission Inventories in China from 1980 to 2005. *Atmos. Environ.* **2007**, *41*, 8963–8979. [CrossRef]
16. Wang, H.; Chen, C.; Huang, C.; Fu, L. On-Road Vehicle Emission Inventory and Its Uncertainty Analysis for Shanghai, China. *Sci. Total Environ.* **2008**, *398*, 60–67. [CrossRef]
17. Lyu, T.; Wang, P.; Gao, Y.; Wang, Y. Research on the Big Data of Traditional Taxi and Online Car-Hailing: A Systematic Review. *J. Traffic Transp. Eng. (Engl. Ed.)* **2021**, *8*, 1–34. [CrossRef]
18. Zhenbo, W.; Jingtong, Z.; Shunfeng, S. Study on the Commuting Characteristics of Urban Residents and Its influencing Fac-tors in China: A case Study of Tianjin. *Urban Stud.* **2016**, *23*, 6. Available online: https://kns.cnki.net/kcms/detail/detail.aspx?FileName=CSFY201605025&DbName=CJFQ2016.,2010.05.16 (accessed on 19 May 2022).
19. Rafiq, R.; McNally, M.G. A Structural Analysis of the Work Tour Behavior of Transit Commuters. *Transp. Res. Part A Policy Pract.* **2022**, *160*, 61–79. [CrossRef]

20. Danting, L.; Lanyi, Z.; Cheng, C.; Rongzu, Q. Spatiotemporal distribution characteristics of the carbon emission from passen-ger cars in urban core region. *J. Fujian Argiculture For. Univ.* **2019**, 664–672. [CrossRef]
21. Nyhan, M.; Sobolevsky, S.; Kang, C.; Robinson, P.; Corti, A.; Szell, M.; Streets, D.; Lu, Z.; Britter, R.; Barrett, S.R.H.; et al. Predicting Vehicular Emissions in High Spatial Resolution Using Pervasively Measured Transportation Data and Microscopic Emissions Model. *Atmos. Environ.* **2016**, *140*, 352–363. [CrossRef]
22. Liu, L.; Tang, L. Revealing the Characteristics of Active Area in the City by Taxi GPS Data a Study of Shenzhen, China. In Proceedings of the 2017 IEEE 2nd International Conference on Big Data Analysis (ICBDA), Beijing, China, 10–12 March 2017; pp. 532–536.
23. Liu, Y.; Kang, C.; Gao, S.; Xiao, Y.; Tian, Y. Understanding Intra-Urban Trip Patterns from Taxi Trajectory Data. *J. Geogr. Syst.* **2012**, *14*, 463–483. [CrossRef]
24. Qian, X.; Ukkusuri, S.V. Spatial Variation of the Urban Taxi Ridership Using GPS Data. *Appl. Geogr.* **2015**, *59*, 31–42. [CrossRef]
25. Shen, J.; Liu, X.; Chen, M. Discovering Spatial and Temporal Patterns from Taxi-Based Floating Car Data: A Case Study from Nanjing. *GIScience Remote Sens.* **2017**, *54*, 617–638. [CrossRef]
26. Tang, J.; Liu, F.; Wang, Y.; Wang, H. Uncovering Urban Human Mobility from Large Scale Taxi GPS Data. *Phys. A Stat. Mech. Its Appl.* **2015**, *438*, 140–153. [CrossRef]
27. Zhao, P.; Kwan, M.-P.; Qin, K. Uncovering the Spatiotemporal Patterns of CO_2 Emissions by Taxis Based on Individuals' Daily Travel. *J. Transp. Geogr.* **2017**, *62*, 122–135. [CrossRef]
28. Xia, C.; Xiang, M.; Fang, K.; Li, Y.; Ye, Y.; Shi, Z.; Liu, J. Spatial-Temporal Distribution of Carbon Emissions by Daily Travel and Its Response to Urban Form: A Case Study of Hangzhou, China. *J. Clean. Prod.* **2020**, *257*, 120797. [CrossRef]
29. Jimenez-Palacios, J.L. Understanding and Quantifying Motor Vehicle Emissions with Vehicle Specific Power and TILDAS Remote Sensing. Ph.D. Thesis, Massachusetts Institute of Technology, Cambridge, MA, USA, 1998.
30. Zhao, T. On-Road Fuel Consumption Algorithm Based on Floating Car Data for Light-Duty Vehicles. Ph.D. Thesis, Beijing Jiaotong University, Beijing, China, 2009; pp. 1–85. Available online: https://kns.cnki.net/KCMS/detail/detail.aspx?dbname=CMFD2010&filename=2009147440.nh (accessed on 19 May 2022).
31. Zhang, J.; Chen, F.; Wang, Z.; Wang, R.; Shi, S. Spatiotemporal Patterns of Carbon Emissions and Taxi Travel Using GPS Data in Beijing. *Energies* **2018**, *11*, 500. [CrossRef]
32. Kanuganti, S.; Sarkar, A.K.; Singh, A.P. Quantifying Accessibility to Health Care Using Two-Step Floating Catchment Area Method (2SFCA): A Case Study in Rajasthan. *Transp. Res. Procedia* **2016**, *17*, 391–399. [CrossRef]
33. Li, Z.; Fan, Z.; Song, Y.; Chai, Y. Assessing Equity in Park Accessibility Using a Travel Behavior-Based G2SFCA Method in Nanjing, China. *J. Transp. Geogr.* **2021**, *96*, 103179. [CrossRef]
34. Zhang, F.; Li, D.; Ahrentzen, S.; Zhang, J. Assessing Spatial Disparities of Accessibility to Community-Based Service Resources for Chinese Older Adults Based on Travel Behavior: A City-Wide Study of Nanjing, China. *Habitat Int.* **2019**, *88*, 101984. [CrossRef]
35. Garren, S.J.; Pinjari, A.R.; Brinkmann, R. Carbon Dioxide Emission Trends in Cars and Light Trucks: A Comparative Analysis of Emissions and Methodologies for Florida's Counties (2000 and 2008). *Energy Policy* **2011**, *39*, 5287–5295. [CrossRef]
36. Li, T.; Jing, P.; Li, L.; Sun, D.; Yan, W. Revealing the Varying Impact of Urban Built Environment on Online Car-Hailing Travel in Spatio-Temporal Dimension: An Exploratory Analysis in Chengdu, China. *Sustainability* **2019**, *11*, 1336. [CrossRef]
37. Nam, D.; Hyun, K.; Kim, H.; Ahn, K.; Jayakrishnan, R. Analysis of Grid Cell–Based Taxi Ridership with Large-Scale GPS Data. *Transp. Res. Rec.* **2016**, *2544*, 131–140. [CrossRef]
38. Kim, K. Exploring the Difference between Ridership Patterns of Subway and Taxi: Case Study in Seoul. *J. Transp. Geogr.* **2018**, *66*, 213–223. [CrossRef]
39. Sun, D.; Ding, X. Spatiotemporal Evolution of Ridesourcing Markets under the New Restriction Policy: A Case Study in Shanghai. *Transp. Res. Part A Policy Pract.* **2019**, *130*, 227–239. [CrossRef]
40. Yang, Z.; Franz, M.L.; Zhu, S.; Mahmoudi, J.; Nasri, A.; Zhang, L. Analysis of Washington, DC Taxi Demand Using GPS and Land-Use Data. *J. Transp. Geogr.* **2018**, *66*, 35–44. [CrossRef]
41. Zhang, X.; Xu, Y.; Tu, W.; Ratti, C. Do Different Datasets Tell the Same Story about Urban Mobility — A Comparative Study of Public Transit and Taxi Usage. *J. Transp. Geogr.* **2018**, *70*, 78–90. [CrossRef]
42. Si, Y.; Guan, H.; Cui, Y. Research on the Choice Behavior of Taxis and Express Services Based on the SEM-Logit Model. *Sustainability* **2019**, *11*, 2974. [CrossRef]
43. Xia, C.; Chen, B. Urban Land-Carbon Nexus Based on Ecological Network Analysis. *Appl. Energy* **2020**, *276*, 115465. [CrossRef]
44. Liu, Y.; Wang, F.; Xiao, Y.; Gao, S. Urban Land Uses and Traffic 'Source-Sink Areas': Evidence from GPS-Enabled Taxi Data in Shanghai. *Landsc. Urban Plan.* **2012**, *106*, 73–87. [CrossRef]
45. Pan, G.; Qi, G.; Wu, Z.; Zhang, D.; Li, S. Land-Use Classification Using Taxi GPS Traces. *IEEE Trans. Intell. Transport. Syst.* **2013**, *14*, 113–123. [CrossRef]
46. Tang, L.; Gao, J.; Ren, C.; Zhang, X.; Yang, X.; Kan, Z. Detecting and Evaluating Urban Clusters with Spatiotemporal Big Data. *Sensors* **2019**, *19*, 461. [CrossRef]
47. Li, Y.; Shen, J.; Xia, C.; Xiang, M.; Cao, Y.; Yang, J. The Impact of Urban Scale on Carbon Metabolism—A Case Study of Hangzhou, China. *J. Clean. Prod.* **2021**, *292*, 126055. [CrossRef]
48. Xia, C.; Li, Y.; Xu, T.; Ye, Y.; Shi, Z.; Peng, Y.; Liu, J. Quantifying the Spatial Patterns of Urban Carbon Metabolism: A Case Study of Hangzhou, China. *Ecol. Indic.* **2018**, *95*, 474–484. [CrossRef]

49. Wu, H.; Fan, H.; Wu, S. Exploring Spatiotemporal Patterns of Long-Distance Taxi Rides in Shanghai. *ISPRS J. Geo-Inf.* **2017**, *6*, 339. [CrossRef]
50. Gaode Developer Platform. Available online: https://lbs.amap.com/ (accessed on 19 May 2022).

land

MDPI

rticle

The Impact of Urbanization Growth Patterns on Carbon Dioxide Emissions: Evidence from Guizhou, West of China

anying Zeng [1,2], Shaohua Wu [1,*], Hua Zhou [3,4] and Min Cheng [1]

1 Institute of Land and Urban–Rural Development, Zhejiang University of Finance and Economics, Hangzhou 310018, China; zengcy@zufe.edu.cn (C.Z.); chengmin@zufe.edu.cn (M.C.)
2 Collaborative Innovation Center for Data Science and Big Data Analysis, Hangzhou 310018, China
3 The Key Laboratory of Carbon Neutralization and Land Space Optimization, Nanjing 210008, China; huazhou19802022@163.com
4 Institute of Land and Resources Survey and Planning of Guizhou Province, Guiyang 550004, China
* Correspondence: shaohua@zufe.edu.cn

Abstract: Little attention has been paid to the impact of future urban expansion patterns on carbon emissions based on the existing urban pattern of a region. This study used the Central Guizhou Urban Agglomeration as the study area, and the relationships between regional urbanization and CO_2 emissions in the study area were analyzed based on historical data. Urban growth patterns were then simulated in four scenarios that focused on the next 15 years, and they were based on the cellular automaton model. In each different scenario, the CO_2 emissions were predicted, and some implications regarding the impact of those emissions were provided. The results showed that as urban land-use intensity increases, CO_2 emissions first increase then decrease; however, the rate of decline for CO_2 emissions is much slower than the rate at which it rises. Moreover, in the next 15 years, urban expansion will lead to a significant increase in CO_2 emissions. The CO_2 emissions were found to be lowest in the spatial agglomeration scenario and highest in the spatial dispersion scenario. The spatial agglomeration scenario was conducive to understanding how CO_2 emissions eventually peak; however, different cities in the study area should adopt different urban expansion patterns. These research results can provide a reference guide for the government with regard to urban planning.

Keywords: urban growth; CO_2 emission; Guizhou province; urban planning

Citation: Zeng, C.; Wu, S.; Zhou, H.; Cheng, M. The Impact of Urbanization Growth Patterns on Carbon Dioxide Emissions: Evidence from Guizhou, West of China. *Land* **2022**, *11*, 1211. https://doi.org/10.3390/land11081211

Academic Editor: Le Yu

Received: 1 July 2022
Accepted: 29 July 2022
Published: 1 August 2022

Publisher's Note: MDPI stays neutral with regard to jurisdictional claims in published maps and institutional affiliations.

1. Introduction

Urban areas account for three percent of the Earth's land area; however, currently, half of the world's population lives in these areas. They are the main areas of economic activity and the main sources of greenhouse gas emissions, accounting for about 85% of total carbon dioxide (CO_2) emissions [1]. With the acceleration of urbanization in recent decades, big cities continue to grow, and small towns develop rapidly. The agglomeration, expansion, and dispersion expansion of urban spaces have accelerated at different levels [2]. Urban growth has increased energy consumption in industry, the transportation sector, and places of residence, resulting in a rapid upsurge of CO_2 emissions [3,4]. Urban growth is an inevitable part of human development, and thus, how to control the resultant CO_2 emissions has become a key problem that needs to be addressed.

Research on urbanization and CO_2 emissions mainly focuses on three themes. The first concerns the empirical relationships found between urbanization levels and CO_2 emissions; results have shown positive, negative, and inverted U-shaped relationships [5–7]. Using the time series data of different countries, many scholars have demonstrated that urbanization increases carbon emissions [8,9]. Several scholars have found that compact cities reduce the energy consumption of residents. The compact planning of urban transit systems was believed to be causing CO_2 emissions to decrease per capita, and a negative correlation was found between per capita CO_2 emissions and population density in the transportatio sector [7]. In addition, an inverted U-shaped relationship between urbanization and carbo emissions has been demonstrated, which means that with the expansion of urban space carbon emissions first increase and then decrease [10,11].

The second topic focuses on evaluating the impact of historical urban expansio patterns on CO_2 emissions [11,12]. For example, Carpio [12] analyzed urban expansio from 1990 to 2019 using satellite images of the Monterrey Metropolitan Area, Mexico, t determine the relationship between the area's expansion and carbon emissions. Li [1 explored the impact that urban expansion had on ecosystem services between 1990 an 2019, taking into account the different rates of urban expansion and city sizes. Numerou studies have demonstrated that the impact of urban growth on CO_2 emissions manifests i three ways [14]. The first is "production", which refers to the process of urbanization as th process of industrialization; the use of fossil fuel energy in industrial production processe is the main source of CO_2 emissions [15,16]. The second is "land", which refers to th process of urbanization as the process of transforming agricultural land into constructio land; this leads to the reduction of carbon storage areas, thus affecting the overall impact c carbon emissions [14,17]. The third is "people", which refers to the process of urbanizatio as the process of population agglomeration. An increase in the urban population leads t an increase in building construction, transportation use, and the energy consumption o urban residents [7,18]. The effects of urban morphology and urban planning on the spatia patterns of CO_2 emissions in cities have also been examined [1,19].

The third topic concerns the consequences of carbon emissions, as revealed by urba expansion simulations of the future, and how they may support urban developmen decisions [20–22]. Some studies have focused on simulating future urban land-use chang and predicting carbon storage losses caused by the expansion of urban fabric into forests farmlands, and other carbon storage areas [23,24]. These studies mainly focus on th impact of land-use changes caused by urban expansion and the resultant CO_2 emissions In addition, some other studies explored simulated urban expansion scenarios with eco environmental constraints, that were based on cellular automaton data. For example Zhang [25] developed a framework, which was based on a modified CA model, in order to simulate a scenario wherein an urban area expands and develops in a 'low-carbon' manne Chen [20] assessed the impact that urban expansion will have on carbon storage areas ir the future, using the belt and road initiative scenario, cropland protection scenario, and ecological protection scenario.

Although the impact of urbanization on carbon emissions has been widely studied further research is still needed. Urban expansion has different spatial patterns, such as spatial agglomerations or spatial dispersions, which have been confirmed by many stud ies [2,26]. The spatial patterns of urban expansion in a region can potentially have a significant impact on the layout of industrial, housing, and transportation infrastructures and could thus result in different levels of energy consumption and energy efficiency [19] Some existing studies have noted the impact of historical urban expansion patterns on CO_2 emissions [11,12,27]; however, the conclusions drawn from studying historical pro cesses may not be applicable to the development of future solutions. The impact of future urban expansion patterns (based on the impact of existing urban expansion patterns) on CO_2 emissions needs to be further studied. At present, studies which simulate future urban expansion scenarios mainly consider the impact that land-use changes or policy interventions have on CO_2 emissions. Little attention has been paid to the impact of future urban expansion patterns on carbon emissions based on the existing urban patterns of a region. In addition, most studies focus on carbon emissions in economically developed regions [4,6,25]; however, underdeveloped regions, especially urbanized agglomerations within those underdeveloped regions, rapidly develop at a faster rate than areas within economically developed regions, and they also have the fastest-growing carbon emission rate in China. Furthermore, with this in mind, analysis focusing on urbanized agglomerations in the West of China is the most logical course of action.

Moreover, this study aims to explore the impact of future urban growth patterns on CO_2 emissions, and to provide guidance for urban planning. First, using an urban agglomeration in the center of the Guizhou Province as a case study, CO_2 emissions were mapped onto a high resolution spatial grid; then, the impact of urbanization on CO_2 emissions in the urban area was analyzed. From this analysis, a relationship between urbanization and CO_2 emissions in the urban area was derived. Four scenarios of different urban growth patterns were simulated using the CA model as a basis, and for each scenario, the CO_2 emissions were predicted for the next 15 years. This study can provide important guidance in the field of urban planning with regard to choosing urban growth patterns.

2. Framework

Generally speaking, urbanization is regarded as a phenomenon that is related to the population concentration in a specific region, and the high-density and intensive land use of human settlements [3,4]. Urbanization is considered to be one of the main causes of increased CO_2 emissions [3,28]. Most studies show that all modes of urban growth cause CO_2 emissions to increase in the initial growth stage; however, with further growth, CO_2 emissions gradually decrease [6,19]. At the same time, urban growth may reduce CO_2 emissions per unit land area due to intensive land use [29]. Exploring the relationship between urbanization and CO_2 emissions (or CO_2 emissions per unit land area) can help us to better understand the impact of urbanization on CO_2 emissions in cities [30].

Urban land growth mainly occurs in three main forms: spatial agglomeration, spatial dispersion, and a synergistic combination of the two [2,31]. Regarding spatial agglomeration, new urban land is mainly concentrated around big cities, causing the urban area to either 'fill in' space or 'spread out'. This urban growth pattern realizes how the agglomeration of population, industry, and transportation occurs in the original big city. Regarding spatial dispersion, new urban land is mainly concentrated around small towns, in different regions, in the form of urban spreading. The dispersion of urban growth means that the population, economy, and transportation network is concentrated on a smaller scale, and growth can occur in multiple spaces, such as multiple small towns or small cities. The synergy between spatial agglomeration and spatial dispersion means that the growth of new urban land is relatively balanced between spatial agglomeration and dispersion modes (Figure 1). For a given region, the choice of different spatial growth modes has an important impact on the urban scale, layout mode, and land-use intensity of each city in the region, thus affecting the total CO_2 emissions and the CO_2 emissions per unit land area [6,19,30].

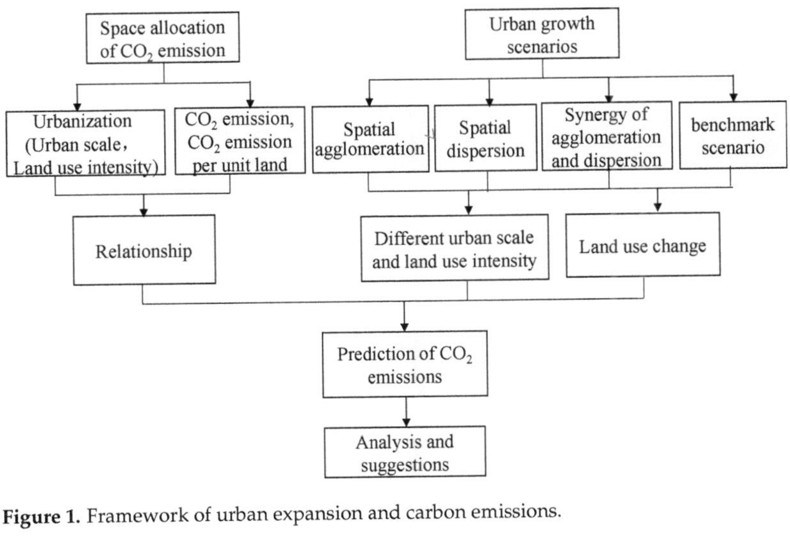

Figure 1. Framework of urban expansion and carbon emissions.

First, this study spatially mapped out CO_2 emissions and obtained their spatial distribution in the study area. The CO_2 emissions of each city in the study area were also obtained. The impact of urbanization—including urban scale and land-use intensity—on the total CO_2 emissions, or CO_2 emissions per unit land area, in urban areas, was explored. Then, four spatial urban growth scenarios were simulated, using the CA model as a basis, in order to analyze potential urban growth and the resultant CO_2 emissions over the next 15 years (to 2035). The CO_2 emissions of the four scenarios were predicted and analyzed (Figure 1). Finally, this paper discusses the impact of spatial urban growth on CO_2 emissions, and provides corresponding suggestions for urban development.

3. Materials and Methods

3.1. Study Area

Most studies focus on the impact that urbanization has on CO_2 emissions in eastern China, or in some of its more developed provinces. Less attention is paid to the less economically developed provinces in western China [4,6,25]; however, these provinces in western China are in the initial stages of rapid urban expansion, and their CO_2 emissions are also rapidly increasing, especially in the Guizhou province, which is experiencing rapid urbanization and industrialization. To balance urban expansion and control of carbon emissions in the future, it is important to explore the impact of urbanization growth patterns on CO_2 emissions, in order to provide support for future urban planning in Guizhou. Thus, we selected Guizhou to be the representative province for western China, as our case study.

The study area is the urban agglomeration in the center of the Guizhou Province, with an area of 53,800 km^2 (Figure 2). It covers 33 counties in six cities (districts) of Guizhou. The Central Guizhou Urban Agglomeration is the key urbanization area in Guizhou, which has great potential for economic development. It is a transportation hub in Southwest China, an important transit area for northwestern and coastal areas, and an important part of the Yangtze River economic belt. The landform of the study area is characterized by plateau–mountainous areas, with an average elevation of about 1100 m; this is consistent with the area's characterization as a place that has "eight mountains, one source of water, and one field." Both the terrain and economy restrict the expansion of the city.

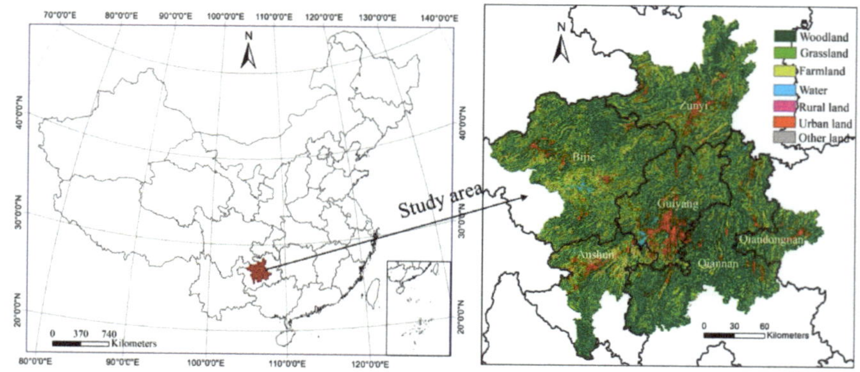

Figure 2. Location of the study area.

With the implementation of the western development strategy of China, the Guizhou economy has ushered in a stage of rapid development. By the end of 2019, the GDP of Guizhou Province reached CNY 1676.934 billion (about USD 248.2 billion, according to the average exchange rate in 2019), with an annual growth rate of 8.3% [32]. The growth rate has been one of the highest in China for nine consecutive years, and it has ranked first for three consecutive years [32]. As the core economic region of the Guizhou Province, the study area has seen increasing rates of industrialization and urbanization, and its energy demands are steadily growing; moreover, its carbon emissions have increased significantly in recent years. The situation regarding energy conservation and emission reduction is grim and worrying.

3.2. Accounting for CO_2 Emissions and Space Allocation

3.2.1. Accounting for CO_2 Emissions

The CO_2 emissions in this study mainly include energy-related and process-related CO_2 emissions. Energy-related CO_2 emissions are emitted during the combustion of fossil fuels, and they can be calculated using Equation (1) [15]:

$$CE_{ij} = FC_{ij} \times NCV_i \times CC_i \times OR_{ij} \quad (1)$$

where CE_{ij} is the CO_2 emissions for fossil fuel type i in socioeconomic sector j. FC_{ij} is the consumption of fossil fuel type i in socioeconomic sector j, obtained from China's energy statistical yearbooks and Guizhou's statistical yearbooks in 2009 and 2019. NCV_i and CC_i are the net calorific value and carbon emissions per unit of net heat generated by fossil fuel i. OR_{ij} is the oxidation rate during fuel combustion. The energy-related CO_2 emissions include 17 fossil fuels in 47 sectors.

Process-related CO_2 emissions mainly refer to the CO_2 emissions from industrial production processes. The main source of process-related CO_2 emissions in China is cement production, accounting for approximately 75% of the total CO_2 emissions produced during industrial production processes [15]; therefore, in this study, the CO_2 emissions from cement production were calculated as follows:

$$CP_k = FC_k \times CF_k \quad (2)$$

where CP_k is the CO_2 emissions from cement production and FC_k is the activity factor of cement production in year k, which was obtained from the official dataset of the National Bureau of Statistics. CF_k is the emission factor for cement production, which was 0.2906 [15].

3.2.2. Spatialization of CO_2 Emissions

The CO_2 emissions of the Guizhou Province were spatially mapped out using a downscaling method and auxiliary data for the years 2009 and 2019. First, this paper established a relationship between the CO_2 emissions of different sectors and their land-use type; then, the CO_2 emissions of different sectors were allocated a space according to their corresponding land-use type. The emissions were allocated a pixel with a resolution of 30 m, and these allocations were based on the corresponding year's auxiliary spatial data for each land-use type [18,33]. The land-use data resolution was 30 m for 2009 and 2019. Land-use types include industrial land, mining land, urban residential land, rural residential land, road traffic land, traffic service facility land, commercial service industry and other service industry land, agricultural facility and hydraulic construction land, woodland, and other lands which produce no carbon emissions. Forty-seven economic sectors were merged into seven sectors, including the sector which controls the production and supply of electric power, the mining sector, transportation service sector, manufacturing industries, the wholesale, retail trade, catering, and other service sectors, the urban residents' energy usage sector, and rural residents' energy usage sector.

CO_2 emissions from different industries were allocated a pixel according to their corresponding land-use types, as specified by the land-use classification system for 2019. Regarding point source data, CO_2 emissions generated by power plants were considered to correspond with the industrial land-use type, and thus, they were allocated as such, in accordance with its power plant coordinates. Data on power plants were obtained from the world power plant database (https://www.wri.org/research/global-database-power-plants (accessed on 13 April 2021)). CO_2 emissions generated by manufacturing processes were considered to correspond with the industrial land-use type, and thus, they were allocated as such; however, this excludes emissions generated from power plant locations. Emissions generated by the combustion of oil fuels in the transportation, storage, post, and telecommunication service sectors were considered to correspond with the road land-use type, an emissions generated by the combustion of non-oil fuels were considered to correspon with the transportation site land-use type; both sets of emissions were allocated accordingl CO_2 emissions generated by the mining sector were considered to correspond with th mining land-use type, and they were allocated accordingly. Emissions generated by th energy use of urban and rural residents were considered to correspond with urban an rural residential land-use types, respectively; both sets of emissions were allocated accord ingly. Emissions generated by the wholesale, retail trade, catering service, and other servic sectors were allocated to the commercial land-use type, and were allocated accordingly.

After the emissions generated by different sectors were matched with their corre sponding land-use type, the emissions were then allocated to a pixel with a resolution o 30 m, based on the corresponding year's auxiliary data, and measured as a weight for eac land-use type [18,33].

$$CE_i^t = \sum_{s=1}^{k} C_{s,j}^t \times \frac{A_i^t}{\sum_{i=1}^{n} A_{i,j}^t} \quad (3$$

where CE_i are the CO_2 emissions at spatial grid location i in year t, and $\sum_{s=1}^{k} C_{s,j}^t$ is the sur of CO_2 emissions for all sectors s in land class j in the study area. A_i^t is the auxiliary dat value for location i in year t, and $\sum_{i=1}^{n} A_{i,j}^t$ is the sum of the auxiliary data for all pixels i i land class j. In this study, the auxiliary data are the corrected global Defense Meteorologica Program—Operational Line-Scan System (DMSP-OLS) Nighttime Lights time (NTL) serie data [34]. The larger the light value of a pixel, the greater the number of emissions tha are allocated to that pixel. To be consistent with the land-use data, the night light data ar resampled to 30 m.

3.3. Impact of Urbanization on CO_2 Emissions from 2009 to 2019

Based on the obtained spatial distribution of CO_2 emissions, the CO_2 emissions generated by the urban land in each county (district) of the study area were extracted, and the impact of urbanization on CO_2 emissions was explored. In this study, urbanization is mainly characterized by two dimensions: urban scale and urban land-use intensity. These dimensions are expressed by the urban area itself and the proportion of urban area to total land area, respectively [35,36]. CO_2 emissions are mainly considered in two ways: in terms of total emissions and CO_2 emissions per unit of urban land area.

For each county, the relationship between urbanization and CO_2 emissions was con structed using statistical models that were mainly based on the spatial distribution of CO_2 emissions and land-use acquisition in 2019. Logarithmic, exponential, primary polynomial quadratic polynomial, and power models were used in this study [37]. The best model was chosen based on its R-squared (R^2) value and Akaike Information Criterion (AIC) [38]. The smaller the AIC value and the higher the R^2 value, the better the model.

3.4. Urban Expansion Simulation and CO_2 Prediction

3.4.1. Cellular Automaton (CA) Model

The cellular automaton (CA) model has been proven to be an effective method for urban spatial simulations of the future [39,40]. It can incorporate the natural and socio-economic characteristics of an urban system to obtain a more reasonable urban simulation. The CA model can simulate different spatial patterns of future urban expansion by changing a few key parameters, such as the distance from towns or big cities. Thus, by using the CA model, it is feasible to perform urban simulations with the goal of simulating different spatial patterns that account for the next 15 years. This study used a CA model to simulate urbanization expansion.

The CA model comprises five parts: cellular, cellular state, transformation rule function, neighborhood, and cell time [39]. The cell is the basic unit of the CA, which is used to describe the attributes of a variable and is represented in a binary form. The neighborhood refers to the surrounding cells of the target cell, and its value is obtained by setting a fixed cell radius. The rule function notes whether the target cell can be expanded into urban land.

In reference to previous studies [27,39] and regional development requirements, the probability (P^{t+1}) of conversion from non-urban cells to urban cells depends on the influence of the following comprehensive factors: the suitability of urban growth, limiting factors, neighborhood influence, and other interferences.

$$P^{t+1}\left(S_{i,j}^{t} \rightarrow urban\right) = D_{i,j}^{t} * N_{i,j}^{t} * L_{i,j}^{t} * \varepsilon_{i,j}^{t} \quad (4)$$

Here, $S_{i,j}^{t}$ is the state of cellular at location (i, j) and time (t). $D_{i,j}^{t}$ is the suitability of urban growth for pixel (i, j) in year t, which refers to the accessibility of roads, towns, and cities in this study. The higher the accessibility, the higher the suitability for urban growth. $N_{i,j}^{t}$ is the neighborhood index, which refers to the proportion of urban or town pixels in the neighborhood. The higher the proportion of urban land in the neighborhood around the pixel, the higher the possibility of converting the pixel into urban land. The radius is set to 3 in this study. $L_{i,j}^{t}$ represent the factors that limit the expansion of urban land, including areas containing water and topographic conditions; the pixels with a water land-use type will not be converted into urban land, and moreover, the greater the size of a slope, the lower the possibility of converting pixels into urban land—pixels with a slope that is greater than 15 degrees will not be converted into urban land. The $\varepsilon_{i,j}^{t}$ term represents random interference.

The threshold for the probability (P^{t+1}) of urban expansion was calculated based on the urban expansion rate of the study area from 2009 to 2019. When the probability of a non-urban cell is greater than the threshold, the cell is converted to urban land. The CA model was constructed using MATLAB 2016a software. The simulation period was from 2019 to 2035.

3.4.2. Scenario Setting

The spatial agglomeration and diffusion of urban space manifests in the different trends of the spatial distribution of the economy and population. The former concerns an agglomeration shifting from a widely spaced area to a relatively narrow, regional space, and the latter concerns the diffusion of a densely packed unit into widely spaced area [16,41]. This study uses four scenarios to simulate the future, with each using a different urban expansion pattern: spatial agglomeration, spatial dispersion, a synergistic combination of agglomeration and dispersion, and a benchmark scenario. Scenario 1 simulates urban spatial agglomeration. In this scenario, new urban land is mainly concentrated around big cities, and these new urban areas are generated by 'filling in' space or 'spreading out'. Scenario 2 simulates urban spatial dispersion. In this scenario, new urban land is mainly concentrated around small towns in different regions, and these new areas are generated by urban spreading. Scenario 3 simulates a synergistic combination of urban spatial agglomeration and dispersion. In this scenario, new urban lands are generated in a relatively balanced manner, via both spatial agglomeration and dispersion. Scenario 4 simulates a benchmark scenario. In this scenario, new urban lands are generated based on urban expansion pattern trends in Guizhou over the past 10 years (2009–2019). According to scenario 4, spatial expansion patterns in the future would emulate how they have existed for the past 10 years.

The different scenarios are realized by changing the suitability parameter in the CA model [40,42]. Scenario 1 is characterized by spatial aggregation, which means that urban growth is more suited to large cities than small towns. In this scenario, the suitability of a cell near a town is one-fifth as suitable as a cell near a city. Scenario 2 is characterized by dispersed urban expansion, meaning that urban growth is less suited to big cities than towns. In this scenario, the CA model demonstrates that the suitability of a cell near a town is five times as suitable as a cell near a city. Scenario 3 is characterized by the balanced development of urban growth, which means that the suitability of areas surrounding small towns is equal to the suitability of areas surrounding big cities. Scenario 4 is a benchmark scenario. In this scenario, the suitability parameters of the CA model are the same as the parameters used for the urbanization expansion that occurred from 2009 to 2019.

3.4.3. Predicted CO_2 Emissions under Different Scenarios

The relationship between urbanization and CO_2 emissions forms the basis for the prediction of CO_2 emissions in different urbanization expansion scenarios [25,35]. In this study, we mainly considered the relationship between the urban area and the CO_2 emissions generated per urban land unit. The main reason for this is that the expansion of urbanization leads directly to an increase in urban land area, which affects the land use intensity, thus finally affecting the efficiency of CO_2 emissions per unit of land. The efficiency of CO_2 emissions per unit of land ultimately affects the total CO_2 emissions produced by differently sized cities.

For the year 2035, the urban area of each county (district) in the study area, in each of the different scenarios, was calculated using the urban expansion results. Then, using the relationship (as it was in 2019) between the urban area and CO_2 emissions per urban land unit for different counties, the CO_2 emissions per urban land unit for each county were predicted for the next 15 years; the total CO_2 emissions generated by urban land use for the whole study area were also calculated [25,43]. The CO_2 emissions' characteristics and the emission trends of the study area, in each of the different scenarios for the next 15 years, were then analyzed.

4. Results

4.1. CO_2 Emissions of the Study Area

The spatial distributions of the CO_2 emissions in the study area, shown at a resolution of 30 m, indicated that the emissions in 2009 were concentrated in the center of the urban areas of each city (Figure S1). In 2019, carbon emission areas showed characteristics of diffusion and dispersion. High carbon emission areas reduced in size, whereas areas that generated a medium amount of carbon emissions increased from 2009 to 2019. As they were affected by urban expansion, many areas changed from being areas that produced no carbon emissions to areas that did produce carbon emissions; this shows how, in recent years, the trend of industrial development and urbanization in the Central Guizhou urban agglomeration has been characterized by a synergistic combination of agglomeration and diffusion.

Carbon dioxide emissions generated by urban land use were calculated based on the spatial distributions of the CO_2 emissions. From 2009 to 2019, Guizhou gradually entered a stage of rapid economic development. This period of urbanization mainly centered on the city, and it expanded to the surrounding areas, notably Guiyang, Zunyi, and Anshun (Figure S2). Due to the expansion of urban areas, carbon emissions increased significantly in Central Guizhou. In 2009, the urban built-up area containing the urban agglomeration was 493.74 km^2, and produced 101.67 million tons of CO_2 emissions, which accounted for 80.9% of total emissions. By 2019, the urban area increased to 937.65 km^2, and CO_2 emissions increased to 133.14 million tons. The carbon emissions of urban areas accounted for 85.4% of total carbon emissions; therefore, controlling CO_2 emissions in urban areas is key to achieving carbon neutralization.

4.2. Impact of Urbanization on CO_2 Emissions from 2009 to 2019

The relationship between urbanization and CO_2 emissions in each county (district) of the study area is shown in Figure 3, and the statistical parameters for each model are shown in Table S1. Results show that urbanization, which is expressed by the proportion of the urban area in relation to the total land area, had a significant effect on CO_2 emissions. The CO_2 emissions per land unit show a significant decreasing logarithmic trend as the urban area increases (Figure 3a). When the urban land area is small, increased urbanization leads to a rapid decline in CO_2 emissions per unit; however, when the city grows larger, with the further expansion of the city, the rate of decline for CO_2 emissions per unit slows down.

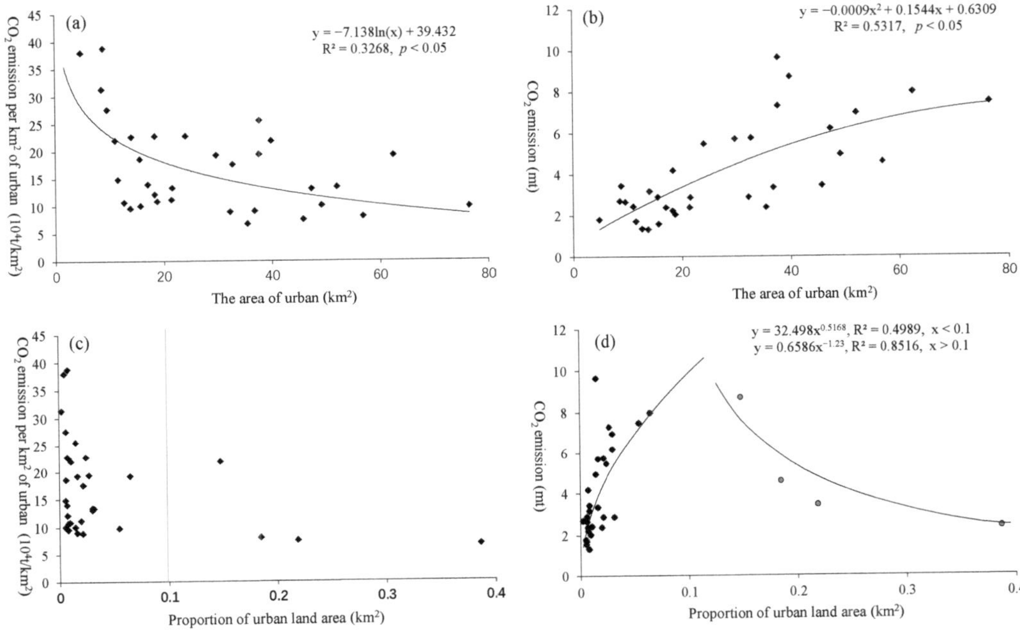

Figure 3. Relationships between CO_2 emissions and urbanization in 2019. Panel (**a**) shows the relationship between the average CO_2 emissions generated by urban land use and the urban area. Panel (**b**) shows the relationship between the total CO_2 emissions generated and the urban area. Panel (**c**) shows the relationship between the average CO_2 emissions generated by urban land use and the proportion of the urban land area. Panel (**d**) shows the relationship between total CO_2 emissions and the proportion of the urban land area.

A scatter diagram of the total CO_2 emissions produced and the area of urban land in each county (district) of the study area is shown in Figure 3b. The results reveal a quadratic relationship between the total urban CO_2 emissions and the urban area—an inverted U-shaped relationship. As the size of the urban area increases, the total amount of urban carbon emissions first increases, then decreases. At present, Guizhou is in the initial stages of urbanization, when CO_2 emissions rapidly increase in accordance with the increasing size of the urban area. Attention should be paid to two key issues in the future: accelerating the development of urbanization and causing CO_2 emissions to peak as soon as possible.

The relationship between the proportion of urban land area and CO_2 emissions are shown in Figure 3c,d. When the urban land-use intensity is relatively low (the proportion of urban land area is less than 10%), the CO_2 emissions show an increasing trend in accordance with the increasing proportion of urban land area. When the proportion of the urban area is greater than 10%, the CO_2 emissions begin to decline exponentially as the proportion of urban area increases. Nevertheless, the rate of decline for CO_2 emissions is slow. The CO_2 emissions per unit of urban land show no obvious trend; however, when the proportion of urban land area is very high (10%), the CO_2 emissions per unit of land are relatively low.

4.3. Urbanization Expansion Results under Different Scenarios

The spatial distribution of urban land in 2035, in accordance with the results of the four scenarios, is shown in Figure 4; all urban land shows a clear expansion over the next 15 years. Moreover, in each different scenario, the distribution of new urban lan is significantly different. In scenario 1, new urban land is mainly concentrated aroun big cities, which emerged due to 'filling' or 'spreading'. In scenario 2, new urban land mainly concentrated around small towns or cities in different regions, emerging as a resu of spreading. In scenario 3, new urban land emerged as a result of the relatively balance relationship between spatial agglomeration and dispersion; the distribution of new urba land in scenario 4 is similar to scenario 3.

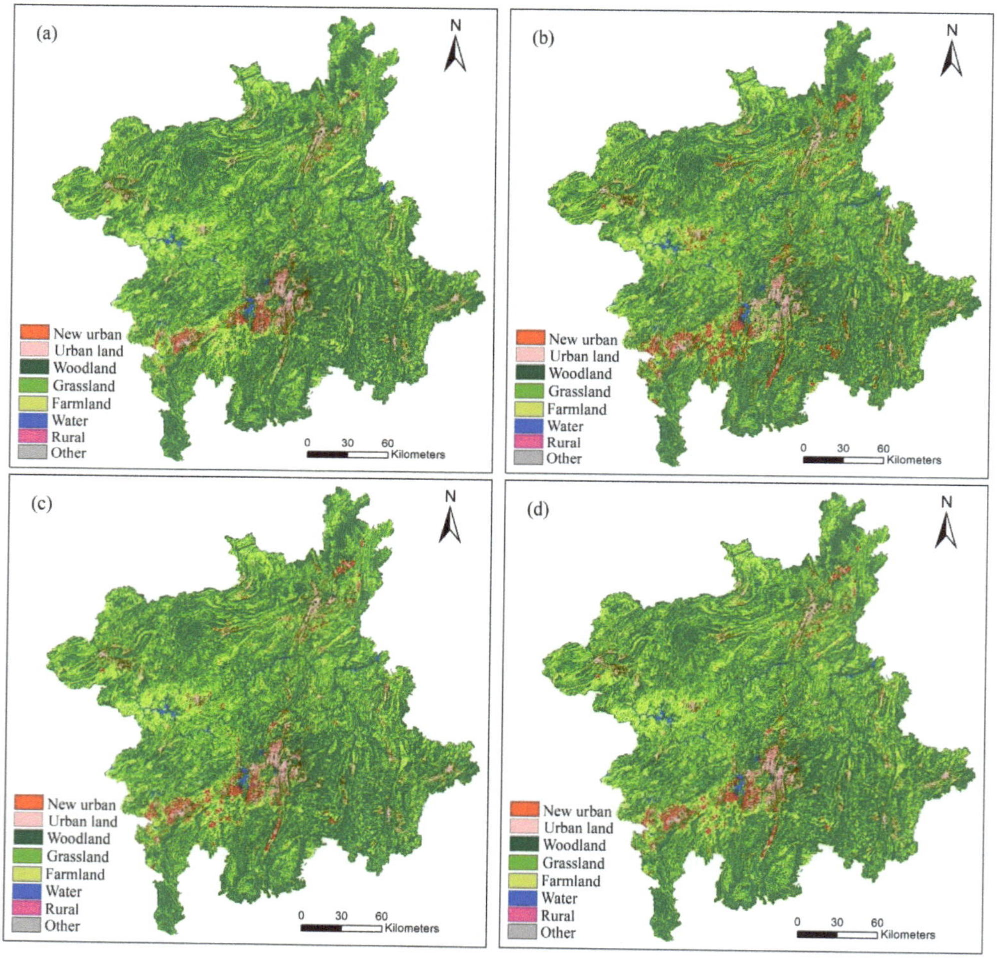

Figure 4. Results of urbanization expansion for 2035, for each different scenario. Panels (**a–d**) show the expansion results for scenario 1 to scenario 4, respectively.

For all scenarios, the spatial patterns of urban land have obvious differences to the spatial pattern of 2019. The increased size of the urban areas in scenarios 1 to 4 are 621 km^2, 1350.24 km^2, 911.02 km^2, and 842.35 km^2, respectively. The new urban area of the decentralized mode (scenario 2) is more than twice that of the agglomeration mode (scenario 1), and 1.5 times that of the benchmark scenario (scenario 4). The new urban area in scenario 3 is similar to that of the benchmark scenario. The urban expansion for the agglomeration scenario (1) is characterized by the manner in which it expands (via 'filling'), and its single-center pattern; therefore, in scenario 1, the probability of urban expansion around small towns is relatively low. In this scenario, urban expansion is mainly concentrated in Guiyang city, the provincial capital, as well as Anshun, and Zunyi city.

The urban expansion in scenario 2 has multi-center characteristics. In addition to Guiyang, obvious urbanization expansion occurs in many spaces, such as small towns. Scenario 3 shows that big cities and small towns both expand rapidly. The spatial distribution of urban land for the benchmark scenario (scenario 4) is similar to that of scenario 3. This means that over the past decade, urbanization in Guizhou developed as a result of a combination of agglomeration and decentralization.

The proportion of other land-use types that were converted to urban land in each of the different scenarios for 2035 are significantly different (Figure S3). More than half of new urban land comes from cultivated land in each of the four scenarios. The occupied proportion of cultivated land is highest under the decentralized expansion pattern, accounting for 64.78% of the total occupied land. The occupied proportions of cultivated land in the other scenarios are all greater than 50%. Forest land is also a common land type that is often converted to new urban land; this was especially the case in scenario 1, with forest land conversions accounting for 28.51% of the total occupied land. In scenario 2, a greater proportion of cultivated land is occupied; however, the proportion of occupied forest land was the lowest of the four scenarios. In addition, the proportion of rural residential land that converted to urban land was lower than 10%. The proportion of other land types that converted to urban land was very small.

4.4. CO_2 Emissions under Different Scenarios

Figure 5 shows the predicted CO_2 emissions in 2035, for each different scenario. Scenario 1 (spatial agglomeration pattern) had the lowest predicted emissions, at 156.29 mt; this demonstrates an increase of 23.15 mt when compared with the emissions generated in 2019. Compared with scenario 2 (urban dispersion expansion pattern), scenario 1 shows a reduction in carbon emissions by about 14%. Compared with scenario 3 and scenario 4, scenario 1 shows a reduction in carbon emissions by about 14.81 mt and 10.58 mt, respectively. This means that by 2035, the CO_2 emissions of the study area will experience its greatest ever reduction if the agglomeration development pattern is adopted. In contrast, scenario 2 generated the highest amount of CO_2 emissions, with a value of 181.62 mt. The emissions of the dispersion model are 14.75 mt higher than that of the baseline scenario. The CO_2 emissions of scenario 3 are 171.12 mt, which is about 4.25 mt higher than scenario 4.

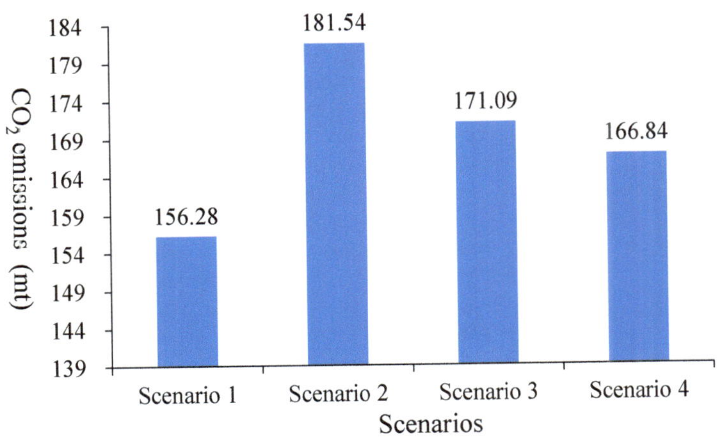

Figure 5. Predicted CO_2 emissions in urban land areas for the year 2035; results from the different scenarios.

When using the CA model to simulate urbanization expansion, the new urban are differs in terms of expansion cycles (years). For each different scenario, the variation i CO_2 emissions, in relation to new urban areas, is shown in Figure 6. The agglomeratio pattern is always better than other patterns in terms of emission reduction when the nev urban area is the same; however, the decentralized pattern of urbanization is not conduciv to reducing carbon emissions, as it persistently produces the highest emissions.

Furthermore, CO_2 emissions and new urban areas have an inverted U-shape relatior ship in all scenarios; in other words, as the new urban area increases, CO_2 emissions firs increase, then decrease. Urban growth, as described by the spatial agglomeration patterr will reach peak CO_2 emissions earlier than other urban growth patterns. In contrast, th spatial dispersion pattern will be last to reach peak carbon emissions. After urbanizatior reaches a certain stage, CO_2 emissions show a downward trend in all scenarios. The result indicate that the spatial agglomeration of urban growth is conducive to reaching peak CO emissions early.

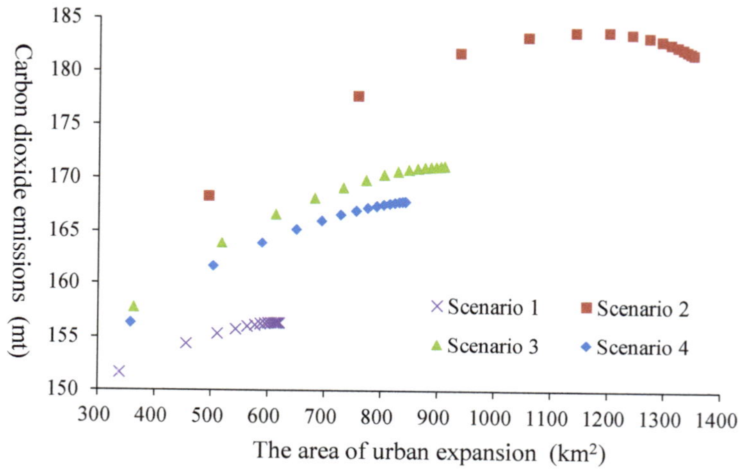

Figure 6. Variation in CO_2 emissions in new urban areas from 2020 to 2035 in each different scenario.

The CO_2 emissions in different regions, using different urbanization expansion scenarios, are shown in Figure 7. Guiyang city, the provincial capital, shows little difference in emissions in each different expansion scenario during the early stages of expansion; however, with regard to the later stages of expansion, scenario 1 is more conducive to reducing carbon emissions, whereas scenario 2 generates the highest amount of emissions. Zunyi is the second largest city in the Guizhou Province. In contrast to Guiyang, the urban expansion in scenario 2, in the Zunyi region, produces a low amount of CO_2 emissions, and the CO_2 emissions in scenario 1 are the highest. In the early stages of urban expansion in Anshun, scenario 1 is more conducive to reducing CO_2 emissions compared with the other scenarios. In the later stages of expansion, as the city expands further, the CO_2 emissions in all scenarios decrease rapidly. Bijie, Qiannan, and Qiandongnan all have the lowest emissions in scenario 1, followed by scenario 3 and scenario 4. The CO_2 emissions in scenario 2 are the highest.

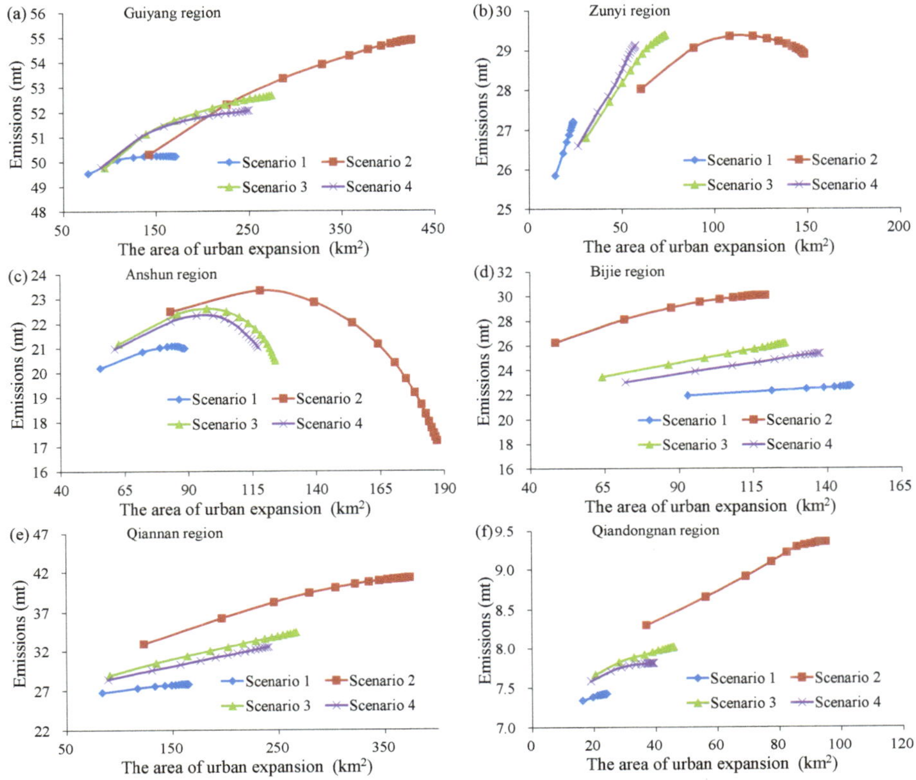

Figure 7. Comparison of CO_2 emissions in different regions for each of the different scenarios from 2020 to 2035. Panel (**a**) is for Guiyang region, panel (**b**) is for Zunyi region, panel (**c**) is for Anshun region, panel (**d**) is for Bijie region, panel (**e**) is for Qiannan region, panel (**f**) is for Qiandongnan region.

5. Discussions and Implications

The world is undergoing a process of rapid urbanization, and people are increasingly concerned about the impact of future urban expansion on carbon emissions. Some studies use different policy interventions to simulate future urban expansion, in order to obtain urban expansion schemes that are more conducive to the reduction of CO_2 emissions [20–24]. Compared with our research, these studies explored urban expansion from a pre-policy intervention perspective. In addition, some studies evaluated the consequences of carbon emissions using future projections of urban expansion, which comprise ex post assessment research [12,44,45]. These studies mainly explored the impact of land-use changes, caused by urban expansion, on CO_2 emissions. Our research focuses on the future impact of different urban spatial expansion patterns on CO_2 emissions based on the existing urban spatial expansion pattern. Different spatial patterns of urban areas mean different layouts of industrial, housing, and transportation networks, thus resulting in different levels of energy consumption and energy efficiency [19]. This study answers the question of which urban expansion pattern should be adopted by the different cities in the study area in order to reduce carbon emissions.

Some existing studies have paid attention to the impact of urban expansion patterns on CO_2 emissions [11,12,27]; however, these studies mainly offer relevant recommendations after assessing the impact of historical urban expansion patterns on CO_2 emissions. First, our study discusses the relationship between regional urbanization and CO_2 emissions based on historical data; then, it explores which future spatial expansion pattern will be more conducive to reducing carbon emissions based on the existing urban pattern. Previous findings indicated that the increased fragmentation of urban forms could result in more carbon emissions, and an agglomeration development pattern of urban land would help reduce carbon emissions [46]. Some other studies also pointed out that greater levels of urban expansion should be encouraged in the developed cities of developing countries to achieve low-carbon development [47]. This is partly similar to our research results. Regarding future scenarios, our results showed that in the study area, the agglomeration pattern of urban land is more conducive to reducing carbon emissions, but different cities in the study area need to adopt different expansion patterns.

This study can provide policy implications for future urban expansion; indeed, this study used four scenarios that included different cities, and it took carbon emission characteristics into account. The results indicated that further agglomeration does not produce a rapid increase in carbon emissions in regions where a certain level of urbanization has been reached, such as in Guiyang city. For medium-sized cities with characteristic industries, such as in Zunyi, urbanization expansion should be guided by industrial agglomeration rather than urban spatial location agglomeration. Cities with low urbanization levels, such as Bijie, Qiannan, and Qiandongnan city, should expand with an aim to achieve spatial agglomeration; however, due to the fact that these cities have the best capacity for carbon storage, emphasizing ecological governance and the development of eco-tourism products may be a better choice for economic development in these cities.

The spatial agglomeration of urban land reduces the need for low-level repeated infrastructure construction, which disrupts residents' lives. This will improve the level of resource sharing and energy efficiency [19]; however, the spatial agglomeration of cities will also increase the traffic pressure and pressures related to resource supplies [47]. Therefore, in urban planning, more attention should be paid to the reasonable adjustment of traffic land to alleviate traffic congestion. In urban planning, the regional area balance between working place and residence, as well as catering space and residence, should be fully considered to reduce the long distances that residents are required to travel in their daily lives; this should improve people's happiness. In addition, urban planning also plays an auxiliary role in assisting other means of carbon emission control, such as the implementation of a fee charged on vehicles that are driven within a particular urban area; such schemes should ensure that there is a reasonable alternative for new energy vehicles. Urban agglomeration maximizes the use of these facilities. When an area of new urban land remains the same, the development of urban spatial agglomeration often increases peak CO_2 emissions; however, it is difficult to considerably reduce carbon emissions through urban planning only. Controlling the consumption of fossil energy and increasing the utilization of non-fossil energy through various means are prerequisites to carbon neutralization in the future.

In the future, land-use change during urban expansion ought to be a factor that is considered in the urban planning process. Most areas in the Guizhou Province are mountainous; these are known as "eight mountains, one source of water, and one field" areas. Indeed, 80% of the land is mountainous, 10% of the area is water, and 10% of the land is farmland. Due to the peculiar terrain of Guizhou, the amount of land available for urban growth is limited. Most rural settlements are scattered in mountainous areas, which are far larger than urban land areas, and are far away from cities. Here, the land that experiences urban growth is mainly composed of agricultural land and forest land. Compared with forest land, Guizhou's agricultural land is limited in terms of helping to facilitate urban growth. Urban expansion agglomeration (Scenario 1) occupies more forest land than the decentralized pattern (Scenario 2), and thus, it is more conducive to maintaining cultivated land areas and securing the food supply for local residents. Spatially dispersed urban growth takes up more cultivated land than woodland, which is conducive to maintaining forests and increasing carbon sequestration in Guizhou. Different regions can choose suitable urban growth modes according to their regional characteristics.

In addition, nighttime light data has been widely used as a proxy for urban issues. For example, it has been used for spatial–temporal dynamic analysis of urban expansion, defining urban boundaries, and estimating the spatial distribution of CO_2 emissions [18,26,33,48]. This paper uses nighttime light data to obtain CO_2 emissions on a pixel scale, and to extract CO_2 emission measurements in urban land areas [18,26]. Based on the CO_2 emissions generated by urban land use in the study area, we explored the relationship between urbanization and CO_2 emissions. The research results show that the relationship between city scale and CO_2 emissions forms an inverted U-shape, which is similar to previous research results [10,11]; however, the relationship between urban land-use intensity and CO_2 emissions is not a strictly inverted U-shaped or linear relationship. With regard to urban land-use intensity, CO_2 emissions increase rapidly when urban land-use intensity is low. After reaching its peak, as urban land-use intensity increases, CO_2 emissions begin to decrease, but the rate of reduction is far slower than the rate of increase. The relationship between urban land-use intensity and CO_2 emissions may therefore be accurately described as erratic, especially when land-use intensity reaches a certain level.

The urban boundary predictions are difficult to directly verify. Although subjective judgment exists, no relevant research has proposed a direct method to verify the rationality of the simulations' boundaries; however, the effectiveness of the parameters of the CA model for the benchmark scenario had been validated. We obtained the optimum parameters for the benchmark scenario by repeatedly simulating urban growth using the 2009 and 2019 land-use data. Other scenarios were implemented by changing the importance of accessibility in towns and cities, and its relation to the suitability of urban growth in the benchmark scenario. Results indicated that the simulated urban spatial distribution conforms to the initial goal of achieving spatial agglomeration or spatial dispersion.

This study aims to explore the impact of future urban spatial expansion patterns on CO_2 emissions. It does not consider whether the spatial agglomeration of different industries will have an important impact on CO_2 emissions. In the future, we could simulate different spatial patterns for the housing, industrial, and commercial sectors, among others; then, the impact of these patterns on CO_2 emissions could be explored. In addition, this study only examines urban agglomeration in Guizhou, an underdeveloped province, as the study area. More studies can be conducted in the future to compare the differences between different urban expansion patterns and their impact on CO_2 emissions in different regions.

6. Conclusions

The spatial patterns of urban expansion have a potentially significant impact on carbon emissions; however, little attention has been paid to the impact of future urban expansion patterns on carbon emissions based on the existing urban expansion pattern. Focusing upon urban agglomeration in Central Guizhou, this study aimed to investigate the impact of urban expansion patterns on the reduction of carbon emissions, and it provided suggestions for low-carbon development. The relationships between regional urbanization and CO_2 emissions in the study area were analyzed based on historical data. Four urban expansion scenarios, including spatial agglomeration, spatial dispersion, a synergistic combination of agglomeration and dispersion, and a benchmark scenario, were simulated; the CO_2 emissions under each scenario were subsequently predicted for the next 15 years.

Regional urbanization had significant effect on CO_2 emissions. The CO_2 emissions per land unit showed a significant decreasing logarithmic trend as the urban area increased; although, the total carbon emissions showed an inverted U-shaped trend as the urban area increased. When the urban land-use intensity was relatively low, CO_2 emissions showed an increasing trend as land-use intensity also increased. When the land-use intensity was greater than 10%, the CO_2 emissions began to decline exponentially with increasing land-use intensity; however, the rate of decline was slow.

In the next 15 years, urban expansion will lead to a significant increase in CO_2 emissions. The CO_2 emissions were lowest in the spatial agglomeration scenario and highest in the spatial dispersion scenario. The spatial agglomeration of urban growth helps to quickly facilitate the peaking of CO_2 emissions in urban areas; however, different cities in the study area can adopt different urban expansion modes. In areas where urbanization has reached a certain level, such as in Guiyang city, further agglomeration did not produce a rapid increase in carbon emissions. For medium-sized cities with characteristic industries, such as in Zunyi, urbanization expansion should be guided by industrial agglomeration, rather than urban spatial agglomeration. The expansion of cities with low urbanization levels such as Bijie, Qiannan, and Qiandongnan city, should develop with the aim of achieving spatial agglomeration; however, due to the fact that these cities have the best capacity for carbon storage, emphasizing ecological governance and the development of eco-tourism products may be a better choice for economic development.

The spatial agglomeration of cities will also increase the traffic pressure and pressures relating to resource supplies. In urban planning, more attention should be paid to the reasonable adjustment of traffic land to alleviate traffic congestion. In urban planning the regional area balance between working place and residence, and catering land and residence, should be fully considered to reduce the length of time that residents must travel long distances, thus improving people's happiness. This research provides an important reference guide for urban planning in the Guizhou province and other similar areas. In the future, more studies could be conducted to compare the impact of urban expansion patterns on CO_2 emissions in different regions, with different urbanization degrees. Moreover, the spatial expansion patterns of different industries and their impact on CO_2 emissions also deserves further study.

Supplementary Materials: The following supporting information can be downloaded at: https://www.mdpi.com/article/10.3390/land11081211/s1, Figure S1: Spatial distribution of CO2 emissions for 2009 (a) and 2019 (b) in the study area; Figure S2: Land use data for 2009 (a) and 2019 (b) in the study area; Figure S3: Proportion of other land-use types converted to urban land in 2035; Table S1 Statistical parameters for each model.

Author Contributions: Conceptualization, C.Z. and S.W.; methodology, C.Z. and M.C.; software C.Z.; validation, H.Z. and S.W.; formal analysis, M.C.; investigation, H.Z.; resources, H.Z.; data curation, C.Z.; writing—original draft preparation, C.Z. and S.W.; writing—review and editing C.Z. and S.W.; visualization, C.Z.; supervision, S.W.; project administration, H.Z. and S.W.; funding acquisition, C.Z. All authors have read and agreed to the published version of the manuscript.

Funding: This study was funded by the National Natural Science Foundation of China (41901062), Natural Science Foundation of Zhejiang Province (LY22D010009), and Postdoctoral Science Foundation of China (2018M642389).

Institutional Review Board Statement: Not applicable.

Informed Consent Statement: Not applicable.

Data Availability Statement: The data presented in this study are available on request from the corresponding author.

Conflicts of Interest: The authors declare no conflict of interest.

References

1. Privitera, R.; Palermo, V.; Martinico, F.; Fichera, A.; Rosa, D.L. Towards lower carbon cities: Urban morphology contribution in climate change adaptation strategies. *Eur. Plan. Stud.* **2018**, *26*, 812–837. [CrossRef]
2. Strano, E.; Simini, F.; Nadai, M.D.; Esch, T.; Marconcini, M. The agglomeration and dispersion dichotomy of human settlements on earth. *Sci. Rep.* **2021**, *11*, 23289. [CrossRef]
3. Ahmad, M.; Akram, W.; Ikram, M.; Shah, A.A.; Jabeen, G. Estimating dynamic interactive linkages among urban agglomeration, economic performance, carbon emissions, and health expenditures across developmental disparities. *Sustain. Prod. Consump.* **2020**, *26*, 239–255. [CrossRef]
4. Zeng, T.; Jin, H.; Geng, Z.; Kang, Z.; Zhang, Z. The Effect of Urban Shrinkage on Carbon Dioxide Emissions Efficiency in Northeast China. *Int. J. Environ. Res.* **2022**, *19*, 5772. [CrossRef] [PubMed]
5. Sun, W.; Huang, C.C. How does urbanization affect carbon emission efficiency? Evidence from China. *J. Clean. Prod.* **2020**, *272*, 122828. [CrossRef]
6. Zhou, Y.; Chen, M.; Tang, Z.; Mei, Z. Urbanization, land use change, and carbon emissions: Quantitative assessments for city-level carbon emissions in beijing-tianjin-hebei region. *Sustain. Cities Soc.* **2021**, *66*, 102701. [CrossRef]

7. Nima, S.; Ahmad, H.; Younes, N. Developing a system dynamics approach for cng vehicles for low-carbon urban transport: A case study. *Int. J. Low-Carbon Technol.* **2022**, *16*, 577–591. [CrossRef]
8. Sadorsky, P. The Effect of Urbanization on CO_2 Emissions in Emerging Economics. *Energ. Econ.* **2014**, *41*, 147–153. [CrossRef]
9. Emir, F.; Philip, L.D.; Sertoglu, K. Assessing the influence of urbanization and energy on carbon emissions of Turkey: Evidence using the new RALS analysis. *Environ. Sci. Pollut. Res.* **2022**, 1–13. [CrossRef] [PubMed]
10. Cheng, L.; Mi, Z.; Sudmant, A.; Coffman, D. Bigger cities better climate? results from an analysis of urban areas in china. *Energ. Econ.* **2022**, *107*, 105872. [CrossRef]
11. Martínez-Zarzoso, I.; Maruotti, A. The Impact of Urbanization on CO2 Emissions: Evidence From Developing Countries. *Energ. Econ.* **2011**, *70*, 1344–1353. [CrossRef]
12. Liu, S.R.; Shen, J.W.; Liu, G.F.; Wu, Y.Z.; Shi, K.F. Exploring the effect of urban spatial development pattern on carbon dioxide emissions in China: A socioeconomic density distribution approach based on remotely sensed nighttime light data. *Comput. Environ. Urban.* **2022**, *96*, 101847. [CrossRef]
13. Carpio, A.; Ponce-Lopez, R.; Lozano-García, D.F. Urban form, land use, and cover change and their impact on carbon emissions in the monterrey metropolitan area, mexico. *Urban. Clim.* **2021**, *39*, 100947. [CrossRef]
14. Li, S.; He, Y.; Xu, H.; Zhu, C.; Wang, K. Impacts of urban expansion forms on ecosystem services in urban agglomerations: A case study of shanghai-hangzhou bay urban agglomeration. *Remote Sens.* **2021**, *13*, 1908. [CrossRef]
15. Anokye, J.; Logah, V.; Opoku, A. Soil carbon stock and emission: Estimates from three land-use systems in Ghana. *Ecol. Process.* **2021**, *10*, 11. [CrossRef]
16. Shan, Y.; Guan, D.; Zheng, H.; Ou, J.; Qiang, Z. China CO_2 emission accounts 1997–2015. *Sci. Data* **2018**, *5*, 170201. [CrossRef]
17. Hari, B.D.; Sameer, A. Greenhouse gas emission reduction options for cities: Finding the "Coincidence of Agendas" between local priorities and climate change mitigation objectives. *J. Forest. Econ.* **2013**, *38*, 100–105. [CrossRef]
18. Mendelsohn, R.; Sohngen, B. The Net Carbon Emissions from Historic Land Use and Land Use Change. *J. Forest. Econ.* **2019**, *34*, 263–283. [CrossRef]
19. Raupach, M.R.; Rayner, P.J.; Paget, M. Regional variations in spatial structure of nightlights, population density and fossil-fuel CO_2 emissions. *Energy Policy* **2010**, *38*, 4756–4764. [CrossRef]
20. Wang, S.H.; Huang, S.L.; Huang, P.J. Can spatial planning really mitigate carbon dioxide emissions in urban areas? A case study in Taipei, Taiwan. *Landsc. Urban Plan.* **2018**, *169*, 22–36. [CrossRef]
21. Chen, Y.; Yue, W.; Liu, X.; Zhang, L.; Chen, Y. Multi-scenario simulation for the consequence of urban expansion on carbon storage: A comparative study in central asian republics. *Land* **2021**, *10*, 608. [CrossRef]
22. Wang, Z.; Zeng, J.; Chen, W.X. Impact of urban expansion on carbon storage under multi-scenario simulations in Wuhan, China. *Environ. Sci. Pollut. Res.* **2022**, *29*, 45507–45526. [CrossRef]
23. Santos, Y.L.F.; Yanai, A.M.; Ramos, C.J.P.; Graca, P.M.L.A.; Veiga, J.A.P.; Correia, F.W.S.; Fearnside, P.M. Amazon deforestation and urban expansion: Simulating future growth in the Manaus Metropolitan Region, Brazil. *J. Environ. Manag.* **2022**, *304*, 114279. [CrossRef]
24. Seto, K.C.; Güneralp, B.; Hutyra, L.R. Global forecasts of urban expansion to 2030 and direct impacts on biodiversity and carbon pools. *Proc. Natl. Acad. Sci. USA* **2012**, *109*, 16083–16088. [CrossRef]
25. Kukkonen, M.O.; Khamis, M.; Muhammad, M.J.; Kyhk, N.; Luoto, M. Modeling direct above-ground carbon loss due to urban expansion in zanzibar city region, tanzania. *Land Use Policy* **2022**, *112*, 105810. [CrossRef]
26. Zhang, Y.; Liu, Y.; Wang, Y.; Liu, D.; Xia, C.; Wang, Z. Urban expansion simulation towards low-carbon development: A case study of Wuhan, China. *Sustain. Cities Soc.* **2020**, *63*, 102455. [CrossRef]
27. Peng, J.; Lin, H.; Chen, Y. Spatiotemporal evolution of urban agglomerations in China during 2000–2012: A nighttime light approach. *Landsc. Ecol.* **2020**, *35*, 421–434. [CrossRef]
28. Dadashpoor, H.; Azizi, P.; Moghadasi, M. Analyzing spatial patterns, driving forces and predicting future growth scenarios for supporting sustainable urban growth: Evidence from Tabriz metropolitan area, Iran. *Sustain. Cities Soc.* **2019**, *47*, 101502. [CrossRef]
29. Cetin, M.; Ecevit, E. Urbanization energy consumption and CO_2 emissions in Sub-Saharan countries: A panel cointegration and causality analysis. *J. Dev. Econ.* **2015**, *3*, 66–76. [CrossRef]
30. Shi, Y.S.; Wang, H.F.; Shi, S.Z. Relationship between social civilization forms and carbon emission intensity: A study of the Shanghai metropolitan area. *J. Clean. Prod.* **2019**, *228*, 1552–1563. [CrossRef]
31. Romero-Lankao, P.; Gunmey, K.R.; Seto, K.C.; Chester, M.; Duren, R.M.; Hughes, S. A critical knowledge pathway to low carbon sustainable futures: Integrated understanding of urbanization, urban areas, and carbon. *Earths Futur.* **2014**, *2*, 515–532. [CrossRef]
32. Liu, F.; Zhang, Z.X.; Shi, L.F.; Zhao, X.L.; Xu, J.Y.; Yi, L.; Liu, B. Urban expansion in China and its spatial-temporal differences over the past four decades. *J. Geogr. Sci.* **2016**, *26*, 1477–1496. [CrossRef]
33. China National Statistics Yearbook; National Bureau of Statistics of China: Beijing, China. 2020. Available online: http://www.stats.gov.cn/ (accessed on 30 June 2022).
34. Oda, T.; Maksyutov, S. A very high-resolution (1 km $\times$ 1 km) global fossil fuel CO_2 emission inventory derived using a point source database and satellite observations of nighttime lights. *Atmos. Chem. Phys.* **2011**, *11*, 543–556. [CrossRef]
35. Li, X.; Zhou, Y.; Zhao, M.; Zhao, X. A harmonized global nighttime light dataset 1992–2018. *Sci. Data* **2020**, *7*, 168. [CrossRef] [PubMed]

36. Chuai, X.; Feng, J. High resolution carbon emissions simulation and spatial heterogeneity analysis based on big data in Nanjing City. *Sci. Total Environ.* **2019**, *686*, 828–837. [CrossRef] [PubMed]
37. Zeng, C.Y.; Yang, L.; Zhu, A.X. The generation of soil Spectral dynamic feedback using Landsat 8 data for digital soil mapping. *Remote Sens.* **2020**, *12*, 1691–1710. [CrossRef]
38. Burnham, K.; Anderson, D. Multimodel inference: Understanding AIC and BIC in model selection. *Sociol. Methods Res.* **2004**, *3*, 261–304. [CrossRef]
39. Li, X.; Zhang, X.; Yeh, A.; Liu, X. Parallel cellular automata for large-scale urban simulation using load-balancing techniques. *Int. J. Geogr. Inf. Sci.* **2010**, *24*, 803–820. [CrossRef]
40. Zhang, Y.; Chang, X.; Liu, Y.F.; Lu, Y.C.; Wang, Y.H.; Liu, Y.L. Urban expansion simulation under constraint of multiple ecosystem services (MESs) based on cellular automata (CA)-Markov model: Scenario analysis and policy implications. *Land Use Policy* **2021**, *108*, 105667. [CrossRef]
41. Sun, C.; Wu, Z.F.; Lv, Z.Q.; Yao, N.; Wei, J.B. Quantifying different types of urban growth and the change dynamic in Guangzhou using multi-temporal remote sensing data. *Int. J. Appl. Earth Obs.* **2013**, *21*, 409–417. [CrossRef]
42. Zhou, L.; Dang, X.; Sun, Q.; Wang, S. Multi-scenario simulation of urban land change in Shanghai by random forest and CA-Markov model. *Sustain. Cities Soc.* **2020**, *55*, 102045. [CrossRef]
43. Echenique, M.; Hargreaves, A.; Mitchell, G.; Namdeo, A. Growing cities sustainably: Does urban form really matter? *J. Am. Plan. Assoc.* **2012**, *78*, 121–137. [CrossRef]
44. Houghton, R.A.; House, J.I.; Pongratz, J. Carbon emissions from land use and land-cover change. *Biogeosciences* **2012**, *9*, 5125–5142. [CrossRef]
45. Arsanjani, J.J.; Helbich, M.; Kainz, W.; Boloorani, A.D. Integration of logistic regression, Markov chain and cellular automata models to simulate urban expansion. *Int. J. Appl. Earth Obs. Geoinf.* **2013**, *21*, 265–275. [CrossRef]
46. Ou, J.P.; Liu, X.P.; Li, X.; Chen, Y.M. Quantifying the relationship between urban forms and carbon emissions using panel data analysis. *Landsc. Ecol.* **2013**, *28*, 1889–1907. [CrossRef]
47. Li, Y.; Shen, J.Y.; Xia, C.Y.; Xiang, M.T.; Cao, Y.; Yang, J.Y. The impact of urban scale on carbon metabolism—A case study of Hangzhou, China. *J. Clean. Prod.* **2021**, *292*, 126055. [CrossRef]
48. Zheng, B.; Cheng, J.; Geng, G.; Wang, X.H.K. Mapping anthropogenic emissions in china at 1 km spatial resolution and its application in air quality modeling. *Sci. Bull.* **2021**, *66*, 612–620. [CrossRef]

land

MDPI

Article

Evaluating the Effects of Renewable Energy Consumption on Carbon Emissions of China's Provinces: Based on Spatial Durbin Model

Yang Sun [1], Mengna Du [2], Leying Wu [2], Changzhe Li [2] and Yulong Chen [2,*]

[1] Institute of Microbial Engineering, School of Life Sciences, Henan University, Kaifeng 475004, China
[2] Key Research Institute of Yellow River Civilization and Sustainable Development & Collaborative Innovation Center on Yellow River Civilization, Henan University, Kaifeng 475001, China
* Correspondence: chenyulong0203@163.com

Abstract: Renewable energy consumption is considered as the main form of energy consumption in the future. The carbon emissions produced by renewable energy can be approximately ignored, and renewable energy is essential for regional sustainable development. In this study, we used the Durbin model with panel data to explore the spatial dependence between renewable energy consumption the and carbon emissions of China's 30 provinces from 1997 to 2017. The results show that: (1) there is a negative spatial correlation between renewable energy consumption and carbon emissions, and "High-Low" areas are mainly concentrated in southern provinces in 1997–2011; (2) the center of gravity of renewable energy consumption moves southwest, which is consistent with the center of gravity of carbon emissions; (3) renewable energy consumption has a significant inhibitory effect on carbon emissions of a local region, but the spatial spillover effect is not significant. Specifically, a 1% increase in renewable energy consumption in a region will reduce carbon emissions by 0.05%. Finally, on the basis of this study, it was proposed to give full play to the advantages of renewable energy in the western region, and further accelerate the development of the renewable energy industry.

Keywords: renewable energy consumption; carbon emissions; spatial correlation; spatial Durbin model

Citation: Sun, Y.; Du, M.; Wu, L.; Li, C.; Chen, Y. Evaluating the Effects of Renewable Energy Consumption on Carbon Emissions of China's Provinces: Based on Spatial Durbin Model. *Land* **2022**, *11*, 1316. https://doi.org/10.3390/land11081316

Academic Editor: Yan Li

Received: 26 June 2022
Accepted: 12 August 2022
Published: 15 August 2022

1. Introduction

The global warming caused by increasing carbon emissions has attracted widespread attention from the international community. Although the COVID-19 pandemic has slowed down global carbon dioxide concentration in 2020, the global temperature will still maintain an upward trend. How to reduce carbon emissions in order to achieve the goal of global temperature control is the current concern of the international community. On account of the vital status of the energy supply system, developing renewable energy has become an important way to deal with increasing energy demand and climate change. As a responsible developing country, China promised to increase its nationally determined contribution, strive to peak in its carbon dioxide emissions by 2030 and achieve carbon neutrality by 2060[1]. To realize this goal, China proposed to build a clean, low-carbon, safe and efficient energy system. The system will control the total amount of fossil energy and improve the utilization of renewable energy to promote the reduction of carbon emissions[2]. In 2020, China's renewable energy development and utilization scale reached 680 million tons of standard coal, equivalent to replacing nearly 1 billion tons of coal [1]. Scientific analysis of the role of renewable energy in carbon emissions is of great significance for achieving carbon peaking and carbon neutrality.

In recent years, scholars have conducted much research on the influencing factors on carbon emissions at different scales by means of econometric models [2–5], structural decomposition methods [6–8], and geographic detectors [9,10]. Most viewpoints include four aspects: (1) economic development will promote the increase of carbon emissions [11],

(2) population size plays a positive role in promoting carbon emissions [12], (3) industrial structure [13,14] is the main factor leading to the growth of carbon emissions,(4) technological progress [15] and low-carbon innovation [16] can effectively reduce carbon emissions. In addition, R&D investment, population density and industrial agglomeration have also been proved to be influencing factors on carbon emissions [17–19]. Due to the short development time of renewable energy and the difficulty in availability of data, there is relatively little literature on the impact of renewable energy consumption on carbon emissions.

In terms of the relationship between renewable energy consumption and carbon emissions, scholars have found that renewable energy consumption has a negative impact on carbon emissions [20–23]. Chen [24] used the Auto-regressive Distributed Lag (ARDL) model to study the impact of renewable energy consumption on carbon emissions in China, and found that renewable energy consumption has a negative impact on carbon emissions, with a relatively significant long-term coefficient and insignificant short-term coefficient. Qiu [25] showed that renewable energy consumption has a significant impact on energy carbon emissions and there was a one-way causality between them. Some scholars believed that there was a two-way causal relationship between renewable energy consumption and carbon emissions [26–28]. Wang and Zhao [29] used the cointegration test, impulse response analysis and variance decomposition to analyze the dynamic relationship between renewable energy consumption, carbon emissions and economic growth in China from 1980 to 2009, and found that renewable energy consumption has a weak impact on carbon emissions.

In general, some researchers pay attention to the relationship between renewable energy consumption and carbon emissions, but most studies assume that each region is an island, and few or no studies have quantified the spatial effect of renewable energy consumption on carbon emissions. In fact, as the First Law of Geography states, everything is related to everything else, but near things are more related than distant things. Where China is concerned, regional carbon emissions have strong spatial and temporal attributes considering the obvious differences in regional development levels, which means that regional energy structure, urbanization level, economic development and technology spillovers all affect carbon emissions of adjacent regions. The purpose of this paper is to develop a spatial panel model to analyze the spatial spillover effect between China's 30 provinces for the period 1997–2017.

The remainder of this study proceeds as follows. Section 2 discusses the methods and data preparation. Section 3 describes spatial characteristics of renewable energy consumption and carbon emissions, and presents the results of the spatial econometric model. Section 4 provides a discussion, with Section 5 providing concluding remarks.

2. Methods and Data

2.1. Spatial Correlation Analysis

The Moran Index, 1948, is one of the most commonly used measures of spatial autocorrelation. The univariate global Moran index(I) was adopted here to measure the spatial autocorrelation of renewable energy consumption and carbon emissions. It is calculated by following formula [30,31]:

$$I = \frac{n\left[\sum_{i=1}^{n}\sum_{j=1}^{n} W_{ij}(x_i - \overline{x})(x_j - \overline{x})\right]}{\left[\sum_{i=1}^{n}\sum_{j=1}^{n} W_{ij}(x_i - \overline{x})^2\right]} \quad (1)$$

where i and j represent the province, I represents the global Moran index, n represents the number of provinces, W_{ij} represents the spatial weight matrix, and x and $\overline{x}$ represent the variables and their mean values. Generally, the value of I ranges between -1 and 1. The value from 0 to 1 means a positive spatial correlation. Otherwise, the value from -1 to 0 means a negative spatial correlation. The value 0 means no spatial autocorrelation.

The bivariate global Moran index indicates the correlation of one variable of a spatial unit to another observed variable of an adjacent spatial unit. This paper used the bivariate global Moran index and the bivariate local Moran index to measure bivariate spatial correlation. The bivariate global spatial autocorrelation (Bivariate Moran's I, I_{kl}) is used to explore the spatial correlation characteristics of two variables, and its expression is [32]:

$$I_{kl} = \frac{\sum_{i=1}^{n}\sum_{j=1}^{n} W_{ij}(x_k^i - \overline{x}_l)(x_k^j - \overline{x}_l)}{S^2\sum_{i=1}^{n}\sum_{j=1}^{n} W_{ij}} \tag{2}$$

The expression for the bivariate local Moran index (I'_{kl}) is [33]:

$$I'_{kl} = \frac{x_k^i - \overline{x}_k}{S_k^2}\sum_{j=1}^{n} W_{ij}\frac{x_l^j - \overline{x}_l}{S_l^2} \tag{3}$$

where x_k^i represents the observed value of k variable on space unit i, x_l^j represent the observed value of l variable on spatial unit j, $\overline{x_k}$ and $\overline{x_l}$ represent the mean values of the observed values of the k and l variables, W_{ij} is the spatial weight matrix, and S_k^2 and S_l^2 represent the variances of the observed values of the k and l variables.

2.2. Standard Deviation Ellipse

The standard deviation ellipse can reflect the overall dominant distribution direction of spatial elements and the degree of dispersion in each direction [34], revealing the spatial distribution characteristics of the dispersion and direction. This study drew the standard deviation ellipse of carbon emissions and renewable energy consumption in China from 1997 to 2017, and analyzed the spatial pattern changes of these two according to their azimuth and standard deviation at the X/Y axis.

Azimuth tan ∝ can be calculated by following expression:

$$\tan\alpha = \frac{\sum_{i=1}^{n}\widetilde{m}_i^2 - \sum_{i=1}^{n}\widetilde{n}_i^2 + \sqrt{\left(\sum_{i=1}^{n}\widetilde{m}_i^2 - \sum_{i=1}^{n}\widetilde{n}_i^2\right)^2 + 4\left(\sum_{i=1}^{n}\widetilde{m}_i\widetilde{n}_i\right)^2}}{2\sum_{i=1}^{n}\widetilde{m}_i\widetilde{n}_i} \tag{4}$$

where m_i and n_i represent the horizontal and vertical coordinates of center of each province, $\widetilde{m}_i$ and $\widetilde{n}_i$ represent the deviations from the center coordinates of each province to the barycentric coordinates.

Standard deviation along the X axis (σ_x) and Y axis (σ_y) are calculated by the following expression:

$$\sigma_x = \sqrt{2}\sqrt{\frac{\sum_{i=1}^{n}(\widetilde{m}_i\cos\alpha - \widetilde{n}_i\sin\alpha)^2}{n}} \tag{5}$$

$$\sigma_y = \sqrt{2}\sqrt{\frac{\sum_{i=1}^{n}(\widetilde{m}_i\sin\alpha + \widetilde{n}_i\cos\alpha)^2}{n}} \tag{6}$$

2.3. Center of Gravity Analysis

The center of gravity model is derived from the concept of center of gravity in physics; the forces in all directions around the center of gravity of the regional space are relatively balanced, which can reveal the degree of spatial equilibrium in the distribution of regional carbon emissions and renewable energy consumption. In order to detect the center of gravity and moving direction of China's carbon emissions and renewable energy consumption, the formulas are as follows [35]:

$$U = \frac{\sum_{i=1}^{n} x_i m_i}{\sum_{i=1}^{n} x_i} \tag{7}$$

$$v = \frac{\sum_{i=1}^{n} x_i n_i}{\sum_{i=1}^{n} x_i} \tag{8}$$

where U and V represent the longitude and latitude of the center of gravity of a certain attribute (per capita carbon emissions and per capita renewable energy consumption) in China, respectively; x_i represents the per capita carbon emissions and per capita renewable energy consumption of each province in China.

2.4. Spatial Econometric Model

The spatial Durbin model can test the influence of the explained variables affected by explanatory variables in local areas and neighboring areas [36,37]. Considering the spatial correlation of dependent and independent variables, this study employed spatial Durbin model to test the influence of renewable energy consumption on carbon emissions. This study takes explanatory variables in natural logarithm to solve the endogenous problem [38]. The model is described as in [39]:

$$\ln(Y_{it}) = \rho\Sigma_{j=1}^{n} W_{ij}\ln(Y_{it}) + \beta_1\ln(REC_{it}) + \beta_2\ln(X_{it}) + \theta_1\Sigma_{j=1}^{n} W_{ij}\ln(REC_{it}) + \theta_2\Sigma_{j=1}^{n} W_{ij}\ln(X_{it}) + \mu_i + \gamma_t + \varepsilon_{it} \quad (9)$$

where Y denotes the dependent variable, W_{ij} indicates spatial weight matrix, X_{it} refers generically to the explanatory variables in the model, $\theta_1\sum_{j=1}^{n} W_{ij}\ln(REC_{it})$ and $\theta_2\sum_{j=1}^{n} W_{ij}\ln(X_{it})$ denote the spatially lagged exogenous regressors. ρ denotes spatial autoregressive parameter representing the intensity of spatial autocorrelation. β denotes fixed unknown associated coefficients. θ_1 and θ_2 indicate the spatial impacts from explanatory variables in neighboring provinces. μ_i, γ_t *and* ϵ_{it} represent space effect, time effect and the theoretical disturbance, respectively.

To analyze the spatial correlation characteristics of renewable energy consumption and carbon emissions for a more comprehensive analysis, this paper established three spatial weight matrices based on the existing literature [40–42]:

$$Queen\ adjacency.W_1 = \begin{cases} W_{ij} = 1\ if\ province\ i\ and\ j\ are\ neighbors \\ W_{ij} = 0\ otherwise \end{cases} \quad (10)$$

$$Inverse\ distance.W_2 = \begin{cases} W_{ij} = \frac{1}{d_{ij}}\ if\ province\ i\ and\ j\ are\ neighbors \\ W_{ij} = 0\ otherwise \end{cases} \quad (11)$$

$$Economic\ distance.W_3 = \begin{cases} W_{ij} = 0\ i = j \\ W_{ij} = \frac{1}{|GDP_i - GDP_j|}\ i \neq j \end{cases} \quad (12)$$

where d_{ij} represents the distance between the centroids of provinces i and j, GDP_i represents the GDP per capital of spatial unit i in 2017.

2.5. Index Selection and Data Source

Due to the lack of data from Tibet, Hong Kong, Macao and Taiwan, 30 provinces and autonomous regions in China are adopted as the research objects. The research period is 1997–2017 due to data limitations on renewable energy consumption. Description of the variables are shown in Table 1.

Table 1. Description of variables.

Variables	Abbreviation	Unit	Mean	Min	Max
carbon emissions	C	ton	6.281	0.130	42.540
economic growth	GDP	yuan	20,006.771	2107.815	93,123.248
energy intensity	EI	10^4 tons of standard coal/10^4 yuan	1.382	0.253	5.147
renewable energy consumption	REC	tons of standard coal	2.267	0.002	3.548
technological progress	PT	pieces	8305.171	5	199,293

(1) **Dependent variable**. Carbon emissions per capita (C), which is calculated by total emissions and population. Provincial total carbon emissions are from the CEAD database (https://www.ceads.net/, accessed on 1 June 2022) and population from the provincial statistical yearbooks.

(2) **Independent variable**. Renewable energy consumption per capita (REC); renewable energy does not emit greenhouse gases. The development of renewable energy (water, wind, light, biomass, etc.) is the main way to deal with climate change and achieve sustainable development. Renewable energy consumption data come from the Energy Economics Data Platform (http://inems2.bit.edu.cn/Home/Menu, accessed on 1 June 2022).

(3) **Control variables**. We set economic growth, quadratic economic growth, energy intensity, and technological progress as control variables. Economic growth is described by per capita *GDP* of each province. Quadratic economic growth is added to verify the EKC curve. Energy intensity is calculated by energy consumption divided by *GDP*, which means the higher the energy intensity, the more energy contained in economic activities. Since technological progress has positive effects on improving production processes and energy utilization efficiency, this study employed the number of patents as the indicator of technological progress. Previous studies found R&D investment would overestimate the actual level of technological progress [16]. Provincial *GDP* is from the statistical yearbooks and treated at a constant price of 1990. Energy consumption and patents are from the China Energy Statistical Yearbook and China Science and Technology Statistical Yearbook, respectively.

3. Result

3.1. Spatial Characteristics of Renewable Energy Consumption and Carbon Emissions

3.1.1. Univariate Spatial Autocorrelation Analysis

To test the spatial autocorrelation before constructing the spatial Durbin model, we firstly examine the spatial dependence among energy consumption and carbon emissions by global Moran'I, which are shown in Table 2.

Table 2. Global Moran'I of energy consumption and carbon emissions from 1997 to 2017.

Year	REC	C	Year	REC	C
1997	0.260 ***	0.361 ***	2008	0.094 *	0.398 ***
1998	0.229 ***	0.326 ***	2009	0.045	0.355 ***
1999	0.214 ***	0.412 ***	2010	0.054	0.393 ***
2000	0.196 ***	0.345 ***	2011	0.169 ***	0.377 ***
2001	0.332 ***	0.364 ***	2012	0.132 **	0.388 ***
2002	0.300 ***	0.348 ***	2013	0.086 **	0.327 ***
2003	0.280 ***	0.333 ***	2014	0.150 ***	0.316 ***
2004	0.085	0.389 ***	2015	0.103 *	0.324 ***
2005	0.050	0.450 ***	2016	0.072	0.327 ***
2006	0.098 *	0.396 ***	2017	0.081	0.308 ***
2007	0.090 *	0.405 ***			

Notes: ***, **, and * indicate the significance of the 1%, 5% and 10% level, respectively.

As shown in Table 2, the Moran'I of carbon emissions are all significant at the 1% level, indicating that there is a positive spatial correlation in carbon emissions, and carbon emissions show spatial agglomeration characteristics rather than randomly distribution in geographic space. Renewable energy consumption shows a random distribution in several years since the total renewable energy consumption increased to a large extent on the basis of a steady increase in a few years. However, the Moran'I of renewable energy consumption in most years is larger than 0 and passes the 10% significance test, which indicates a positive spatial correlation, i.e., renewable energy consumption presents a state of agglomeration distribution in geographic space.

3.1.2. Bivariate Spatial Correlation Analysis

Considering the spatial interaction of regional renewable energy consumption and carbon emissions in adjacent provinces, we calculated the bivariate global Moran'I of 30 provinces from 1997 to 2017 using provincial renewable energy consumption as the first variable and carbon emissions as the second variable. Results are shown in Table 3. On the whole, the Moran index ranks from −0.239 to −0.076 in the study period, and most years passed the 10% significance test, which indicates that REC has a negative spatial spillover effect on carbon emissions, and REC in one area has a negative effect on carbon emissions in adjacent areas.

Table 3. Bivariate global Moran index from 1997 to 2017.

Year	Value	Year	Value	Year	Value
1997	−0.128 *	2004	−0.220 ***	2011	−0.124 *
1998	−0.107	2005	−0.206 ***	2012	−0.121 *
1999	−0.102	2006	−0.217 ***	2013	−0.076
2000	−0.098	2007	−0.177 **	2014	−0.089
2001	−0.155 *	2008	−0.259 ***	2015	−0.128 *
2002	−0.191 **	2009	−0.183 **	2016	−0.134 *
2003	−0.239 ***	2010	−0.136 **	2017	−0.161 **

Notes: ***, **, and * indicate the significance of the 1%, 5% and 10% level, respectively.

To intuitively observe the spatial relationship and distribution between renewable energy consumption and carbon emissions in China, we drew bivariate local spatial LISA agglomeration maps of 1997, 2004, 2011, and 2017 (Figure 1). The cluster types are "High-High", "Low-Low", "High-Low" and "Low-High", while "High-Low" type indicates provinces with high renewable energy consumption surrounded by low carbon emissions provinces.

Figure 1 shows that China's renewable energy consumption-carbon emissions agglomeration type is mainly "High-Low", which fluctuated during the study period. There are three provinces (Guangdong, Hunan, Guizhou) belonging to "High-Low" in 1997 and seven provinces (Guangdong, Hunan, Guizhou, Hubei, Guangxi, Sichuan, Yunnan) in 2004, which means that the utilization rate of renewable energy in these provinces is increasing, from high energy consumption–high emission to low energy consumption–low emission. On the whole, the use of renewable energy has a certain negative effect on the carbon emissions of surrounding provinces. However, due to policies such as the development of the western region and the rise of the central region, the carbon emissions of the central and western regions increased dramatically, resulting in a downward trend in the "High-Low" regions from 2004 to 2017.

"Low-High" areas move from North China to Northeast China, and then to North China and Northwest China from the spatial perspective. The number of "Low-High" areas increases from three provinces (Inner Mongolia, Jilin and Hebei) in 1997 to six provinces (Heilongjiang, Jilin, Liaoning, Hebei, Shanxi and Ningxia) in 2011, and reduces to four (Gansu, Shanxi, Shanxi, Hebei) in 2017. As China's heavy industrial bases, three northeast provinces gradually enter the "Low-High" region from 1997 to 2011, along with small proportion of renewable energy in the energy consumption structure. During 2011–2017, the Northeast region increased the use of renewable energy and did not belong to the "Low-High" areas.

The number of provinces belonging to "High-High" area is the smallest. Inner Mongolia has been in the "High-High" area since 2011, owing to substantial fossil energy and an energy-intensive development pattern. The "Low-Low" area has a small change range, and is situated in the southern provinces during the study period.

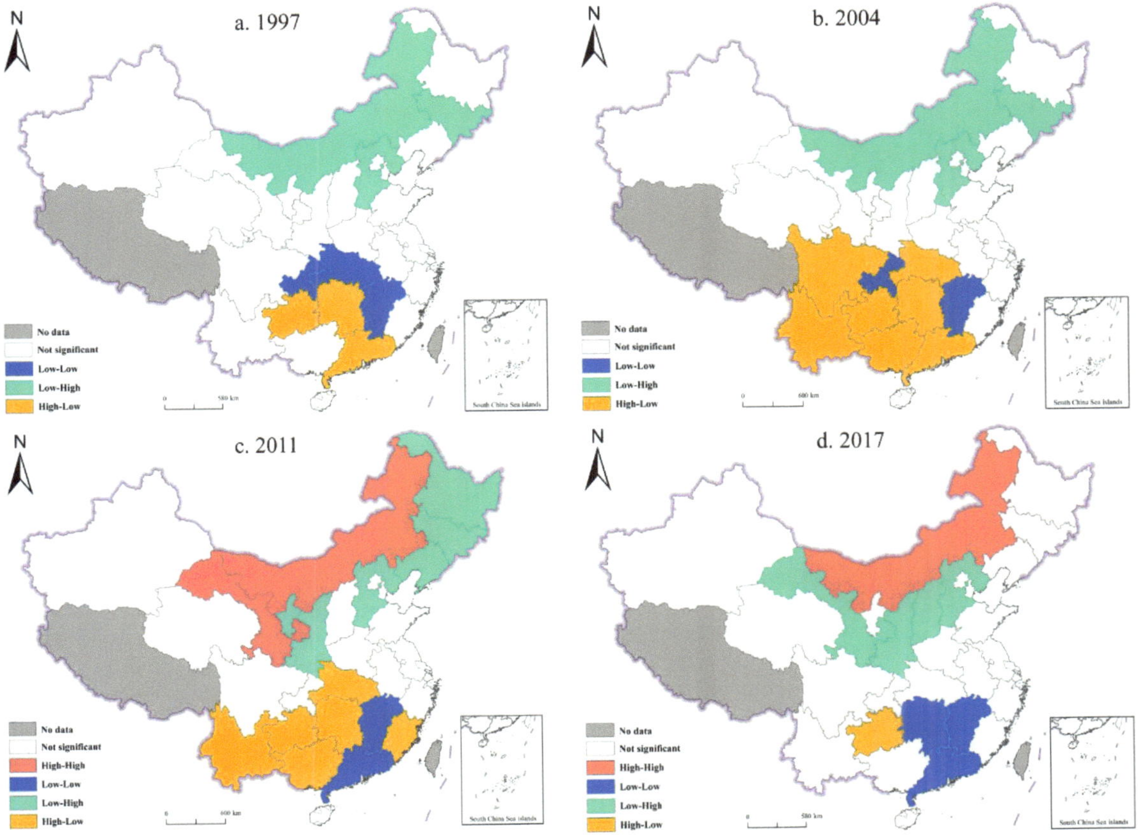

Figure 1. Spatiotemporal agglomeration of renewable energy consumption-carbon emissions (Notes: This map is based on the standard map with the approval number of GS (2020)4632).

3.1.3. Standard Deviation Ellipse and Center of Gravity Evolution

As shown in Figure 2, the standard deviation ellipses of carbon emissions are characterized by developing along the east-west (*X*) axis and spreading along the north-south (*Y*) axis. Their spatial distributions show a "northeast-southwest" trend on the whole. During the study period, the coverage area of the carbon emissions standard deviation ellipse gradually decreased, from 3,721,807 km^2 (in 1997) to 3,619,345 km^2 (in 2017), indicating that the growth rate of carbon emissions inside the standard deviation ellipse is greater than that outside. Compared with the eastern provinces, carbon emissions in the western provinces grew faster, mainly due to the abounding natural resources, and high proportion of heavy industry in the industry in western regions (except for Yunnan, Guangxi). The standard deviation along the *Y*-axis decreased during 1997–2017, while it fluctuated in the *X*-axis direction, indicating that the spatial distribution of carbon emissions tends to be concentrated in the north-south direction. The main reason may be that the growth rate of carbon emissions of provinces such as Sichuan, Guizhou and Hunan decreased through replacing fossil energy gradually by clean energy.

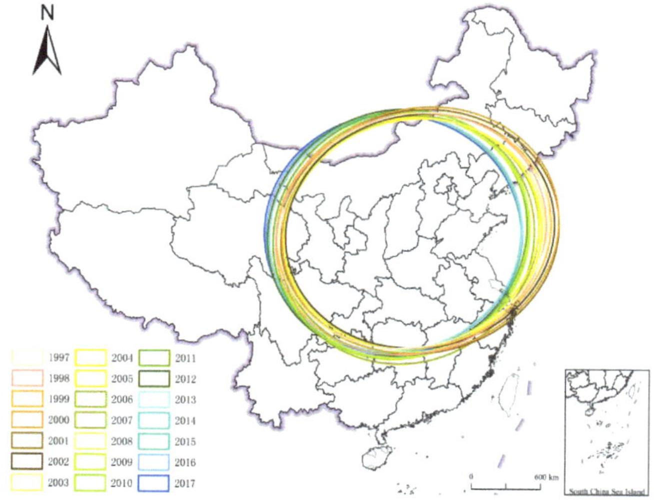

Figure 2. Standard deviation ellipse of per capita carbon emissions.

China's renewable energy consumption showed a "northwest-southeast" spatial distribution pattern during 1991–2007 (Figure 3). The azimuths of the standard deviation ellipses were large and fluctuated with time, indicating that the spatial pattern of renewable energy consumption was in an unstable state. In 2009, the standard deviation ellipse increased significantly along the *Y*-axis, indicating that the spatial distribution of renewable energy along the *Y*-axis tends to be scattered. The main reason is that the national renewable energy consumption level in 2009 had risen to a certain extent, which also echoes the aforementioned—that the spatial autocorrelation of renewable energy did not pass the significance test in 2009. From 1999 to 2003, the standard deviation ellipse gradually shortened in the direction of the *X*-axis, indicating a shrinking trend of renewable energy in the east-west direction; while from 2004 to 2014, the fluctuation of the standard deviation of the *Y*-axis was small, which means there was no obvious contraction or expansion in the north-south direction.

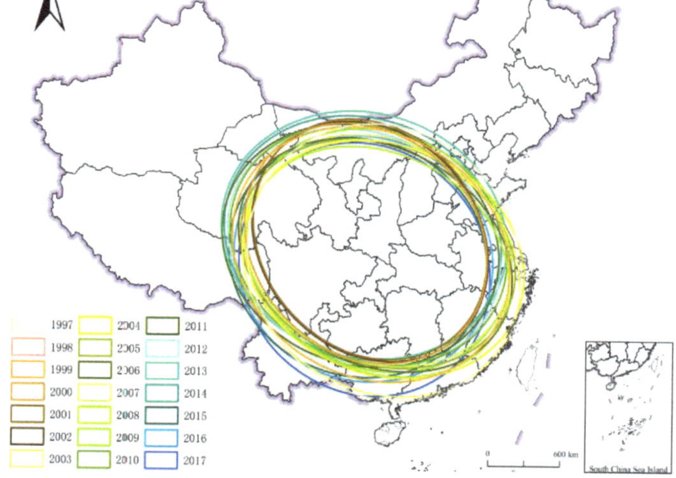

Figure 3. Standard deviation ellipse of per capita renewable energy consumption.

It can be seen from Figure 4 that the center of gravity of China's per capita carbon emissions is mainly distributed longitudinally between 111.28 and 114.06° E and latitudinally between 35.73 and 36.41° N during 1997–2017. From 1997 to 2005, the center of gravity of China's per capita carbon emissions generally shifted to the southwest, from 1998 (113.43° E, 36.41° N) to 2005 (113.15° E, 36.22° N). After 2005, the center of per capita carbon emissions showed a trend of moving to the southwest firstly and then to the northwest. Changes of carbon emissions at latitude is more obvious than that at longitude, indicating that China's carbon emissions show large changes in the east-west direction. This is directly related to the adjustment of the national strategy, which is consistent with the industrial gradient transfer effect caused by China's eastern, central and western development strategies, indicating that structural adjustment will directly lead to changes in carbon emissions.

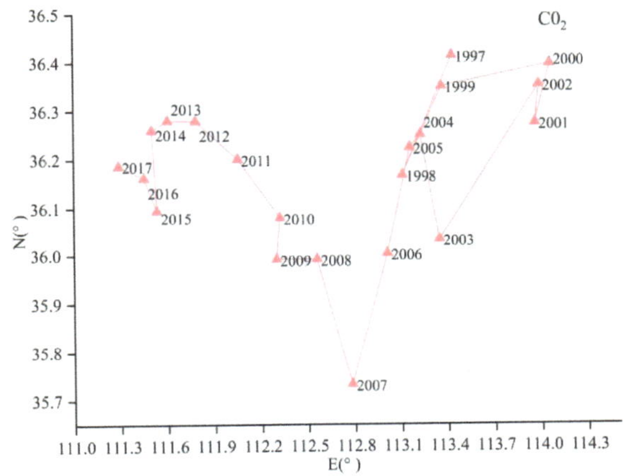

Figure 4. The evolution of the center of gravity of per capita carbon emissions.

During the research period, the center of gravity of China's per capita renewable energy consumption was between 106.71–109.06° E, 31.02–33.47° N, and showed a relatively large offset in most years (Figure 5). The moving trend of this center of gravity is mainly divided into two stages: the center of gravity of per capita renewable energy consumption fluctuated and moved to the northwest from 1997 to 2013; and moved to the south from 2013 to 2017. The moving tendency to the south is mainly due to the increased utilization of hydro, wind and biomass energy in the southern regions (Guangdong, Guangxi, Yunnan, Guizhou).

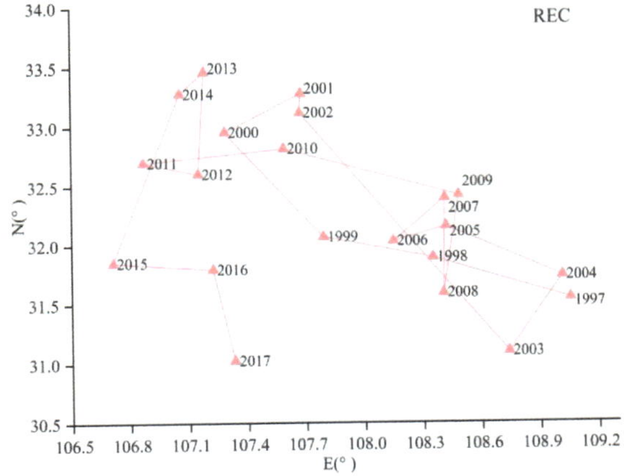

Figure 5. The evolution of the center of gravity of per capita renewable energy consumption.

Form the analysis above, we confirm that there is an obvious spatial agglomeration between renewable energy consumption and carbon emissions, and their distribution within regions is not balanced. The characteristics of "High-Low" and "Low-High" agglomeration indicate that renewable energy consumption has a spatial spillover effect on carbon emissions. The center of gravity of these two variables generally shows the same direction shifting, indicating that renewable energy has a significant impact on carbon emissions. In order to fully understand the impact of renewable energy consumption on carbon emissions, the spatial econometric model is employed for further analysis.

3.2. Result of Spatial Econometric Estimation

Considering the spatial incorporation of renewable energy consumption and carbon emissions, three classical specifications with panel data are tested: spatial autoregressive model (SAR), spatial error model (SEM) and spatial Durbin model (SDM). The specification of the Queen adjacency (W1) was adopted as spatial weight matrix when carrying out these tests. First, Lagrange multiplier (LM) tests were adopted to determine which model is appropriate [43]. LM tests consist of the error dependence test (LM-error) and the test for endogenous spatially lagged dependent variables (LM-lag). Results showed that both the LM-error (LM = 18.17, p = 0.000) and the LM-lag (LM = 81.11, p = 0.000) test passed the 1% significance level, indicating the necessary need for spatial auto-correlation of disturbances [44]. The robust LM-error and robust LM-lag test were used since both tests above are significant. The robust LM-error (LM = 11.05, p = 0.000) and robust LM-lag (73.99, p = 0.000) test are both significant, indicating that using SAR or SEM singly would have deviations, and therefore SDM is appropriate [38]. Additionally, we also employed the LR-test to testify whether the SDM would not be simplified to the SAR and SEM [45]. Results of the LR-test (in Table 4) indicate that SDM is appropriate here. Second, several tests are employed to decide the form of SDM. The Hausman test is used to make sure the fixed effects or the random effects are suitable [46]. The results (chi-square = 46.35, p-value = 0.000) show that the optimal model is a panel with fixed effects. Generally, we see the spatial fixed effect, the temporal fixed effect and spatial-temporal-fixed effect in SDM. Bayesian information criterion (BIC) and Akaike information criterion (AIC) are employed to determine which effect is appropriate [47]. The lowest AIC and BIC indicators indicate that spatial-temporal-fixed effect is preferred in this study.

Table 4. SDM model analysis results.

Variable	Spatial Fixed Effect	Temporal Fixed Effect	Spatial-Temporal-Fixed Effect
LnGDP	5.621 ***	2.438 ***	5.450 ***
LnGDP^2	−0.191 ***	−0.088 ***	−0.183 ***
LnEI	0.479 ***	0.897 ***	0.508 ***
LnREC	−0.050 ***	−0.075 ***	−0.051 ***
LnPT	−0.170 ***	0.025	−0.154 ***
W * LnGDP	−4.766 ***	1.419	−2.862 ***
W * LnGDP^2	0.214 ***	−0.056	0.136 ***
W * LnEI	0.487 ***	0.449 ***	0.980 ***
W * LnREC	−0.030	−0.018	−0.074
W * LnPT	0.163 **	−0.046	0.147 **
LR test-spatial lag	52.43 ***	33.59 ***	40.05 ***
LR test-spatial error	38.34 ***	41.40 ***	41.50 ***
AIC	307.948	537.501	227.350
BIC	361.296	590.850	280.698

Notes: ***, **, and * indicate significance of the 1%, 5% and 10% level, respectively.

Analysis in Section 3.1 shows that the renewable energy consumption and carbon emissions of China's 30 provinces fluctuated with time, and the unit root test is necessary to verify the stationarity of variables. The panel unit root test has advantages of cross-sectional correlations, rather than unit root tests based on univariate time series or cross-sectional data [38]. Three unit-root tests were used in this paper, including the Levin-Lin-Chu (LLC) test, the Im-Pesaran-Shin (IPS) test, and the Breitung test. It can be seen from the test results in Table 5 that all variables show a stable trend at the 1% significance level after the first-order difference.

Table 5. Panel unit root test.

Variable		LLC	IPS	Breitung
Ln	level	−7.351 ***	−5.866 ***	−4.098 ***
	First difference level	−6.410 ***	−12.686 ***	−13.061 ***
LnGDP	level	−1.991 **	4.267	7.025
	First difference level	−3.462 ***	−6.156 ***	−3.389 ***
LnGDP^2	level	−1.718 **	5.351	7.634
	First difference level	−3.446 ***	−5.961 ***	−3.080 ***
LnEI	level	−2.408 ***	−1.197	1.203
	First difference level	−7.321 ***	−11.843 ***	−10.933 ***
LnREC	level	−2.857 ***	−4.518 ***	−1.411 *
	First difference level	−8.345 ***	−13.302 ***	−11.720 ***
LnPT	level	0.128	−0.451	3.599
	First difference level	−5.398 ***	−11.506 ***	−7.992 ***

Notes: ***, **, and * indicate the significance of the 1%, 5% and 10% level, respectively.

Based on the stationarity test, it is found that the variables after the logarithmic first-order difference are single integral of the same order. The cointegration test is adopted to examine the long-term equilibrium relationship between variables [38,39]. Passing the cointegration test means there is a long-term stable equilibrium relationship between variables, and the variance regression residuals are stable. The Pedroni test and Kao test are used to verify whether the regression result is accurate. Results of both tests (Table 6) reject the null hypothesis at the 1% significance level, indicating that there is a co-integration relationship between panel data and a long-term stable relationship between variables.

Table 6. Panel cointegration test.

Test		Statistic
	Modified Phillips-Perron t	5.485 *
Pedroni	Phillips-Perron t	−8.404 *
	Augmented Dickey-Fuller t	−9.157 *
	Modified Dickey-Fuller t	−5.977 *
	Dickey-Fuller t	−5.269 *
Kao	Augmented Dickey-Fuller t	−5.897 *
	Unadjusted modified Dickey-Fuller t	−8.121 *
	Unadjusted Dickey-Fuller t	−5.993 *

Notes: * indicates the significance of the 1% level.

The Spatial Durbin model contains the total effects of independent variables on dependent variables, both the direct and indirect effects. The direct effect is the influence of the explanatory variable on the carbon emissions of the local province; the indirect effect is the influence of the explanatory variable of other neighboring provinces on carbon emissions in a local province, and the total effects are the sum of direct effects and indirect effects. Results of SDM based on three weight matrices are shown in Tables 7–9.

According to the estimated results in Table 7, the direct effect of REC is −0.050, which means REC is significantly negative for local carbon emissions. This is verified by other studies [20,24,29]. Renewable energy consumption will restrain the growth of local carbon emissions to a certain extent, which means that, with the increase of wind, hydro, solar, nuclear and other renewable energy increase, local carbon emissions will gradually decrease. The indirect effect of REC is −0.064, while not passing the significance test. REC of neighboring regions may have negative effect on local carbon emission while the spillover effect of renewable energy consumption is not obvious. One possible reason

may be the administrative barriers in forms of energy consumption such as electricity [48
It is worth mentioning that the effect of REC is smaller than other explanatory variable
which is consistent with previous studies [49,50]. One possible reason is that renewabl
energy accounts for a small proportion in China's energy structure, which is still dominate
by fossil energy [51], and the impact of renewable energy policies on carbon emission
has a lagging effect [52]. Another possible reason may be insufficient renewable energ
storage technology, since there is abundant wind-generated electricity in China [53]. Th
total effect of REC is −0.114 with 5% significance, indicating the increase in renewabl
energy consumption has an inhibitory effect on the growth of carbon emissions. From th
specific path of the impact of renewable energy on carbon emissions, the most obviou
effect on carbon emission reduction is the proportion of renewable energy consumptior
and renewable energy can inhibit carbon emissions through energy efficiency [54]. T
realize the reduction goal of carbon emissions, it is necessary to effectively increase th
proportion of clean renewable energy in the energy consumption side through incentiv
policies and economic means, such as the tiered electricity price and the peak-valley time
of-use electricity price system, and take multiple measures to promote the consumption o
renewable energy. On this basis, carbon emission reduction can be achieved.

Table 7. Effects of variables on carbon emissions based on W1.

Variable	Excluding REC Variables (M1)			Including REC Variables (M2)		
	Direct Effects	Indirect Effects	Total Effects	Direct Effects	Indirect Effects	Total Effects
$\text{Ln}(GDP)$	5.669 ***	−3.115 ***	2.554 ***	5.516 ***	−3.051 ***	2.465 ***
$\text{Ln}(GDP^2)$	−0.197 ***	0.132 ***	−0.065 **	−0.186 ***	0.141 ***	−0.045
$\text{Ln}(EI)$	0.497 ***	1.030 ***	1.527 ***	0.507 ***	0.901 ***	1.408 ***
$\text{Ln}(PT)$	−0.166 ***	0.131 *	−0.035	−0.154 ***	0.151 **	−0.004
$\text{Ln}(REC)$	–	–	–	−0.050 ***	−0.064	−0.114 **

Notes: ***, **, and * indicate the significance of the 1%, 5% and 10% level, respectively.

Table 8. Effects of variables on carbon emissions based on W2.

Variable	M1			M2		
	Direct Effects	Indirect Effects	Total Effects	Direct Effects	Indirect Effects	Total Effects
$\text{Ln}(GDP)$	3.502 ***	1.223	4.725 ***	3.582 ***	0.755	4.338 ***
$\text{Ln}(GDP^2)$	−0.091 ***	−0.088 *	−0.179 ***	−0.092 ***	−0.066	−0.158 ***
$\text{Ln}(EI)$	0.437 ***	0.725 ***	1.162 ***	0.476 ***	0.627 **	1.103 ***
$\text{Ln}(PT)$	−0.163 ***	0.091	−0.073	−0.150 ***	0.102	−0.048
$\text{Ln}(REC)$	–	–	–	−0.057 ***	−0.019	−0.076

Notes: ***, **, and * indicate the significance of the 1%, 5% and 10% level, respectively.

Table 9. Effects of variables on carbon emissions based on W3.

Variable	M1			M2		
	Direct Effects	Indirect Effects	Total Effects	Direct Effects	Indirect Effects	Total Effects
$\text{Ln}(GDP)$	3.470 ***	−0.272	3.197 ***	3.693 ***	−0.531	3.162 ***
$\text{Ln}(GDP^2)$	−0.092 ***	0.027	−0.065 *	−0.101 ***	0.034	−0.067 *
$\text{Ln}(EI)$	0.580 ***	0.830 ***	1.411 ***	0.600 ***	0.721 ***	1.322 ***
$\text{Ln}(PT)$	−0.202 ***	0.068	−0.134 **	−0.196 ***	0.095	−0.101
$\text{Ln}(REC)$	–	–	–	−0.053 ***	−0.010	−0.062

Notes: ***, **, and * indicate the significance of the 1%, 5% and 10% level, respectively.

As shown in Table 7, the direct effect of *GDP* is positive significantly, indicating that *GDP* has a promoting effect on local carbon emissions, while the indirect effect of *GDP* is negative significantly, indicating that *GDP* has an inhibitory effect on carbon emissions of adjacent regions. As for quadratic economic growth, the relationship between GDP^2 and carbon emissions shows an inverted U-shaped relationship [55,56], indicating that when *GDP* reaches a certain level, carbon emissions will decline with economic growth. The effects of EI are significantly positive, with high energy intensity economic growth during 1997–2017, which is consistent with previous studies [57,58]. The direct effect of PT is −0.154 indicating that technological progress has an obvious inhibitory effect on local carbon emissions. In other words, China's technological innovation is still in the commercialization stage on the whole, and its role is still mainly limited to local regions, while the spatial spillover effect of PT is positive at the 5% significance level, indicating that increase on the number of patents in local region may lead to an increase in carbon emissions of adjacent areas. This means that technological benefits are more likely to happen in local regions, a situation also found in a previous study [59]. One possible reason is that when technological innovation promotes the reduction of local carbon emissions, low-tech industries would be transferred to adjacent areas with low environmental regulations, resulting in increased carbon emissions in other regions.

4. Discussion

4.1. Results Based SDM without REC

A previous study [22] found that, when adding renewable energy consumption in the ARDL model, the inversed U-shaped EKC hypothesis is supported in the long run, while this did not happen without renewable energy consumption. Thus, we constructed a spatial Durbin model (SDM) without renewable energy consumption (M1) to verify whether REC affected other variables. Our results show the inversed U-shaped EKC hypothesis is both supported under these two types of SDM, which is contrary to Chen et al. [22]. A possible reason may be due to the fact that the variables and data selected in the empirical analysis are different. This result is similar to previous studies that verified an EKC curve in China [60,61].

Comparing the effects of other variables in M1 and M2, under the W1 weight matrix, the elasticity coefficient of economic growth on local carbon emissions in M2 is lower than that in M1, weakening the promotion effect of economic growth on local carbon emissions and the inhibition effect on carbon emissions of adjacent provinces and regions. The negative effects of PT on local carbon emissions slightly increased, while the positive effects on adjacent regions' carbon emissions slightly decreased in M1 compared with that in M2. As for EI, the direct and indirect effects on carbon emissions decreased less in M1 compared with that in M2.

In general, in the case of three spatial weight matrices, results in M1 are slightly changed, indicating that the results in Section 3.2 are reliable.

4.2. Robustness Test

Since the spatial weight matrix is a substantial part of the spatial Durbin model, different spatial weight matrices need to be employed for verifying the stability of our empirical results. We examined the spatial weight matrix based on inverse distance and economic distance, which are described in Section 2.4. The results of robustness tests are listed in Tables 7–9. The estimations of renewable energy consumption, *GDP*, GDP^2, EI and PT of both M1 and M2 are similar in coefficients and significance under three spatial weight matrices. The estimated results of all variables are not significantly changed by the spatial weight matrix. In a word, the estimations are robust.

4.3. Policy Recommendations

Renewable energy consumption will become the main energy consumption in future considering increasingly serious climate change [62], and China need to increase the proportion of renewable energy in its energy structure. On the one hand, this energy structure should be improved gradually by establishing an energy price system with fossil energy tax and renewable energy subsidies. On the other hand, energy transformation should be promoted through diverse ways, such as taking advantage of the abundant reserves of renewable energy such as hydropower, wind energy and biomass energy in the western region to build a large-scale renewable energy base, or accelerating the construction of large-scale wind power photovoltaic base projects focusing on desert areas.

Regional cooperation in energy transition should be promoted, due to the diverse development levels of different provinces in China. Since the center of gravity of renewable energy consumption and carbon emissions is moving southwest, western provinces with advantages in substantial renewable energy resources and a developing economy need establish a long-term and stable partnership of cooperation with provinces in coastal regions which have leading technologies and a developed economy, to realize advantages of complementarity and to promote common development. An effective compensation mechanism based on a market-oriented economy, such as a carbon emission trading market will realize this common development.

Investment and subsidies in technological innovation for renewable energy consumption should be increased. The level of key technology development in China lags behind developed countries on the whole [63]. As for provinces with abundant natural reserves such as Yunnan and Guizhou, it is necessary to increase research on energy alternative technologies [18]. As for provinces with renewable energy consumption, such as Qinghai and Xizang, it is necessary to improve renewable energy utilization by developing new energy storage technology.

4.4. Research Deficiencies and Prospects

The limitation of this study is that it mainly focuses on the impact of renewable energy consumption on carbon emissions, with less attention to other factors and the role of co-occurring factors that would influence aggregate emissions. Some socio-economic factors, such as environmental regulation and industrial structure, are neglected. In fact, the environmental regulations formulated by government or the market could promote energy efficiency [64], and industrial structure optimization is essential to carbon performance [65]. Future research can focus on the effects of renewable energy consumption on these socio-economic factors and the role of co-occurring factors on carbon emissions. Besides, when using the number of patents to represent technological progress, this may overestimate the actual level of technological progress, and further research can focus on the representation of technological progress. In addition, this study constructed a research framework from a spatial perspective to clarify the relationship between carbon emissions, renewable energy consumption and other influencing factors. Due to the limitation on renewable energy consumption data, the research was analyzed at provincial scale. In fact, the prefecture-level city is considered as the core of low-carbon development [66]. Furthermore, adjacent cites will have more impact on space through the liquidity of elements, such as capital and labor [67]. Future research can focus on the effects of renewable energy consumption on carbon emissions at the prefecture level or the country level.

5. Conclusions

From the provincial perspective, this study analyzes the spatial correlation and spatial distribution pattern of renewable energy consumption and carbon emissions of China's 30 provinces from 1997 to 2017, and explores the impact of renewable energy consumption on carbon emissions based on the spatial-temporal-fixed SDM. Conclusions are as follows.

First, carbon emissions and renewable energy consumption presented statistically significant spatial autocorrelations at the provincial level from 1997 to 2017. Bivariate global Moran'I values confirmed the spatial negative correlation between renewable energy consumption and carbon emissions in most years. According to bivariate local spatial LISA agglomeration maps, negative spatial correlation was the main spatial association among provinces exhibiting spatial clustering. "High-Low" areas are mainly concentrated in southern provinces, while "Low-High" areas are mainly concentrated in northeastern provinces. Second, the center of gravity of carbon emissions was located in Shanxi province and presented a "northeast-southwest" trend on the whole. The center of gravity of renewable energy consumption was unstable, and showed a northwest trend during

1997–2013, and a southerly trend during 2013–2017. The center of gravity for renewable energy consumption moved southwest in the same direction as carbon emissions.

Third, results of the spatial Durbin model indicate that renewable energy consumption has a significant inhibitory effect on the growth of carbon emissions of a local region. Specifically, a 1% increase in renewable energy consumption in a region will reduce carbon emissions by 0.05% in that region. Renewable energy consumption has not played a significant role in reducing carbon emissions as seen by the statistically smaller effect, along its small proportion of the total energy structure of China. Form the spatial perspective, renewable energy consumption also has negative effect on carbon emissions of neighboring provinces, but this negative effect is not significant statistically.

Author Contributions: Conceptualization—Y.S. and Y.C.; methodology—Y.S. and M.D.; software—M.D.; formal analysis—Y.S.; data curation, M.D. and C.L.; writing—original draft preparation, M.D. and L.W.; writing—Y.S. and Y.C. All authors have read and agreed to the published version of the manuscript.

Funding: This research was funded by The National Natural Science Foundation of China, grant number 41901239; Humanities and Social Sciences Research Project of the Ministry of Education, grant number 20YJC790017. Young Elite Scientists Sponsorship Program by Henan Association for Science and Technology, grant number 2022HYTP027. GEIGC Science and Technology Project SGGEIG00JYJS2100056.

Institutional Review Board Statement: Not applicable.

Informed Consent Statement: Not applicable.

Data Availability Statement: The data that support the findings of this study are available from the corresponding author upon reasonable request.

Conflicts of Interest: The authors declare no conflict of interest.

Notes

1. Chinese Government Network. Xi Jinping delivers an important speech at the general debate of the 75th UN General Assembly. (2020-09-22) [2020-11-20].
2. Grasp the key period and window period of carbon peak in the 14th five-year plan. China Electric Power News, 2021-03-23(1).

References

1. Yang, S.M.; Shen, C.L.; Wang, F. Exploring Breakthrough in Carbon Peak and Carbon Neutrality and Promoting High-Quality Development of China's Petrochemical Industry. *Strategy Investig.* **2022**, *Z3*, 20–22.
2. Polloni-Silva, E.; Silveira, N.; Ferraz, D.; de Mello, D.S.; Moralles, H.F. The drivers of energy-related CO_2 emissions in Brazil: A regional application of the STIRPAT model. Environmental science and pollution. *Res. Int.* **2021**, *28*, 51745–51762.
3. Cheng, Y.; Wang, Z.; Zhang, S.; Ye, X.; Jiang, H. Spatial econometric analysis of carbon emission intensity and its driving factors from energy consumption in China. *J. Geogr. Sci.* **2013**, *68*, 1418–1431.
4. Lin, X.Q.; Bian, Y.; Wang, D. Spatiotemporal evolution characteristics and influencing factors of industrial carbon emission efficiency in Beijing-Tianjin-Hebei region. *Econ. Geogr.* **2021**, *41*, 187–195.
5. Su, K.; Chen, Y.H.; Fan, S.S.; Zhang, M.M. Influencing factors and reduction mechanism of carbon emissions at city-range: An empirical study on Fujian province. *China Environ. Sci.* **2019**, *39*, 859–867.
6. Yan, Y.M.; Wang, Z.; Wu, L.Y.; Liu, C.X. Analysis of the determinants of carbon emission intensity on regional differences. *Acta Sci. Circumstantiae* **2016**, *36*, 3436–3444.
7. Peng, L.L.; Li, N.; Zheng, Z.Y.; Li, F.; Wang, Z. Spatial-temporal heterogeneity of carbon emissions and influencing factors on household consumption of China. *China Environ. Sci.* **2021**, *41*, 463–472.
8. Zhao, J.C.; Liu, Q.Q. Examining the driving factors of urban residential carbon intensity using the LMDI method: Evidence from China's county-level cities. *Int. J. Environ. Res. Public Health* **2021**, *18*, 3929. [CrossRef]
9. Han, Y.Y.; Pi, H.J.; Shi, Z.N. Study on measurement and influencing factors of industrial CO_2 emission in Beijing-Tianjin-Hebei Region. *World Reg. Stud.* **2020**, *29*, 140–147.
10. Fang, Y.B.; Wang, L.M.; Mou, C.F. Determinants of spatial disparities of petroleum terminal utilization carbon emissions in China. *Resour. Sci.* **2017**, *39*, 2233–2246.
11. Deng, J.X.; Liu, X.; Wang, Z. Characteristics analysis and factor decomposition based on the regional difference changes in China's CO_2 emission. *J. Nat. Resour.* **2014**, *29*, 189–200.

12. Wang, Y.; He, Y.F. Spatiotemporal dynamics and influencing factors of provincial carbon emissions in China. *World Reg. Stu* **2020**, *29*, 512–522.
13. Mo, H.B.; Wang, S.J. Spatio-temporal evolution and spatial effect mechanism of carbon emission at county level in the Yello River Basin. *Sci. Geogr. Sin.* **2021**, *41*, 1324–1335.
14. Huang, L.L.; Wang, Y.; Zhang, C.; Huang, Y.M. A spatial-temporal decomposition analysis of CO_2 emissions in Fujian Southeas triangle region. *China Environ. Sci.* **2020**, *40*, 2312–2320.
15. Ma, X.J.; Chen, R.M.; Dong, B.Y.; Niu, X.Q. Factor decomposition and decoupling effect of China's industrial carbon emission *China Environ. Sci.* **2019**, *39*, 3549–3557.
16. Lu, N.; Wang, W.; Wang, M.; Zhang, C.; Lu, H. Breakthrough low-carbon technology innovation and carbon emissions: Direc and spatial spillover effect. *China Popul. Resour. Environ.* **2019**, *29*, 30–39.
17. Xu, L.; Du, H.R.; Zhang, X.L. Driving forces of carbon dioxide emissions in China's cities: An empirical analysis based on th geodetector method. *J. Clean. Prod.* **2021**, *287*, 125169. [CrossRef]
18. Wang, S.J.; Xie, Z.H.; Wang, Z.H. The spatiotemporal pattern evolution and influencing factors of CO_2 emissions at the count level of China. *Acta Geogr. Sin.* **2021**, *76*, 3103–3118.
19. Zhang, H.M.; Yuan, P.F.; Zhu, Z.S. City population size, industrial agglomeration and CO_2 emission in Chinese prefectures. *Chin Environ. Sci.* **2021**, *41*, 2459–2470.
20. Jia, J.; Lei, J.; Chen, C.; Song, X.; Zhong, Y. Contribution of renewable energy consumption to CO_2 Emission mitigation: comparative analysis from a global geographic perspective. *Sustainability* **2021**, *13*, 3853. [CrossRef]
21. Naz, S.; Sultan, R.; Zaman, K.; Aldakhil, A.M.; Nassani, A.A.; Abro, M.M.Q. Moderating and mediating role of renewable energ consumption, FDI inflows, and economic growth on carbon dioxide emissions: Evidence from robust least square estimato *Environ. Sci. Pollut. Res. Int.* **2019**, *26*, 2806–2819. [CrossRef] [PubMed]
22. Chen, Y.L.; Wang, Z.; Zhong, Z.Q. CO_2 emissions, economic growth, renewable and non-renewable energy production anc foreign trade in China. *Renew. Energy* **2019**, *131*, 208–216. [CrossRef]
23. Zhang, M.; Yang, Z.; Liu, L.; Zhou, D. Impact of renewable energy investment on carbon emissions in China—An empirical stud using a nonparametric additive regression model. *Sci. Total Environ.* **2021**, *785*, 147109. [CrossRef] [PubMed]
24. Chen, Y.; Wang, Z.; Zhong, Z. Renewable energy consumption, foreign trade, carbon emissions in China. *Soft Sci.* **2017**, *31*, 49–52
25. Qiu, Q.Q.; Li, S.; Xia, Q. Renewable energy consumption-energy-related CO_2 emission nexus for China. *Ecol. Econ.* **2017**, *33* 19–23.
26. Musah, M.; Kong, Y.; Mensah, I.A.; Antwi, S.K.; Donkor, M. The link between carbon emissions, renewable energy consumption and economic growth: A heterogeneous panel evidence from West Africa. *Environ. Sci. Pollut. Res.* **2020**, *27*, 28867–28889 [CrossRef]
27. Radmehr, R.; Henneberry, S.R.; Shayanmehr, S. Renewable energy consumption, CO_2 emissions, and economic growth nexus: A simultaneity spatial modeling analysis of EU countries. *Struct. Chang. Econ. Dyn.* **2021**, *57*, 13–27. [CrossRef]
28. Oppong, A.; Jie, M.; Acheampong, K.N.; Sakyi, M.A. Variations in the environment, energy and macroeconomic interdependencies and related renewable energy transition policies based on sensitive categorization of countries in Africa. *J. Clean. Prod.* **2020** *255*, 119777. [CrossRef]
29. Wang, L.; Zhao, T. Dynamic relationship among consumption of renewable energy, economic growth and carbon emission in China. *Technol. Econ.* **2013**, *32*, 99–104.
30. Cliff, A.D.; Ord, J.K. *Spatial Processes: Models & Applications*; Taylor & Francis: Abingdon, UK, 1981.
31. Zhao, M.; Lv, L.; Wu, J.; Wang, S.; Zhang, N.; Bai, Z.; Luo, H. Total factor productivity of high coal-consuming industries and provincial coal consumption: Based on the dynamic spatial Durbin model. *Energy* **2022**, *251*, 123917. [CrossRef]
32. Anselin, L.; Syabri, I.; Smirnov, O. Visualizing multivariate spatial correlation with dynamically linked windows. In Proceedings of the CSISS Workshop on New Tools for Spatial Data Analysis, Santa Barbara, CA, USA, 10–11 May 2002.
33. Zhang, C.; Gao, J.; Ding, P.Y. Impact of haze on the intention to visit China of overseas tourists: Based on destination image and risk perception theories. *Tour. Trib.* **2017**, *32*, 58–67.
34. Feng, Z.Y.; Shi, R.H. Spatio-temporal features and the association of ground-level PM2.5 concentration and its emission in China *J. Geo-Inf. Sci.* **2021**, *23*, 1221–1230.
35. Song, Y.; Zhang, M. Research on the gravity movement and mitigation potential of Asia's carbon dioxide emissions. *Energy* **2019** *170*, 31–39. [CrossRef]
36. Lee, L.F.; Yu, J. Identification of spatial Durbin panel models. *J. Appl. Econom.* **2015**, *31*, 133–162. [CrossRef]
37. Lesage, J.P.; Pace, R.K. Spatial econometric modeling of origin-destination flow. *Journal of Regional Science* **2010**, *48*, 941–967 [CrossRef]
38. Chica-Olmo, J.; Sari-Hassoun, S.; Moya-Fernández, P. Spatial relationship between economic growth and renewable energy consumption in 26 European countries. *Energy Econ.* **2020**, *92*, 104962. [CrossRef]
39. Li, Z.; Zhou, Y.; Li, K.; Xiao, H.; Cai, Y. The spatial effects of city-level water-energy nexus: A case study of Hebei Province, China *J. Clean. Prod.* **2021**, *310*, 127497. [CrossRef]
40. Wang, Z.B.; Liang, L.W.; Wang, X.J. Spatio-temporal evolution patterns and influencing factors of PM2.5 in Chinese urban agglomerations. *Acta Geogr. Sin.* **2019**, *74*, 2614–2630.

41. Ye, Z.; Zhang LN, X. The spatial spillover effect of pollution and emission reduction in industrial structure adjustment: Estimation based on spatial Durbin model. *Ecol. Econ.* **2021**, *37*, 178–184.
42. Yang, J.L.; Chen, Z.Z. Environmental regulation, new-type urbanization and regional economic growth: An empirical study based on the spatial Durbin model of 30 provinces in China. *Ecol. Econ.* **2021**, *37*, 165–171.
43. Breusch, T.S.; Pagan, A.R. The lagrange multiplier test and its applications to model specification in econometrics. *Rev. Econ. Stud.* **1980**, *47*, 239–253. [CrossRef]
44. Anselin, L.; Rey, S. Properties of tests for spatial dependence in linear regression models. *Geogr. Anal.* **1991**, *23*, 112–131. [CrossRef]
45. Elhorst, J.P. Linear spatial dependence models for cross-section data. In *Spatial econometrics*; Springer: Berlin, Heidelberg, 2014.
46. Hausman, J.A. Specification tests in econometrics. *Econometrica* **1978**, *46*, 1251–1271. [CrossRef]
47. Kopczewska, K.; Kudła, J.; Walczyk, K. Strategy of spatial panel estimation: Spatial spillovers between taxation and economic growth. *Appl. Spat. Anal. Policy* **2017**, *10*, 77–102. [CrossRef]
48. Ding, Z.L. Report of the law enforcement inspection team of the Standing Committee of the National People's Congress on inspecting the implementation of the renewable energy law of the people's Republic of China. In Proceedings of the 15th meeting of the Standing Committee of the 13th National People's Congress, Beijing, China, 14 December 2019.
49. Wang, L.; Zhao, T. The dynamic relationship between China's renewable energy consumption, economic growth and carbon emissions. *Technoeconomics* **2013**, *11*, 99–104.
50. Wang, L.P. Research on the dynamic relationship between China's renewable energy consumption and carbon emissions based on ARDL model. *Resour. Policy* **2022**, *77*, 102764. [CrossRef]
51. Lin, B.Q.; Moubarak, M. Renewable energy consumption—Economic growth nexus for China. *Renew. Sustain. Energy Rev.* **2014**, *40*, 111–117. [CrossRef]
52. Fan, Q.F.; Xing, X.F. Research on the impact of renewable energy policies on carbon emissions. *East China Econ. Manag.* **2022**, *36*, 1–11.
53. Dong, C.; Qi, Y.; Dong, W.; Lu, X.; Liu, T.; Qian, S. Decomposing driving factors for wind curtailment under economic new normal in China. *Appl. Energy* **2018**, *217*, 178–188. [CrossRef]
54. Zhao, Z.Y.; Ma, X. The path and impact of renewable energy power on carbon emissions: Mediating effect test based on interprovincial data. *East China Econ. Manag.* **2022**, *36*, 65–72.
55. Wang, S.J.; Huang, Y.Y. Spatial spillover effect and driving forces of carbon emission intensity at the city level in China. *Acta Geogr. Sin.* **2019**, *74*, 1131–1148. [CrossRef]
56. Chen, Z.; Wu, S.; Ma, W.; Liu, X.; Cai, B.; Liu, J.; Jia, X.; Zhang, M.; Chen, Y.; Xu, L.; et al. Driving forces of carbon dioxide emission for China's cities: Empirical analysis based on extended STIRPAT Model. *China Popul. Resour. Environ.* **2018**, *28*, 45–54.
57. Wang, X.G. The Analysis of Spatiotemporal Dynamics of Carbon Dioxide Emissions from Energy Consumption at Different Scales in China. Ph.D. Thesis, The Northwest A&F University, Yangling, China, 2019.
58. Zhou, J.P. Panel data analysis on relation between per capital *GDP*, energy intensity and carbon emissions. *Environ. Sci. Manag.* **2013**, *38*, 165–167.
59. Li, J.Y.; Li, S.S. Energy investment, economic growth and carbon emissions in China-empirical analysis based on spatial Durbin model. *Energy Policy* **2020**, *140*, 111425. [CrossRef]
60. Ummalla, M.; Samal, A. The impact of natural gas and renewable energy consumption on CO_2 emissions and economic growth in two major emerging market economies. *Environ. Sci. Pollut. Res.* **2019**, *26*, 20893–20907. [CrossRef] [PubMed]
61. Jiang, Q.Q.; Khattak, S.I.; Rahman, Z.U. Measuring the simultaneous effects of electricity consumption and production on carbon dioxide emissions (CO_2e) in China: New evidence from an EKC-based assessment. *Energy* **2021**, *229*, 120616. [CrossRef]
62. Destek, M.A.; Manga, M. Technological innovation, financialization, and ecological footprint: Evidence from BEM economies. *Environ. Sci. Pollut. Res.* **2021**, *28*, 21991–22001. [CrossRef]
63. UNEP. Global Trends in Renewable Energy Investment. Available online: https://www.fs-unep-centre.org/global-trends-in-renewable-energy-investment-2020/ (accessed on 10 June 2020).
64. Guo, R.; Yuan, Y.J. Different types of environmental regulations and heterogeneous influence on energy efficiency in the industrial sector: Evidence from Chinese provincial data. *Energy Policy* **2020**, *145*, 111747. [CrossRef]
65. Chen, Y.; Wang, M.; Feng, C.; Zhou, H.; Wang, K. Total factor energy efficiency in Chinese manufacturing industry under industry and regional heterogeneities. *Resour. Conserv. Recycl.* **2020**, *168*, 105255. [CrossRef]
66. Gu, C.; Tan, Z.; Liu, W.; Yu, T.; Han, Q.; Liu, H.; Dai, Y.; Liu, Z.; Zheng, S. A study on climate change, carbon emissions and low-carbon city planning. *Urban Plan. Forum* **2009**, *3*, 38–45.
67. Xu, B.; Lin, B. Does the high–tech industry consistently reduce CO_2 emissions? results from nonparametric additive regression model. *Environ. Impact Assess. Rev.* **2017**, *63*, 44–58. [CrossRef]

land

MDPI

Article

Urbanization Influences CO_2 Emissions in the Pearl River Delta: A Perspective of the "Space of Flows"

Yabo Zhao [1,2], Ruiyang Chen [2], Tong Sun [2], Ying Yang [3], Shifa Ma [2], Dixiang Xie [2], Xiwen Zhang [2] and Yunnan Cai [2,*]

1 Key Laboratory of Urban Land Resources Monitoring and Simulation, Ministry of Natural Resources, Shenzhen 518034, China
2 School of Architecture and Urban Planning, Guangdong University of Technology, Guangzhou 510090, China
3 Data Management Center of Planning and Natural Resources of Shenzhen, Shenzhen 518034, China
* Correspondence: caiyunnan@gdut.edu.cn; Tel.: +86-020-3762-7563

Abstract: As the largest carbon emitter in the world, China is facing increasing challenge to reduce CO_2 emissions. Given this issue, exploring the influencing factors is of great significance for scientific low-carbon emission policymaking. Although previous literature has explored the effects of urbanization on CO_2 emissions, the impact of the space of flow on urban carbon emissions have been less explored. Due to the increasing connection between cities, its impact on urban carbon emissions cannot be ignored. Thus, this paper takes the space of flows into account as an aspect of urbanization to supplement the existing literature and empirically examines the multiple effects of urbanization on CO_2 emissions in the Pearl River Delta (PRD) urban agglomeration. By using a STIRPAT model, statistical data, and web crawler data, we examined impacts of different types of urbanization on CO_2 emissions. Our empirical results show that: (1) Within the PRD urban agglomeration, urban linkage intensity is strongly connected to urban socioeconomic growth, establishing a geographical structure with Guangzhou and Shenzhen as the double core. (2) Our results show that urbanization exerts two opposite effects on CO_2 emissions: positively connects carbon emissions with population urbanization, integrated urban linkage flow, and energy intensity, whereas economic urbanization and social urbanization are shown to be negatively correlated. However, spatial urbanization has no significant positive effect on urban CO_2 emissions. (3) It is worth noting that urban linkage flows are the second most important factor affecting urban carbon emissions after economic urbanization. Our study could formulate effective planning suggestions for future CO_2 emission reduction paths and development modes in the PRD.

Keywords: space of flows; urbanization; carbon emissions; spatial economic models; Pearl River Delta urban agglomeration

Citation: Zhao, Y.; Chen, R.; Sun, T.; Yang, Y.; Ma, S.; Xie, D.; Zhang, X.; Cai, Y. Urbanization Influences CO_2 Emissions in the Pearl River Delta: A Perspective of the "Space of Flows". *Land* **2022**, *11*, 1373. https://doi.org/10.3390/land11081373

Academic Editor: Yan Li

Received: 7 July 2022
Accepted: 11 August 2022
Published: 22 August 2022

Publisher's Note: MDPI stays neutral with regard to jurisdictional claims in published maps and institutional affiliations.

1. Background of the Study

In promoting soaring energy consumption levels, processes of rapid urbanization raise a range of environmental concerns, not least with respect to the radical increases that have been witnessed in CO_2 emissions globally. Increasing levels of atmospheric greenhouse gases—particularly CO_2—constitute the major impact factor in global warming, which now poses a great threat to nature and the planet [1]. The issue of global warming has attracted widespread attention around the world.

Since 2006, China has been the world's largest carbon emitter and is considered a region for international carbon reduction efforts [2]. According to the "Statistical Review of World Energy 2021," which is published by Beyond Petroleum (BP), China's carbon emissions increased from 7.71 billion tons in 2009 to 9.90 billion tons in 2020, an increase of 30.7% which places the country first in the world in terms of emissions growth [3]. In order to demonstrate its determination to achieve coordinated eco-environment and economic development, the Chinese Government has committed to reductions to the extent that "CO_2 emissions per unit of GDP in 2020 will be reduced by 40–45% compared to 2005" [4]. It is not only emissions that have risen dramatically in recent decades: China's urbanization level has grown rapidly, increasing from 17.9% in 1978 to 63.89% in 2020, with an average annual growth rate of more than 1%. Rapid urbanization of this kind brings with it a series of problems, including an intensification of production and living activities that has resulted in increased fossil energy consumption and thus increased carbon emissions [5,6]. Recognizing that urbanization constitutes the core unit of regional economic and social development, the question of how to achieve carbon emission reduction while ensuring continued, albeit low-carbon, urban development has become a focal issue in China. The relationship between urbanization and CO_2 emissions has been extensively discussed by scholars around the world [7]. Previous findings have shown that urbanization has different degrees and types of impacts on energy consumption and CO_2 emissions through mechanisms that operate at different scales and in different regions [8–11]. The impact of urbanization on CO_2 emissions also varies according to its rate and quality [12,13]. Thus, further study of urban carbon emissions is imperative to meeting this challenge: in fact, if China is to meet its carbon reduction targets, the mechanism that links urbanization to carbon emissions must be uncovered. This knowledge would also be invaluable in guiding international responses and cooperation on matters of climate change.

The issue of urban carbon emissions has been extensively discussed by scholars, and this research has taken a range of different directions. The existing literature that addresses Chinese emissions can be divided into a number of categories, including studies which have addressed, variously: the impact of urbanization on carbon emissions [14,15], the drivers and mechanisms of urban carbon emissions [16,17], the analysis and prediction of carbon emission scenarios [18,19], the path of carbon emission reduction and policy proposals [20,21], the effect of energy consumption and type on carbon emissions in China [22,23], the externalities problems from carbon emission [24], and regional differences in carbon emission efficiency between Chinese cities [25,26].

When considering this body of work, we note that whilst identifying the driving mechanisms and factors behind carbon emissions is key to formulating emission reduction policies and implementing scenario simulations, no uniform conclusions have yet been reached. This is not to say that studies have not addressed the relationship between different drivers and carbon emissions—for example, we note that Solarin et al. [27] found a stable long-term equilibrium relationship between the level of urbanization and carbon emissions, and previous research has shown that the driving factors of carbon emissions include population structure [28], GDP [29], industrialization [30], energy consumption [31], urbanization level [32], and land-use change [33]. The study methods employed include least order squares [34], the STIRPAT model [35], the LMDI model [36], vector autoregressive models [37], geographically weighted regression models [38], and multiplier effects [39]. The action pathway of carbon emission increases and the countermeasure of carbon emission reductions have both been explored, among others, by Tan et al. [40], who predicted the CO_2 reduction potential of Chongqing and analyzed its low-carbon transition path, proposing that improving technology, increasing energy efficiency, optimizing energy input structures, and adjusting the industrial structure are the most effective strategies in achieving Chongqing's carbon emission reduction target. Furthermore, Zhang et al. [41] empirically investigated the impact of fiscal decentralization on environmental functional mechanisms using panel data from 29 Chinese provinces; these scholars found that the impact of fiscal decentralization on environmental policies significantly varies across geographic regions and municipalities, as a result recommending that improvements be made to the institutional environment by incorporating the reduction of emissions into the local government performance assessment system.

The relationship between urbanization and carbon emissions in urban agglomerations is becoming a hot topic in urban geography research. For example, Wang et al. [42] found that urbanization affects CO_2 emissions in two, albeit opposite, ways: while land urbanization, economic urbanization, and social urbanization were shown to exert positive effects on emissions, population urbanization had a negative effect on CO_2 emissions. It is worth noting that the industrial structure and social consumption were found by these scholars to have exerted a positive but not significant effect on CO_2 emissions in the Pearl River Delta. The significant positive effect of energy intensity, on the other hand, suggests that lower energy efficiency tends to increase emissions. In contrast, Zhang et al. [43] empirically measured a city-level "total factor carbon emission efficiency index" (TCEI) at different stages of urbanization using dynamic panel data for four typical urban agglomerations in China from 2006 to 2016. They found the TCEI of these urban agglomerations at different urbanization stages to increase year by year, indicating that increasing economic development and the urbanization level will help to achieve low carbon development in specific urban agglomerations. The optimization of the industrial structure and making improvements to the ecological will in turn help to curb carbon emissions.

While scholars have also made in-depth studies on the relationship between urbanization and carbon emissions, there remain some disagreements on the nature of this relationship, and views tend to conform to three distinct types, each of which respectively support the presence of a driving effect, an inhibiting effect, or uncertainty regarding the effect of urbanization on carbon emissions. Sadorsky et al. [44] analyzed over 80 countries in the last 20 years in order to determine the influence of demographic factors on pollution emissions, pointing out that the level of carbon emissions in cities would increase with increases in their urbanization level. Tang et al. [45] found the carbon emission trading system on land use transition would affect China's CO_2 emissions peak. In contrast to the above studies, Martinez-Zarzoso et al. [46], using urban population data for 88 developing countries over the last 30 years, found an "inverted-U" relationship between urban population density and carbon emissions. Poumanyvong et al. [7] selected 99 countries and regions around the world, classifying them by income level; these scholars found that the impact of urbanization on carbon emissions varied by income. While they were able to show that for countries in the middle-income range, urbanization increases carbon emissions significantly, they were unable to confirm the exact relationship between urbanization and carbon emissions for the sample of low-income countries.

The flow space was originally proposed by Castells as a space of flow interactions [47]. It can reveal the connection of flow between regions. At present, most of the research on flow space is focused on regional spatial structure, theoretical research, and the application of "flow" data. Zhong et al., constructed a weighted directed graph based on tourism flow data, which can reveal the spatial structure of urban movement [48]. Roth et al. use personal mobility data from the London Underground to reveal the structure and organization of the city; the results of their study show that intra-city mobility patterns are quantitatively highly heterogeneous and that there is a polycentric structure consisting of large flows [49]. Shen et al., believe that flow space is a space built around the flow of elements such as human, logistics, capital, technology, and information flow, which is supported by information technology-based network flow lines and rapid traffic flow lines to create a purposeful, repeated, and programmable dynamic movement [50]. Based on the above research, this paper considers flow space as a networked dynamic space which is built around the exchange of people, logistics, and information flow between cities with information network and transportation network as the carrier.

Previous studies on the mechanism linking urbanization and carbon emissions have accomplished much and led to a rich range of results, which provide the theoretical basis and practical experience for this paper. These existing studies, however, have tended to ignore the interconnections between cities within urban agglomerations, instead treating urban agglomerations as static systems. Spaces of flows is a notion first introduced by Manuel Castells, who defined it as the material organization of time-sharing social practices that work through flows [51]. As China's cities step into a new stage in their urbanization, the rapid development of modern transportation, communication technology, and other technologies make the connection between cities increasingly close—the resulting flows of people, goods, and information between cities will also have an impact on urban carbon emissions. Therefore, this paper selects the Pearl River Delta (PRD) urban agglomeration, with its strong inter-city linkages, especially the high-intensity, high-velocity, and high-frequency flow of multiple elements within it [52], as a research case and adds the indicator of "comprehensive urban linkage flow." Establishing indicators across four dimensions, the study aims to explore the relationship between urbanization and carbon emissions and identify the mechanism at work in the impact of new urban development on carbon emissions.

The remainder of this article is organized according to the following structure. In Section 2, the study area, the data source and processing, and the methodology are introduced. In Section 3, the results of urban linkage intensity and the regression results of STIRPAT model are analyzed. Finally, the conclusions and policy implications are presented in Section 4.

2. Materials and Methods

2.1. Study Area

The PRD urban agglomeration forms the study area addressed in this paper. The region is located in the south of China and is one of the most dynamic economic regions in the Asia-Pacific region (Figure 1). The PRD urban agglomeration includes nine cities—Dongguan, Foshan, Guangzhou, Huizhou, Jiangmen, Shenzhen, Zhaoqing, Zhongshan, and Zhuhai.

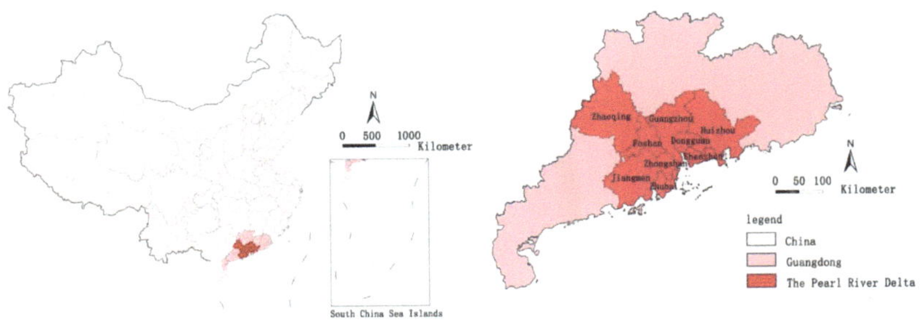

Figure 1. Study area: PRD urban agglomeration.

The PRD urban agglomeration was one of the regions established in the early stages of China's reform and opening-up policy and is an important economic center. It covers an area of about 55,368.7 km^2—accounting for less than one third of the province's area; in 2019, it gathered 53.35% of the province's population and 79.67% of its GDP. The PRD urban agglomeration holds an important position in China's national economic and social development as well as in the overall reform and opening up of the country.

2.2. Data Source and Processing

The data used in this study primarily took the forms of inter-city flow data, urbanization index data, and carbon emission data.

2.2.1. Urban Flow Data

City Passenger Flow

Since passenger flows between cities mainly occur through roads and railroads, we chose to use the passenger frequency of high-speed rail, ordinary trains, and intercity buses in order to calculate population movement between cities in the PRD. The data was mainly obtained by web crawling the bus schedules published on the 12306 websites (https://www.12306.cn (accessed on 7 June 2019)) and the passenger stations in each city using the software Python. We selected the average daily frequency of high-speed trains, trains, ordinary trains, and intercity buses between cities in 2019 to represent the traffic flow data between cities. Passenger trips were converted into passenger flow data. Since different modes of transportation have different carrying capacities, different passenger capacities were assigned to them. Passenger cars are standardized to carry 50 people, and ordinary trains are standardized to carry 1 soft sleeper, 8 hard sleepers, and 8 hard-seat cars in total carrying 1200 people. Dynamic and high-speed trains are made up of a first-class car comprising of 4 sections and a second-class car comprising of 5 sections, carrying 600 people as the standard [53].

Traffic volume was defined using the following formula:

$$M_{ij} = 50A_{ij} + 1200B_{ij} + 600C_{ij} + 600D_{ij} \quad (1)$$

$$R_{ij} = (M_{ij} + M_{ji})/2 \quad (2)$$

In the formula, M_{ij} represents the connectivity of the city passenger flow network from city i to city j; R_{ij} represents the connectivity of the city passenger flow network between city i and city j; and A_{ij}, B_{ij}, C_{ij}, and D_{ij} denotes the daily average intercity bus frequency, ordinary train frequency, moving train (subway) frequency, and high-speed rail frequency from city i to city j, respectively. The results of this calculation were used for the final construction of the city's passenger flow linkage matrix (Table 1). The connection matrix is a matrix that represents the strength of the linkage between cities, with larger numbers in the matrix representing stronger linkages between two cities.

Table 1. PRD city people flow connection matrix 2019.

	DongGuan	FoShan	GuangZhou	HuiZhou	JJiangMen	ShenZhen	ZhaoQing	ZhongShan	ZhuHai
DongGuan		15,000	237,550	101,350	3150	211,800	6850	6250	8500
FoShan			235,550	4650	7800	105,400	360,800	9600	7450
GuangZhou				66,100	151,150	664,300	337,050	148,000	209,600
HuiZhou					1500	146,350	1600	2750	2600
JiangMen						3500	1500	1000	13,200
ShenZhen							214,100	13,050	5800
ZhaoQing								2700	3650
ZhongShan									217,550
ZhuHai									

Logistics Flow

Logistics flow is the process of moving an object entity from one location to another. In the flow space, data from express sites can be used to calculate the level of logistics between regions, representing the frequency of flow of items. In this paper, 12 mainstream logistics companies—SF, CaiBird, Best, Debonair, Annen, China Tong, and Shunxin—were selected as research subjects. We obtained logistics flow data mainly by using Python to crawl the logistics routes between cities in the PRD urban agglomeration in 2019 using the Alibaba logistics network (https://x56.1688.com/56home/home.htm (accessed on 7 June 2019)). Drawing on the method proposed by Taylor for studying the world city network, the score value of a logistics outlet in city i, denoted by X_{ia}, is defined as the importance of a logistics outlet in the whole network [54]. The calculation of the logistics linkage is expressed as:

$$R_{ij,a} = X_{ia} \times X_{ja} \quad (3)$$

R_{ij} denotes the degree of logistics connection between two cities, city i and city j:

$$R_{ij} = \sum_{j} R_{ij,a} \quad (4)$$

Finally, the logistics linkage matrix of the PRD urban agglomeration was constructed (Table 2).

Table 2. PRD city logistics linkage matrix 2019.

	DongGuan	**FoShan**	**GuangZhou**	**HuiZhou**	**JiangMen**	**ShenZhen**	**ZhaoQing**	**ZhongShan**	**ZhuHai**
DongGuan		362,451	1,354,832	101,586	241,344	1,127,103	118,841	112,980	84,404
FoShan			222,935	44,334	55709	299,405	64,878	18,159	22,410
GuangZhou				93,767	386,066	1,240,088	128,623	21,202	68801
HuiZhou					93,767	114,535	33,800	27,795	21,704
JiangMen						117,300	57,563	10,359	31,075
ShenZhen							52,260	22,088	28,170
ZhaoQing								8436	24,352
ZhongShan									5245
ZhuHai									

Information Flow

With the development of communication technology, the internet has become an important communication channel between cities. In fact, the flow of information between cities within an urban agglomeration occurs mainly through the internet. This can be represented in a number of ways, but we used the Baidu index city search volume, which is able to visually show the information exchange status between cities. In this paper, the average daily search volume of Baidu users among cities in the PRD in 2019 was obtained using the keywords of each city in the PRD in the Baidu index interface. A city information connection flow matrix was constructed from the data [55]. The formula for calculating the information linkage flow between cities is as follows:

$$R_{ij} = A_b + B_a \tag{5}$$

A_b in the formula is the search volume of city Baidu users to city A; B_a is the search volume of city A Baidu users to city B; and R_{ij} is the information flow network connection degree between city i and city j. We used the results of this calculation in order to construct the city information flow connection matrix (Table 3).

Table 3. PRD city information flow linkage matrix 2019.

	DongGuan	**FoShan**	**GuangZhou**	**HuiZhou**	**JJiangMen**	**ShenZhen**	**ZhaoQing**	**ZhongShan**	**ZhuHai**
DongGuan		74,784	299,088	89,585	28,776	550,232	29,190	28,356	43,920
FoShan			433,782	46,200	39,579	173,038	36,894	49,644	76,440
GuangZhou				159,104	100,128	923,520	96,084	93,840	146,256
HuiZhou					20,574	339,735	18,522	18,810	23,040
JiangMen						62,037	22,969	18,207	23,490
ShenZhen							61,204	64,944	113,928
ZhaoQing								12,936	15,295
ZhongShan									38,100
ZhuHai									

Flow Data Integration Processing

We then normalized the three sets of data from the flow contact matrices described above using the following equation:

$$X = (x - Min)/(Max - Min) \tag{6}$$

In the formula, x is the original data; Min is the minimum value of the matrix data; Max is the maximum value of the matrix data; and X is the normalized data. The standardized X indicates the degree of inter-city linkage—the higher the value, the stronger the inter-city linkage.

Summation of normalized data was used to express the intensity of urban linkage flow.

$$R_i = \sum_j r_{ij} \quad (7)$$

In the formula, R_i denotes the strength of linkage flow of city i and R_{ij} is the degree of network connectivity between city i and city j. In order to study the spatial structure of integrated flow urban agglomerations, with reference to Liu [56], all factor flows were considered equally important with a weight of 0.33.

2.2.2. Urbanization Index Data

Drawing on the results of the existing comprehensive measurement index system [9,12,25,52,57,58], and following the principles of systematicity, completeness, validity, scientificity, and operability, it was summarized into four aspects from the connotation of the concept of urbanization: (1) Population urbanization indicators mainly reflect the process of population concentration in urban areas, including the proportion of urban population, population density, and the number of people employed in secondary and tertiary industries. (2) Economic urbanization indicators mainly reflect the transformation of economic structure into non-agricultural, including GDP per capita, the proportion of output value of secondary and tertiary industries, and the proportion of research and development investment in regional GDP. (3) Social urbanization indicators mainly reflect the change of lifestyle, including total social consumption, per capita electricity consumption and the proportion of college students. (4) Spatial urbanization index. mainly reflects the process of change of geographical landscape. Specifically, it includes the proportion of built-up area and paved road area per capita. These four indicators are interrelated and distinct, aiming to reflect the comprehensive level of urbanization process (Table 4).

Table 4. Urbanization index system table.

Tier 1 Indicators	Secondary Indicators	Reference
Population urbanization	Urban population to total population	[9]
	Population density	[58]
	Percentage of population in secondary and tertiary industries	[52]
Economic urbanization	The proportion of output value on secondary and tertiary industries to regional GDP	[9]
	GDP per capita	[58]
	Research and development investment as a percentage of regional GDP	[25]
Social urbanization	Total social consumption	[25]
	Urban electricity consumption per capita	[57]
	University student rate	[57]
Spatial urbanization	Area share of built-up area	[57]
	Paved road area per capita	[25]
Urban integrated linkage intensity	City passenger flow	[52]
	Logistics flow	[52]
	Information flow	[52]
Control variables	Energy consumption intensity	[12]

All data on new urbanization that we used in the evaluation system were taken from the Statistical Yearbook of Guangdong Province 2019 and the Statistical Yearbook of Urban Construction of China.

Each variable is described in Table 5.

Table 5. Results of descriptive statistics for each variable.

Variables	Unit	Minimum Value	Maximum Value	Average Value	Standard Deviation
Carbon dioxide emissions	million tons	16.1	67.6	36.3	15.0
Urban population ratio	%	48.6	100.0	82.2	15.5
Population density	person/km^2	281	6484	1870.8	1747.6
Percentage of population in secondary and tertiary industries	%	0.549	0.998	0.881	0.144
The proportion of output value of secondary and tertiary industries	%	99.9	82.8	95.9	5.2
GDP per capita	Yuan/Person	53,936	203,489	119,858.7	48,495.5
R&D investment ratio	%	4.93	1.11	2.55	0.97
Total social consumption	Billion Yuan	996.3	9551.6	3664.2	3193.3
Urban electricity consumption per capita	kWh/person	6363.4	12,752.6	9017.4	1969.3
Percentage of college students	Name/10,000 people	45.1	745.4	244.5	254.8
Area share of built-up area	%	20.0	60.0	38.4	14.3
Paved road area per capita	km^2/person	8.5	24.2	17.4	4.7
Energy consumption intensity	million tons of standard coal	407.6	2379.9	1186.0	659.0

2.2.3. Carbon Emissions Data

The carbon emission data refer to the study of Wang et al. [59], which accounted for the carbon emission of energy consumption with urban energy consumption data according to the IPCC [60,61] greenhouse gas emission inventory, followed by obtaining the nighttime lighting data within the city and counting the total nighttime lighting values based on the DMSP/OLS nighttime lighting images, and then fitting and analyzing them with the corresponding carbon emission statistics, which were simulated and inverted from the nighttime lighting images.

2.3. Methodology

2.3.1. The Entropy Method

Based on the index system of urbanization, the present paper constructs a comprehensive index system of urban carbon emissions. This was done by combining urban linkage flow and urban energy consumption intensity. The index system can be used to analyze the influence of different dimensional indicators on carbon emissions.

In consideration of the different importance of different dimensional indicators in the system, we selected the entropy method in order to calculate the weight index of each indicator. The entropy value method was used to calculate the entropy weight based on the information entropy of each dimensional index. Finally, the weights of each index were corrected using entropy weights, allowing more accurate and objective weighting results to be obtained [58].

The entropy value method is calculated as shown below:

In evaluating the six dimensions of population urbanization, economic urbanization, social urbanization, spatial urbanization, energy consumption intensity, and comprehensive urban linkage flow intensity, each dimensional index U_k was calculated as follows:

$$U_j = \sum_{j=1}^{m} X_{kj} \frac{g_{kj}}{\sum_{i=1}^{n} g_{ki}} \tag{8}$$

In formula (8), X_{kj} denotes the normalized indicator data and g_j denotes the differentiation index of indicator j at that time. Accordingly, the weights of each dimension of the carbon emission index system were able to be obtained (Table 6).

Table 6. Weights of each dimension of the carbon emission index system.

Dimensionality	Population Urbanization	Economic Urbanization	Social Urbanization	Spatial Urbanization	Energy Consumption Intensity	Comprehensive City Connection Degree
Weights	0.220	0.211	0.220	0.131	0.104	0.114

2.3.2. STIRPAT Extension Model

York et al., originally developed the Stochastic Impacts by Regression on Population Affluence, Technology (STIRPAT) [62] model, an extension of the IPAT model, in order to introduce random variation variables based on the previous model and in this way overcome the shortcomings of the original model.

The formula is as follows:

$$I = aP^b A^c T^d e \quad (9)$$

In practice, it is usually used in its logarithmic form:

$$\ln I = \ln a + b \ln P + c \ln A + d \ln T + e \quad (10)$$

In the formula, b, c, and d denote the elasticity coefficients of the population, affluence and technology level, respectively, and e is the random error term.

Based on the STIRPAT model, in this study, we analyzed the effects of different urbanization and integrated urban connectivity on carbon emissions. When the calculated urbanization index variables for each dimension are substituted into the extended model the model becomes:

$$\ln I = \mathrm{a} + \beta_1 \ln PU + \beta_2 \ln EU + \beta_3 \ln SU + \beta_4 \ln LU + \beta_5 \ln UC + \beta_6 \ln EN \quad (11)$$

In the formula, I denotes total carbon emission, PU denotes population urbanization, EU denotes economic urbanization, SU denotes social urbanization, LU denotes spatial urbanization, UC denotes urban integrated correlation, and EN denotes urban energy consumption intensity.

3. Results

3.1. *Urban Linkage Intensity Analysis*

After processing and normalizing the linkage intensity data for each city within the PRD urban agglomeration, we were able to obtain the city integrated linkage flow intensity (Figure 2).

Generally speaking, the core cities in the PRD urban agglomerations (e.g., Guangzhou, Dongguan, Shenzhen, etc.) have relatively consistent linkage intensities in relation to city passenger flows, logistics flows, and information flows. However, differences can be identified in the intensity of linkages between different cities located in peripheral areas. Guangzhou and Shenzhen are the top two cities in the urban agglomeration from the perspective of the intensity of city-integrated linkages, followed by Dongguan and Foshan. Among these cities, Guangzhou displayed the highest integrated linkage flow intensity and Jiangmen the lowest; the value of the former is 9.03 times higher that of the latter. Guangzhou and Shenzhen together form the center of gravity of the national economy as well as a world trade center and an international integrated transportation hub. They have a dominant position in terms of information interaction and transportation links. These two cities form a dual-core network linkage structure in the PRD urban agglomeration. At the same time, different types of mobility within the urban agglomeration exhibit different structural characteristics:

(1) The population movement between Guangzhou and Shenzhen is the closest in network formed by passenger flow linkages between cities. Guangzhou and Shenzhen are closely linked to each other in terms of economy and information. The traffic flow between them far exceeds that between other cities, e.g., the average frequency of

the high-speed train between the two cities is 6 shifts/hour. Guangzhou and Shenzhen form the most important urban passenger connection axis within the urban agglomeration, and they account for 18.8% of the total passenger flow in the PRD urban agglomeration.

(2) In the network formed by logistics flow linkages between cities, Dongguan-Guangzhou is shown to be the strongest linkage. In terms of the strength of logistics flow linkage strength of individual cities, Dongguan has the highest value. The reason for this lies in the city's e-tailing industry, which drove the development of the city's economy and networks, opening up a variety of available service channels for the logistics flow industry. As an important commodity production base in the PRD urban agglomeration, Dongguan has thus witnessed rapid development in urban logistics. The logistics flow link between Dongguan and Guangzhou alone accounts for 10.4% of the total volume of the PRD urban agglomeration.

(3) In the information flow linkage network, Shenzhen-Guangzhou, Shenzhen-Dongguan, and Guangzhou-Foshan have the three highest inter-city information flow intensities; together, they account for 43% of the total regional intensity. From a geospatial location perspective, the information exchange links within the PRD urban agglomeration tend to cluster along the Pearl River Estuary. With its high level of economic vitality, Shenzhen ranks first in terms of overall information linkage intensity—the city's rapid development has allowed Shenzhen to gradually surpass Guangzhou in terms of information linkage flow intensity. Shenzhen is the core city of the PRD urban agglomeration, with an increasing number of internet users and increasing degree of informatization. Topics about Shenzhen often trigger discussions among internet users in different cities on the Internet. Hot topics, such as soaring house prices and technological development in Shenzhen, push internet users in different cities to learn about Shenzhen through online channels.

(4) The intensity of linkage flows in the PRD urban agglomeration can be divided into three tiers (Figure 3). Guangzhou and Shenzhen occupy the core position and in the three tiers of urban passenger flow, logistics flow, and information flow division, both cities belong to the first tier, although Shenzhen's position is relatively lower than that of Guangzhou. Dongguan and Foshan are in the second tier of the city linkage flow hierarchy. Dongguan is ranked in the first tier of the logistics linkage hierarchy. The cities in the urban tier of all connected flows are basically the same. In addition, cities place within different tiers for the different flow types. For example, Zhaoqing has stronger passenger flow linkages and is in the second tier, while in terms of logistics flow and information flow, it is located within the third tier. Jiangmen has relatively stronger logistics flow connections, which also reflects the diversity of development and the difference of relative advantages of different city flow types. From the perspective of the integrated flow of the cities studied, the PRD urban agglomeration shows a distribution pattern of "two cores—multiple centers."

Compared to Dongguan and Foshan, Zhaoqing, and Jiangmen are further away from the urban agglomeration center. They receive relatively less radiation from the core of the urban agglomeration.

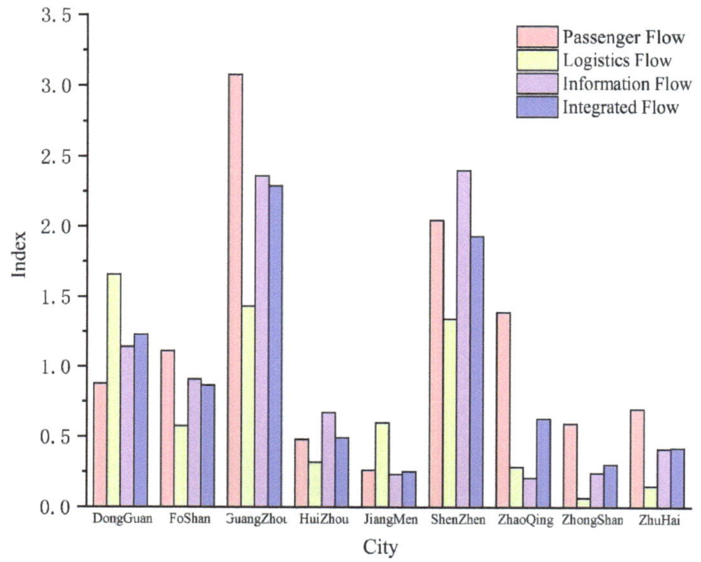

Figure 2. Histogram of urban linkage flow intensity in the PRD urban agglomeration.

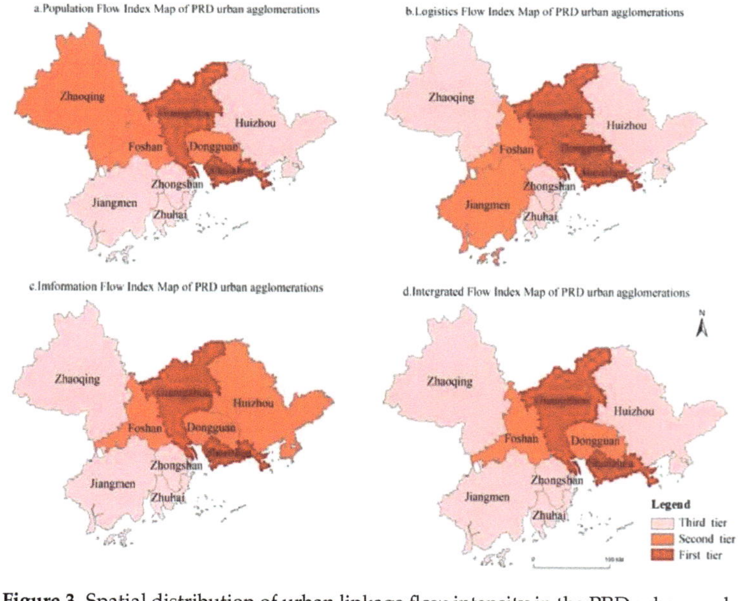

Figure 3. Spatial distribution of urban linkage flow intensity in the PRD urban agglomeration.

3.2. Analysis of Regression Results of STIRPAT Model

The regression results of the STIRPAT model showed that the model had a decidable coefficient R^2 of 0.999 and the Durbin–Watson value of 2.263. The Durbin–Watson value refers to a test statistic proposed by Durbin and Watson for the autocorrelation of diagnostic model residuals. The value of this statistic is between 0 and 4, and the closer it is to 2, the more the model tends not to have autocorrelation. All variables passed the 5% significance test. This shows that the model is statistically significant and has a good fit. The VIF values of each variable were below 10, indicating that no multicollinearity existed among them (Table 7).

Table 7. Model regression results.

Variables	Regression Coefficient	Std.Error	Significance	VIF
CONSTANT	3.136	0.100	0.000	
Ln(PU)	1.185	0.064	0.002	6.503
Ln(EU)	−1.207	0.175	0.038	9.811
Ln(SU)	−0.275	0.028	0.013	2.309
Ln(LU)	0.265	0.083	0.033	5.731
Ln(NYXF)	0.415	0.042	0.004	1.697
LXD	1.140	0.116	0.003	9.967
F Value	13.683		0.028	
R2	0.999			

(1) Population urbanization was found to exert a significant positive effect on urban carbon emissions. For every 1% increase in population urbanization, urban carbon emissions were shown to increase by 1.185%; population urbanization is thus an important factor in increases in carbon emissions. As one of the regions with the highest level of development in the country, the PRD urban agglomeration has absorbed a large number of foreign populations. According to data published by the National Health and Wellness Commission, the PRD urban agglomeration had a resident population of 59.985 million in 2016, with 33.5053 million local household residents and 26.4797 million migrants, which is one of the largest influxes of migrants in the country. At the same time, the large number of migrants keeps the population density high and drives urban development, the latter of which in turns stimulates the consumption of logistics goods like fuel and cement, which in turn has led to increases in carbon emissions. Compared with other cities, megacities within urban agglomerations have higher urbanization levels, higher population densities, and more diverse consumption patterns. Urban residents can afford higher consumption levels, and many families have purchased private cars due to their improved living standards, all of which increase CO_2 emissions from transportation trips. In addition, the limited size of the built-up areas of megacities and the rapid population growth rates are not matched by corresponding levels of urban facilities and environmental population affordability (such as urban park green space, sewage treatment plant, sanitation facilities, etc.). These reasons further contribute to the deterioration of the carbon emission situation in cities. As a result, cities should consider measures to relieve urban population pressure (such as optimize the layout of urban population, guide the population to the peripheral areas of the PRD urban agglomeration orderly transfer, fully liberalize the restrictions on settlement in established towns and small cities, liberalize the restrictions on settlement in medium-sized cities in an organized manner, reasonably determine the conditions for settlement in large cities, and strictly control the size of the population of megacities), develop and improve urban facilities, and build new urban areas.

(2) Economic urbanization is the most important factor that affects urban carbon emissions, and a negative correlation exists between the two. Our results show that for every 1% increase in the economic urbanization index, there is a corresponding 1.207% reduction in urban carbon emissions. This is the largest regression variable among the five independent variables. Moreover, the economic urbanization index passed the 1% significance test, indicating that this variable is the important factor affecting urban carbon emissions. With the national goals of "carbon peaking in 2030" and "carbon neutral in 2060," the PRD urban agglomeration is facing double pressures in the form of economic development and energy consumption control. Therefore, cities in the PRD urban agglomeration have increased their investment in research and development and strengthened their policy support for new energy technology enterprises, such as photovoltaic and hydrogen-based energy. Moreover, the PRD urban agglomeration has economic advantages, market potential, and talent structure, and these advantages have attracted many large enterprises and foreign businesspeople to collaborate and start businesses. These collaborations have helped introduce advanced pollution prevention technologies and management methods,

and they have also promoted the transformation and upgrading of enterprises. At the same time, these energy technology advances have also driven carbon emission reduction in cities within the agglomeration and improved the energy utilization efficiency and clean energy capabilities of the cities.

For example, most cities have switched from previously fuel-driven to electric buses. In December 2020, the electrification rate of urban buses in Guangdong Province reached 93.5%. Among the cities, the bus electrification rates of Guangzhou, Shenzhen, and Zhuhai have reached 100%, effectively optimizing these cities' energy consumption structures. While cities are investing in research and development in energy technology, they are also targeting higher pollution emission standards in some heavy industries. The impact of traditional high-polluting industries on carbon emissions is being limited, which indirectly promotes the reduction of carbon dioxide emissions in cities and the implementation and effectiveness of national emission reduction initiatives. Finally, the operation of the carbon trading market should be promoted, the carbon emission statistics and accounting system should be improved, and more binding regulations should be introduced for key emitters in different industries to better play the role of energy conservation and emission reduction. Although total carbon emissions have slowed in the past two years, there is still a long way to go to achieve the goal of "carbon neutrality" by 2060. The PRD urban agglomeration should continue to encourage and support the research and development of clean energy and energy storage technologies, optimize the overall industrial structure of the urban agglomeration (such as promote the transformation and development of the PRD region with a single industrial structure, use national policies to promote counterpart cooperation with the central cities in urban agglomeration, and accelerate the joint construction of industrial parks, factor flow, and resource sharing), guide the gradual replacement of traditional energy with clean energy, and promote low-carbon urban development.

(3) Social urbanization has a negative effect on urban carbon emissions. Specifically, for every 1% increase in the social urbanization index, urban carbon emissions are reduced by 0.275%. Since the number of college students in cities has increased in recent years, the overall quality of the urban population has also improved. In addition, with the publicity and encouragement of environmental behaviors by urban media, residents are gradually establishing basic environmental concepts such as green travel and water conservation. People are gradually aware of the importance of establishing a more harmonious coexistence between human beings and nature.

Cities in the PRD urban agglomeration have also proposed policies to promote energy conservation and emission reduction in recent years. In December 2013, the PRD urban agglomeration officially launched a carbon emissions trading scheme. Heavy chemical industries with high energy consumption—such as steel, petrochemicals, electricity, and cement—were included in the scope of the carbon trading market. At the same time, enterprises with serious pollution emissions were required to carry out rectification measures. Market mechanisms are thus being employed to effectively control greenhouse gas emissions. In addition, Guangdong Province decided to implement the "Emission Limits and Measurement Methods for Light Duty Vehicles (China Stage 6)" in 2018, which strictly limits the pol-luting exhaust emissions from vehicle production. Some cities even have introduced policies to restrict the travel of motorcycles and other means of transport with serious exhaust emissions, as well as to strengthen the control of energy consumption standards for the production that enterprises produce. All these measures have effectively slowed down the contribution of total social energy consumption and per capita electricity consumption to carbon emissions. In this way, urban carbon emissions are effectively curbed.

(4) Spatial urbanization is positively correlated with carbon emissions. For every 1% increase in the spatial urbanization index, the corresponding increase in urban carbon emissions is 0.265%. During the pre-city construction period, the population was heavily concentrated as well as the industries were rapidly upgraded and expanded. To meet the needs of accelerated urbanization, commercial development was built substantially within

the city and the scale of urban roads was expanded. High levels of energy consumption are known to cause environmental stress, destroy urban greenery, and worsen the carbon emission situation of cities. Compared to other indicators, however, spatial urbanization was found to exert the least impact on carbon emissions, which can be attributed to the following reasons: First, the improvement of the environment by park green space. While the extent of urban development is expanding in cities within the PRD, the scale of green spaces is also being expanded and the construction of green spaces accelerated, in line with the greening of urban parks; these initiatives have a positive effect on the mitigation of carbon emissions in cities. Second, the expansion of urban built-up areas and traffic roads in urban agglomerations has slowed, which has allowed the environmental impacts of spatial urbanization to also slow in recent years as the urbanization rate has increased. Third, the regional implementation of building energy efficiency standards and energy limit standards for building material products have been raised. With the improvement of technology and techniques, green building materials are gradually put into use. The state has improved building energy efficiency standards and encouraged the construction of low-energy buildings. The energy consumption generated in urban construction has been reduced and the energy-saving capacity of buildings has been improved—for these reasons, the rate of increase of carbon emissions generated by the expansion of urban development has slowed.

(5) Our results indicate that a positive correlation exists between the intensity of urban integrated linkage flows and carbon emissions. In fact, every 1% increase in this index is shown to increase the city's carbon emissions by 1.140%. Although the economic growth rate of cities in the PRD has slowed in recent years, the economic cooperation and development links between cities continue to gain in strength. The formation of Guangzhou-Foshan-Zhuhai, Shenzhen-Dongguan-Huizhou, and Zhuhai-Zhongshan-Jiangmen metropolitan areas is gradually establishing a new urban cluster development pattern with a clear division of labor and complementary functions. The construction of several intercity transportation arteries, such as the Guangzhou-Zhuhai Intercity Railway and the Guangzhou-Foshan Ring Road, has shortened transportation travel time and reduced the cost of factor movement. This has had a long-term impact on optimizing urban functions and smoothing economic links between these cities, but the tailpipe emissions generated by transportation have exacerbated urban carbon emissions. It is worth noting that urban integrated linkage flows within the PRD urban agglomeration are mostly in the direction of Guangzhou, Shenzhen, Dongguan, and other developed cities in the region, rather than the relatively backward economic industries cities in the outer layer of the urban agglomeration. In the long run, this will be detrimental to the future development of the urban agglomeration as a whole. As the core of the PRD urban agglomeration, Guangzhou and Shenzhen should actively use their industrial and location advantages to play a guiding role. These cities should accelerate the economic and logistics links with the cities at the edge of the urban agglomeration, drive the backward economic industries cities to accelerate their development, alleviate the situation of excessive concentration of resource elements in the core cities, and promote the coordinated development of carbon emission reduction and economy.

(6) Urban energy consumption intensity was found to positively correlated with carbon emissions and emerges as one of the main factors affecting carbon emissions in urban agglomerations. For every 1% increase in urban energy consumption intensity, urban carbon emissions are shown to increase by 0.415%. For a long time, China's energy consumption structure was coal-fired, making the energy structure inefficient. According to data published by the National Bureau of Statistics in 2020, the power generation structure of Guangdong Province is still dominated and thermal power generation, and thermal power generation accounts for 71.38% of the energy grid. To break the traditional coal-fired energy structure, ease the "double control pressure on energy consumption," and achieve the "double carbon goal," Guangdong Province should improve the energy efficiency of traditional heavy industries and accelerate the optimization and upgrading

of industrial structures, such as metal smelting, construction logistics, cement, and othe heavy chemical industries. However, these industries cannot be replaced in the short tern because of the huge demand that still exists in the current market. Therefore, on the on hand, we should target high energy-consuming and high-emission projects, strengthen th supervision work of relevant departments, while controlling high-energy-consuming an high-emission enterprises, replacing and shutting down excess capacity, and improve th pollution emission standards of high-energy-consuming and high-emission enterprise On the other hand, we suggest that more attention be paid to research and developmen investment in the direction of strategic future industries such as green materials, advance high-end equipment manufacturing, and new energy by driving technological innovatior improving energy utilization efficiency, and reducing the energy consumption of product to promote the transformation of high-energy-consumption industries to low-carbon de velopment. At the same time, the PRD urban agglomeration should optimize the energ consumption structure within the urban agglomeration. A gap still exists between th environmental quality of the PRD region and other international urban agglomeratior To achieve a green development model and build ecological power systems as soon a possible, it is necessary to adhere to the green and low-carbon development of energy in th future, even in the face of rapid economic growth, by establishing an energy consumptior system centered on new energy to promote research and development into clean energ technologies such as hydropower, nuclear power, and wind power. Furthermore, status quo energy consumption—dominated by coal-fired thermal power generation—will need to be controlled if cities are to reduce their carbon emissions.

4. Conclusions and Policy Implications

4.1. Conclusions

Cities are complex systems composed of multiple elements, and changes in population growth, logistics flows, the economic structure, development patterns, and other factors within a city will cause changes in the scale, intensity, and impact of these factors on carbon emissions [63]. The present study has constructed a comprehensive index system for urban carbon emissions based on a STIRPAT model, combining integrated urban linkage flows and a 4-dimensional urbanization index. The results of this research provide a number of important findings.

Primarily, the PRD urban agglomeration has formed a network linkage structure which places Guangzhou and Shenzhen at its core—these two cities are among the top urban agglomerations in terms of population, logistics, information flow, and other factor flows. In contrast, Jiangmen, Zhaoqing, and other cities located in the peripheral areas of the PRD urban agglomeration find themselves at a lower level in terms of the intensity of their linkage flows. Guangzhou has the highest integrated linkage flow intensity—a level 9.03 times higher than Jiangmen, which has the lowest flow intensity. However, cities are also located at quite different levels in terms of different categories of flow types, and the development structure of the linked flows in each city is diverse and differentiated in terms of relative advantages. For example, Zhaoqing has a strong, second-tier position in terms of passenger flow linkages, while it is third tier in terms of logistics and information flows; despite having low flow intensity, Jiangmen has a relatively strong logistics linkage. It is worth noting that Dongguan, as a new first-tier city and the production base of goods in the PRD urban agglomeration, is at the top level of the urban agglomeration in terms of the strength of its logistics links; it also performs well in terms of the strength of its population and information flows linkages.

Secondly, the regression analysis that we conducted using a STIRPAT model revealed that population urbanization, spatial urbanization, and urban integrated linkage flows all correlated positively with urban carbon emissions, while economic urbanization and social urbanization showed a negative correlation. Economic urbanization was found to be the most significant and important factor affecting urban carbon emissions, with every 1% increase in this variable reducing urban carbon emissions by 1.207% accordingly.

Population urbanization exerted the second highest influence, increasing urban carbon emissions by 1.185% for every 1% increase. The spatial urbanization index exerted the least impact, with a regression coefficient of only 0.265%.

Based on these analysis results, the following insights can be drawn:

First, the government clearly has a guiding role to play in reducing carbon emissions, e.g., through actions such as continuing to increase economic investment in research and development [64]; accelerating the transformation of the city's economic structure to low-energy-consuming industries; developing new energy enterprises in cities, focusing on solar energy, water energy, wind energy, and other renewable energy sources; improving the city's energy consumption structure [65,66]; and allowing renewable energy to gradually replace local traditional high-energy-consuming production capacity [67]. Such measures would allow the slowing of growth in urban carbon emissions while achieving sustainable development goals and protecting the ecology of urban areas [68].

Second, our results emphasize the importance of improvements to the infrastructure of cities, through measures such as the provision of sanitation facilities and sewage treatment plants, which are appropriately scaled for the population of a given area or improving the affordability of living expenses for the populations of cities. In 2011, Cai Fang [69] pointed out that China's demographic dividend would be lost around 2015, and that the total population would experience negative growth in 2025. If this is true, this also means that China's total population will reach its peak in coming years, and the huge population size of the PRD urban agglomeration, one of China's major population clusters, poses a serious challenge to the development of a low-carbon economy. Therefore, controlling the external population of cities, curbing the high levels of energy consumption caused by urban overpopulation, and mitigating the impact of population growth on carbon emissions constitute important priorities [70].

Third, knowledge about saving energy and green environmental protection must be popularized. There are many migrant workers in the PRD urban agglomeration, who, whilst they are less educated and less aware of environmental protection, could in the future play a more positive role in promoting carbon emission reduction in the city. In this regard, municipalities should increase the public dissemination of environmental protection knowledge, improve the legal systems around environmental protection, and raise environmental awareness amongst urban residents. While controlling the outflow of highly skilled personnel to overseas, China should formulate relevant policies to attract high-end talent for improving the structure of external population, thereby effectively reducing energy consumption and the energy consumption ratio per unit of GDP [71].

Fourth, strengthen the regulatory work of urban carbon emissions. For high energy-consuming and high-emission projects, promote local pilot projects, study and improve the double control degree of energy consumption, establish a long-term carbon emission management evaluation system, incorporate it into the environmental assessment work, and strengthen the supervision of pollution emissions by ecological-related departments. At the same time, improve management effectiveness and curb blind development. Control high energy-consuming and high-emission enterprises (such as coal power, petrochemical, chemical, steel, non-ferrous metal smelting, building materials, and other industries), replace the shutdown of excess capacity, improve the pollution emission standards of double-high industries, and expand the regional environmental capacity.

Fifth, promote the smooth operation of the carbon trading market and carbon market links to improve the construction of the green financial system. According to the 2019 Guangdong Province Ecological and Environmental Status Bulletin released by the Department of Ecology and Environment of Guangdong Province, the province's carbon market ranked first in the country with a cumulative turnover of 139 million tons of carbon emission allowances and a turnover of 2.715 billion yuan. Obviously, the operation of the carbon trading market has provided an effective method to promote carbon emission reduction in Guangdong Province. However, there are still problems in the current carbon trading market, such as the statistical accounting system needs to be improved, the quality of carbon emission data varies; the market access rules of the carbon pilot are of low standard and the type of entering enterprises is single. Accordingly, we should continue to improve the system of the carbon trading market and introduce more binding regulations for key emitters in different industries to better guide the carbon emission reduction work of key emitters. At the same time, we should encourage the innovative development of green financial products in many aspects, increase policy support and improve standards, and provide more flexible financing methods for green industries to achieve the purpose of carbon emission reduction [72].

Last but not least, developed cities have a role in strengthening linkages between core cities and edge cities within urban agglomerations. This is particularly true of the PRD urban agglomeration, where measures must be taken in order to break the pattern of strong central cities and weak peripheral cities, and rather promote the coordinated development of the cities in the PRD urban agglomeration, the optimization of regional development and the formation of a complementary pattern of advantages. The outline of the PRD Reform and Development Plan is put forward as a core policy instrument which could be used to build a polycentric spatial structure of "three major metropolitan areas" and make the development of the three major metropolitan areas of "Guangzhou-Foshan-Zhaoqing," "Shenzhen-Dongguan-Huizhou," and "Zhuhai-Zhongshan- Jiangmen" the focus of future economic and social development within the agglomeration. The definition of these three metropolitan areas could work to combine the social characteristics and advantages of different cities in the area, with each area carrying out a differentiated division of economic industries based on that area's complementary industrial advantages [73]. In order to develop sustainable transport links, intercity rail lines need to be expanded and adjusted according to the traffic flow between different cities and future urban development needs, optimizing the planning and construction of inter-city transportation networks and strengthening transportation links. In this way, each city, metropolitan area, urban agglomeration, and the region as a whole might better utilize the advantages of metropolitan area cooperation, promote the integration and interactive development between regional cities [74], establish an urban development pattern with complementary functions, realize differential industrial construction, enhance economic cooperation, drive the construction and development of science and technology and economic quality in marginal cities, and relieve the pressure of core cities in terms of population and land [75].

4.2. Limitations and Future Direction of the Research

In this study, we used a STIRPAT model to construct an urban carbon emission index system by combining the comprehensive urban linkage flows and a four-dimensional urbanization index, taking into account the methods and finding of existing studies. In this way, we conducted an in-depth study of the factors influencing urban carbon emissions. However, due to some constraints, this study still has several limitations.

First, in terms of research data, we acknowledge that the data accuracy is still not high enough—for example, the data of city passenger flows in urban linkage flows ignore the travel of private cars, and the logistics data lack the consideration of large material operation problems. In addition, the research data are only collected for one year in 2019. In the following study, we need to further improve the accuracy, representativeness, and rigor of the data by obtaining supplementary data, data from multiple channels, and data from a longer time series; such measures would improve the richness of the study. Secondly, in terms of the study area, the selected area is limited to the Pearl River Delta urban agglomeration in Guangdong Province, China, and we will consider expanding the study area in the next study so that the conclusions can have a broader reference value. Third, in terms of indicator selection, the secondary indicators selected in this study were uniformly grouped into one level for uniform calculation. In the subsequent study, we will consider calculating the more detailed secondary indicators separately to obtain more accurate results. Finally, the policy recommendations presented at the end of the article are too general due to various objective factors in the study. In future research, we will focus more on more targeted policy recommendations.

Author Contributions: Y.Z. and Y.C. conceived and designed the research; Y.Z., R.C., T.S., Y.Y., S.M. and X.Z. performed the methodology and wrote the original draft; Y.Z., R.C., T.S., Y.Y., S.M., D.X. and Y.C. finished the final manuscript. All authors have read and agreed to the published version of the manuscript.

Funding: This work was supported by the Project Supported by the Open Fund of Key Laboratory of Urban Land Resources Monitoring and Simulation, Ministry of Natural Resources (Grant No. KF-2020-05-010), Natural Science Foundation of Guangdong Province (Grant No. 2019A1515011653), National Natural Science Foundation of China (Grant No. 51978173; Grant No. 42101186), and Science and Technology Program of Guangzhou (Grant No. 201904010465).

Institutional Review Board Statement: Not applicable.

Informed Consent Statement: Not applicable.

Data Availability Statement: Not applicable.

Acknowledgments: The authors' deepest gratitude goes to the anonymous reviewers and editors for their careful work and thoughtful suggestions that have helped improve this paper substantially.

Conflicts of Interest: The authors declare no conflict of interest.

References

1. Friedlingstein, P.; Andrew, R.; Rogelj, J.; Peters, G.; Canadell, J.; Knutti, R.; Luderer, G.; Raupach, M.; Schaeffer, M.; van Vuuren, D.P.; et al. Persistent growth of CO_2 emissions and implications for reaching climate targets. *Nat. Geosci.* **2014**, *7*, 709–715. [CrossRef]
2. Wang, H.; Jin, Y.; Hong, X.; Tian, F.; Wu, J.; Nie, X. Integrating IPAT and CLU Mondo models to assess the impact of carbon peak on land use. *Land* **2022**, *11*, 573. [CrossRef]
3. British Petroleum (BP). Statistical Review of World Energy 2021. Available online: https://www.bp.com/statisticalreview (accessed on 12 September 2021).
4. Zhang, X.; Geng, Y.; Shao, S.; Dong, H.; Wu, R.; Yao, T. How to achieve China's CO_2 emission reduction targets by provincial efforts?—An analysis based on generalized Divisia index and dynamic scenario simulation. *Renew. Sust. Energy Rev.* **2020**, *127*, 109892. [CrossRef]
5. Yang, J.; Chen, B. Using LMDI method to analyze the change of industrial CO_2 emission from energy use in Chongqing. *Front. Earth Sci.* **2011**, *5*, 103–109. [CrossRef]
6. Huang, Y.; Raza, S.M.F.; Hanif, I.; Alharthi, M.; Abbas, Q.; Zain-ul-Abidin, S. The role of forest resources, mineral resources, and oil extraction in economic progress of developing Asian economies. *Resour. Policy* **2020**, *69*, 101878. [CrossRef]
7. Poumanyvong, P.; Kaneko, S. Does urbanization lead to less energy use and lower CO_2 emissions? A cross-country analysis. *Ecol. Econ.* **2010**, *70*, 434–444. [CrossRef]
8. Cole, M.A.; Neumayer, E. Examining the impact of demographic factors on air pollution. *Popul. Environ.* **2004**, *26*, 5–21. [CrossRef]
9. Zhang, C.; Lin, Y. Panel estimation for urbanization, energy consumption and CO_2 emissions: A regional analysis in China. *Energ. Policy* **2012**, *49*, 488–498. [CrossRef]
10. Zi, C.; Jie, W.; Hong-Bo, C. CO_2 emissions and urbanization correlation in China based on threshold analysis. *Ecol. Indic.* **2016**, *61*, 193–201. [CrossRef]
11. Ji, X.; Chen, B. Assessing the energy-saving effect of urbanization in China based on stochastic impacts by regression on population, affluence and technology (STIRPAT) model. *J. Clean. Prod.* **2017**, *163*, S306–S314. [CrossRef]
12. Du, L.; Wei, C.; Cai, S. Economic development and carbon dioxide emissions in China: Provincial panel data analysis. *China Econ. Rev.* **2012**, *23*, 371–384. [CrossRef]
13. He, Z.; Xu, S.; Shen, W.; Long, R.; Chen, H. Impact of urbanization on energy related CO_2 emission at different development levels: Regional difference in China based on panel estimation. *J. Clean. Prod.* **2017**, *140*, 1719–1730. [CrossRef]
14. Dogan, E.; Turkekul, B. CO_2 emissions, real output, energy consumption, trade, urbanization and financial development: Testing the EKC hypothesis for the USA. *Environ. Sci. Pollut. Res.* **2016**, *23*, 1203–1213. [CrossRef]
15. Farhani, S.; Ozturk, I. Causal relationship between CO_2 emissions, real GDP, energy consumption, financial development, trade openness, and urbanization in Tunisia. *Environ. Sci. Pollut. Res.* **2015**, *22*, 15663–15676. [CrossRef]
16. Nguyen, D.; Huynh, T.; Nasir, M. Carbon emissions determinants and forecasting: Evidence from G6 countries. *J. Environ. Manag.* **2021**, *285*, 111988. [CrossRef]
17. Shen, L.; Wu, Y.; Lou, Y.; Zeng, D.; Shuai, C.; Song, X. What drives the carbon emission in the Chinese cities? —A case of pilot low carbon city of Beijing. *J. Clean. Prod.* **2018**, *174*, 343–354. [CrossRef]

18. Qiao, W.; Lu, H.; Zhou, G.; Azimi, M.; Yang, Q.; Tian, W. A hybrid algorithm for carbon dioxide emissions forecasting based o improved lion swarm optimizer. *J. Clean. Prod.* **2020**, *244*, 118612. [CrossRef]
19. Nathaniel, S.; Khan, S. The nexus between urbanization, renewable energy, trade, and ecological footprint in ASEAN countries. *Clean. Prod.* **2020**, *272*, 122709. [CrossRef]
20. Adebayo, T.; Udemba, E.; Ahmed, Z.; Kirikkaleli, D. Determinants of consumption-based carbon emissions in Chile: A application of non-linear ARDL. *Environ. Sci. Pollut. Res.* **2021**, *28*, 43908–43922. [CrossRef]
21. Yuan, J.; Ng, S. Emission reduction measures ranking under uncertainty. *Appl. Energy* **2017**, *188*, 270–279. [CrossRef]
22. Khan, M.; Teng, J.; Khan, M. Effect of energy consumption and economic growth on carbon dioxide emissions in Pakistan wit dynamic ARDL simulations approach. *Environ. Sci. Pollut. Res.* **2019**, *26*, 23480–23490. [CrossRef]
23. De Oliveira-De Jesus, P. Effect of generation capacity factors on carbon emission intensity of electricity of Latin America & th Caribbean, a temporal IDA-LMDI analysis. *Renew. Sust. Energ. Rev.* **2019**, *101*, 516–526.
24. Waldo, S.; Jensen, F.; Nielsen, M.; Ellefsen, H.; Hallgrimsson, J.; Hammarlund, C.; Hermansen, O.; Isaksen, J. Regulating Multipl Externalities: The case of Nordic Fisheries. *Mar. Resour. Econ.* **2016**, *31*, 233–257. [CrossRef]
25. Wang, S.; Liu, X. China's city-level energy-related CO_2 emissions: Spatiotemporal patterns and driving forces. *Appl. Energy* **2017** *200*, 204–214. [CrossRef]
26. Wang, K.; Wei, Y. China's regional industrial energy efficiency and carbon emissions abatement costs. *Appl. Energy* **2014**, *130* 617–631. [CrossRef]
27. Solarin, S.; Nathaniel, S.; Bekun, F.; Okunola, A.; Alhassan, A. Towards achieving environmental sustainability: Environmenta quality versus economic growth in a developing economy on ecological footprint via dynamic simulations of ARDL. *Environ. Sc Pollut. Res.* **2021**, *28*, 17942–17959. [CrossRef]
28. Huo, T.; Li, X.; Cai, W.; Zuo, J.; Jia, F.; Wei, H. Exploring the impact of urbanization on urban building carbon emissions in China Evidence from a provincial panel data model. *Sustain. Cities Soc.* **2020**, *56*, 102068. [CrossRef]
29. Nasir, M.; Canh, N.; Le, T. Environmental degradation & role of financialization, economic development, industrialization and trade liberalization. *J. Environ. Manag.* **2021**, *277*, 111471.
30. Dong, F.; Wang, Y.; Su, B.; Hua, Y.; Zhang, Y. The process of peak CO_2 emissions in developed economies: A perspective o industrialization and urbanization. *Resour. Conserv. Recycl.* **2019**, *141*, 61–75. [CrossRef]
31. Naz, S.; Sultan, R.; Zaman, K.; Aldakhil, A.; Nassani, A.; Abro, M. Moderating and mediating role of renewable energy consumption, FDI inflows, and economic growth on carbon dioxide emissions: Evidence from robust least square estimato *Environ. Sci. Pollut. Res.* **2019**, *26*, 2806–2819. [CrossRef] [PubMed]
32. Pang, Q.; Zhou, W.; Zhao, T.; Zhang, L. Impact of urbanization and industrial structure on carbon emissions: Evidence from Huaihe River Eco-Economic Zone. *Land* **2021**, *10*, 1130. [CrossRef]
33. Searchinger, T.; Wirsenius, S.; Beringer, T.; Dumas, P. Assessing the efficiency of changes in land use for mitigating climate change *Nature* **2019**, *565*, E9. [CrossRef] [PubMed]
34. Al-Mulali, U.; Sheau-Ting, L.; Ozturk, I. The global move toward Internet shopping and its influence on pollution: An empirical analysis. *Environ. Impact Assess. Rev.* **2015**, *22*, 9717–9727. [CrossRef]
35. Shahbaz, M.; Loganathan, N.; Muzaffar, A.; Ahmed, K.; Jabran, M. How urbanization affects CO_2 emissions of STIRPAT model in Malaysia? The application. *Renew. Sust. Energ. Rev.* **2016**, *57*, 83–93. [CrossRef]
36. Zhao, Y.; Ma, S.; Fan, J.; Cai, Y. Examining the effects of land use on carbon emissions: Evidence from Pearl River Delta. *Int. J. Environ. Res. Public Health* **2021**, *18*, 3623. [CrossRef]
37. Zeng, S.; Nan, X.; Liu, C.; Chen, J. The response of the Beijing carbon emissions allowance price (BJC) to macroeconomic and energy price indices. *Energy Policy* **2017**, *106*, 111–121. [CrossRef]
38. Qin, H.; Huang, Q.; Zhang, Z.; Lu, Y.; Li, M.; Xu, L.; Chen, Z. Carbon dioxide emission driving factors analysis and policy implications of Chinese cities: Combining geographically weighted regression with two-step cluster. *Sci. Total Environ.* **2019**, *684*, 413–424. [CrossRef]
39. Zhang, Y.; Zhang, K. The linkage of CO_2 emissions for China, EU, and USA: Evidence from the regional and sectoral analyses. *Environ. Sci. Pollut. Res.* **2018**, *25*, 20179–20192. [CrossRef]
40. Tan, X.; Dong, L.; Chen, D.; Gu, B.; Zeng, Y. China's regional CO_2 emissions reduction potential: A study of Chongqing city. *Appl. Energy* **2016**, *162*, 1345–1354. [CrossRef]
41. Zhang, K.; Zhang, Z.; Liang, Q. An empirical analysis of the green paradox in China: From the perspective of fiscal decentralization. *Energy Policy* **2017**, *103*, 203–211. [CrossRef]
42. Wang, S.; Zeng, J.; Huang, Y.; Shi, C.; Zhan, P. The effects of urbanization on CO_2 emissions in the Pearl River Delta: A comprehensive assessment and panel data analysis. *Appl. Energy* **2018**, *228*, 1693–1706. [CrossRef]
43. Zhang, F.; Jin, G.; Li, J.; Wang, C.; Xu, N. Study on dynamic total factor carbon emission efficiency in China's urban agglomerations. *Sustainability* **2020**, *12*, 2675. [CrossRef]
44. Sadorsky, P. The effect of urbanization on CO_2 emissions in emerging economies. *Energ. Econ.* **2014**, *41*, 147–153. [CrossRef]
45. Tang, Y.; Yang, Y.; Xu, H. The impact of China carbon emission trading system on land use transition: A macroscopic economic perspective. *Land* **2022**, *11*, 41. [CrossRef]
46. Martinez-Zarzoso, I.; Maruotti, A. The impact of urbanization on CO_2 emissions: Evidence from developing countries. *Ecol. Econ.* **2011**, *70*, 1344–1353. [CrossRef]

47. Castells, M. High technology, economic restructuring and the urban regional process in the United States. *High Technol. Spaces Soc.* **1985**, *28*, 11–40.
48. Zhong, C.; Arisona, S.M.; Huang, X.; Batty, M.; Schmitt, G. Detecting the dynamics of urban structure through spatial network analysis. *Int. J. Geogr. Inf. Sci.* **2015**, *28*, 2178–2199. [CrossRef]
49. Roth, C.; Kang, S.M.; Batty, M.; Barthelemy, M. Structure of urban movements: Polycentric activity and entangled hierarchical flows. *PLoS ONE* **2011**, *6*, e15923. [CrossRef]
50. Shen, L.; Gu, Z. Integration of regional space of flows and construction of global urban network. *Sci. Geol. Sin.* **2009**, *29*, 787–793. (In Chinese)
51. Manuel, C. *The Informational City: Information Technology, Economic Restructuring and the Urban-Regional Progress*; Blackwell: Oxford, UK; Cambridge, MA, USA, 1989.
52. Wang, S.; Gao, S.; Wang, Y. Spatial structure of the urban agglomeration based on space of flows: The study of the Pearl River Delta. *Geogr. Res.* **2019**, *38*, 1849–1861. (In Chinese)
53. Xia, C.; Zhang, A.; Wang, H.; Zhang, B.; Zhang, Y. Bidirectional urban flows in rapidly urbanizing metropolitan areas and their macro and micro impacts on urban growth: A case study of the Yangtze River middle reaches megalopolis, China. *Land Use Policy* **2019**, *82*, 158–168. [CrossRef]
54. Li, W.; Wu, Q.; Wu, K.; Wang, Y.; Zhang, H.; Zhou, Z. Spatial structure of the urban agglomeration based on space of flows: The study of the Pearl River Delta. *Area Res. Dev.* **2021**, *40*, 20–26. (In Chinese)
55. Zong, H.; Hao, L.; Dai, J. Study of urban network structure in Chengdu-Chongqing economic circle based on Baidu Index. *J. Southwest Univ.* **2021**, *44*, 36–45 (In Chinese)
56. Liu, C.; Zeng, J. The calculating method about the comprehensive transport accessibility and its correlation with economic development at county level: The statistical analysis of 79 counties in Hubei Province. *Geogr. Res.* **2011**, *30*, 2209–2221. (In Chinese)
57. Chen, M.; Lu, D.; Zhang, H. Comprehensive evaluation and the driving factors of China's urbanization. *Acta. Geogr. Sin.* **2009**, *64*, 387–398. (In Chinese)
58. Zhao, Y.; Wang, S.; Zhou, C. Understanding the relation between urbanization and the eco-environment in China's Yangtze River Delta using an improved EKC model and coupling analysis. *Sci. Total Environ.* **2016**, *571*, 862–875. [CrossRef] [PubMed]
59. Wang, Y.; Li, G. Mapping urban CO_2 emissions using DMSP/OLS 'city lights' satellite data in China. *Environ. Plann. A* **2017**, *49*, 248–251. [CrossRef]
60. IPCC. *IPCC Guidelines for National Greenhouse Gas Inventories*; IGES: Kanagawa, Japan, 2006.
61. Shan, Y.; Huang, Q.; Guan, D.; Hubacek, H. China CO_2 emission accounts 2016–2017. *Sci. Data* **2020**, *7*, 54. [CrossRef]
62. York, R.; Rose, E.; Dietz, T. STIRPAT, IPAT and IMPACT: Analystic tools for unpacking the draving forces of environmental impacts. *Ecol. Econ.* **2003**, *46*, 351–365. [CrossRef]
63. He, J.; Wang, S.; Liu, Y.; Ma, H.; Liu, Q. Examining the relationship between urbanization and the eco-environment using a coupling analysis: Case study of Shanghai, China. *Ecol. Indic.* **2017**, *77*, 185–193. [CrossRef]
64. Shao, S.; Yang, L.; Gan, C.; Cao, J.; Geng, Y.; Guan, D. Using an extended LMDI model to explore techno-economic drivers of energy-related industrial CO_2 emission changes: A case study for Shanghai (China). *Renew. Sust. Energy. Rev.* **2016**, *55*, 516–536. [CrossRef]
65. Hanif, I.; Raza, S.M.F.; Gago-de-Santos, P.; Abbas, Q. Fossil fuels, foreign direct investment, and economic growth have triggered CO_2 emissions in emerging Asian economies: Some empirical evidence. *Energy* **2019**, *171*, 493–501. [CrossRef]
66. Leng, C.; Zain-ul-Abidin, S.; Majeed, W.; Raza, S.M.F.; Ahmad, I. The non-linear relationship between carbon dioxide emissions, financial development and energy consumption in developing European and Central Asian economies. *Environ. Sci. Pollut. Res.* **2021**, *28*, 63330–63345.
67. Batool, Z.; Raza, S.M.F.; Ali, S.; Ul Abidin, S.Z. ICT, renewable energy, financial development, and CO_2 emissions in developing countries of East and South Asia. *Environ. Sci. Pollut. Res.* **2022**, *29*, 35025–35035. [CrossRef]
68. Zhang, X.; Zhao, X.; Jiang, Z.; Shao, S. How to achieve the 2030 CO_2 emission-reduction targets for China's industrial sector: Retrospective decomposition and prospective trajectories. *Glob. Environ. Chang.* **2017**, *44*, 83–97. [CrossRef]
69. Wang, H.; Yang, Q. Seventy years of population dynamics and the current aging challenge: A literature review and policy analysis. *J. Macro-Qual. Res.* **2019**, *7*, 30–54. (In Chinese)
70. Zhang, Y.; Liu, Z.; Zhang, H.; Tan, T. The impact of economic growth, industrial structure and urbanization on carbon emission intensity in China. *Nat. Hazards* **2014**, *73*, 579–595. [CrossRef]
71. Madlener, R.; Sunak, Y. Impacts of urbanization on urban structures and energy demand: What can we learn for urban energy planning and urbanization management? *Sustain. Cities Soc.* **2011**, *1*, 45–53. [CrossRef]
72. Jiang, Y.; Batool, Z.; Raza, S.M.F.; Haseeb, M.; Ali, S.; Ul Abidin, S.Z. Analyzing the Asymmetric Effect of Renewable Energy Consumption on Environment in STIRPAT-Kaya-EKC Framework: A NARDL Approach for China. *Int. J. Env. Res. Public Health* **2022**, *19*, 7100. [CrossRef]
73. Liang, L.; Wang, Z.; Li, J. The effect of urbanization on environmental pollution in rapidly developing urban agglomerations. *J. Clean. Prod.* **2019**, *237*, 117649. [CrossRef]

74. He, D.; Liu, H.; He, K.; Meng, F.; Jiang, Y.; Wang, M.; Zhou, J.; Calthorpe, P.; Guo, J.; Yao, Z.; et al. Energy use of, and CO_2 emissions from China's urban passenger transportation sector—Carbon mitigation scenarios upon the transportation mode choices. *Transp. Res. Part-Policy Pract.* **2013**, *53*, 53–67. [CrossRef]
75. Fan, F.; Lei, Y. Decomposition analysis of energy-related carbon emissions from the transportation sector in Beijing. *Transp. Res Part-Transp. Environ.* **2016**, *42*, 135–145. [CrossRef]

land

Article

Carbon Metabolism in Urban "Production–Living–Ecological" Space Based on Ecological Network Analysis

Xinhui Feng [1], Yan Li [1,*], Lu Zhang [2], Chuyu Xia [3], Er Yu [1] and Jiayu Yang [1]

1 School of Public Affairs, Institute of Land Science and Property Management, Zhejiang University, Hangzhou 310058, China
2 Eco-Environmental Science Research & Design Institute of Zhejiang Province, Hangzhou 310007, China
3 Faculty of Architecture, Civil and Transportation Engineering, Beijing University of Technology, Beijing 100124, China
* Correspondence: liyan522@zju.edu.cn

Abstract: To understand the changing pattern of urban carbon metabolism from the perspective of urban "production–living–ecological" (PLE) space, taking Suzhou City as an example, this study constructed a carbon metabolic network model in urban PLE space, analyzed the changes of horizontal carbon flow, and evaluated the comprehensive effect of the PLE space changes using the ecological network analysis method. The results showed that the total carbon sequestration showed a fluctuating change of increasing and then decreasing, while the total carbon emissions grew dramatically. Production spaces were the key nodes for the generation of horizontal carbon flow. The exploitation relationship was the dominant ecological relationship in the network, the mutualism relationship was abundant from 2005 to 2010 and gradually decreased from 2010 to 2018, and the frequency of competition relationship appeared gradually increased. The ecological network hierarchy evolved from an irregular shape dominated by primary consumers in 2000–2005 to a pyramidal shape dominated by producers in 2010–2018 at the driving weight end, and the pull weight showed a declining trend, with pull weight of producers increasing from 1.72% to 24.33%. The results can provide a theoretical basis for planning adjustments to the city's PLE space structure to achieve low-carbon goals.

Keywords: urban carbon metabolism; ecological network analysis; ecological relationship; "production-living-ecological" space

Citation: Feng, X.; Li, Y.; Zhang, L.; Xia, C.; Yu, E.; Yang, J. Carbon Metabolism in Urban 'Production–Living–Ecological" Space Based on Ecological Network Analysis. *Land* **2022**, *11*, 1445. https://doi.org/10.3390/land11091445

Academic Editor: Marko Scholze

Received: 7 July 2022
Accepted: 25 August 2022
Published: 31 August 2022

1. Introduction

Urban areas, which account for about 2% of the global land area, contribute nearly 75% of global carbon emissions and are one of the major causes of global warming [1]. Land use change in the urban "production–living–ecological" (PLE) space is an important source of urban carbon emissions [2]. Land use change, as the dual body of both carbon emissions and sequestration, directly transforms or indirectly affects the carbon emission and sequestration processes between the terrestrial ecosystem and the atmosphere by changing the land use type, function, and structure [3,4]. Therefore, it is necessary to understand the linkage between urban carbon and the change of PLE space to identify the critical path of carbon emissions reduction through land use adjustment in spatial planning [5]. By constructing a carbon metabolic network model in urban PLE space based on land use change, we can simulate and account for the carbon metabolism flow in cities and analyze the ecological relationship and hierarchical structure in the process of carbon metabolism, which will firstly help to reflect the interrelationship and coordination between the urban economy, environment, and resources [6]. Secondly, it can measure the changes in carbon metabolism flow, metabolism anomalies, and the rationality of urban PLE space structure, to provide a scientific reference for regulating the process of urban carbon metabolism through the optimization of territorial space, and for building a more low-carbon and healthy territorial space pattern.

Urban metabolism refers to the flow of substances and energy through the whole urban system, which is analogous to the process of biological metabolism. The study of it provides an efficient way to understand sustainable urban development [7,8]. In recent years, various studies on the flow processes of single elements or substances such as water [9], nitrogen [10], and energy [11] in cities based on the concept of urban metabolism have been conducted. As climate change, mainly characterized by global warming, has become a worldwide environmental issue [12], carbon as a key element of urban metabolism has gradually attracted the attention of researchers. Ye et al. [13] explored the impacts of building characteristics and residential energy consumption on urban carbon emissions with linear models. Kennedy et al. [14] quantitatively measured carbon emissions from industrial activities such as transportation, electricity production, and coal combustion based on the data sets of comprehensive energy consumption in major cities globally. Li et al. [15] investigated the carbon emissions efficiency of various industrial sectors in Guangdong Province, China and the carbon transfer between them based on the input–output analysis, to explore reasonable energy conservation and emissions reduction strategies. As for the methodology of urban carbon metabolism research, the ecological network analysis (ENA) method can determine the flow direction and intensity of ecological elements in the network, which is a mainstream way to quantify the horizontal carbon flow and its ecological impact [16]. For example, Zhang et al. identified the energy flow process between 28 socio-economic sectors—such as industry and transportation—constructed a network model of Beijing City, and analyzed its carbon footprint. The study reported that the sectors of social services, consumption, and heavy industry were the key nodes [17]. And they constructed a network model of the carbon metabolism process based on land transfer, noting the transformation between transportation and agriculture as the key path of the network [18].

However, most of previous studies have focused on the carbon flow in different industrial sectors of the urban socio-economic system, but few have investigated the influence mechanism of the transition of urban PLE space on carbon metabolism. In addition, there is a lack of detailed studies on the ecological relationship and hierarchical structure among different carbon metabolism subjects. In addition, many studies on ecological network analysis and the simulation of carbon metabolic processes focus on the influence of socio-economic activities on urban carbon metabolic processes, and are less concerned with differences in the attributes of the natural environment or various natural subjects. There are also a few studies that attempt to distinguish different attributes within the environment in the urban carbon metabolism process. For instance, Zhang et al. [18] and Xia et al. [19] included natural metabolic agents with different attributes (woodlands, grasslands, etc.) into the ecological network model for the first time in their analysis and simulation of a carbon metabolism network in Beijing City. However, studies on the participation of natural subjects in carbon metabolic processes are still relatively lacking, and the spatialized representation of carbon metabolic processes is still insufficient.

The classification of urban land functions from the perspectives of PLE is coincident with the concept of the "three pillars" (economic, social, and ecological) of international sustainable development [20,21], which has been recognized by many researchers in recent years [22,23]. The PLE space classification of land use is in accordance with the requirements of territorial spatial planning [24], fully reflecting the PLE dominant function of land use, which is a scientific way to optimize the exploitation of material resources and urban spatial structure [25]. The construction of the urban carbon metabolic network from the perspective of PLE space not only takes into account the influence of socio-economic activities on the urban carbon metabolic process, but also incorporates natural subjects with different attributes into the analysis of the carbon metabolic process, which can make up for the deficiencies of previous studies in this regard. Moreover, the ENA method can deeply explore the changes in the ecological relationship and hierarchical structure of different carbon metabolic subjects due to the transition of PLE space. It is more conducive to policymakers carrying out land planning from a socio-economic macro perspective to promote urban carbon neutrality, taking into account the overall characteristics of the region at the same time.

With the release of the outline of the integrated regional development in the Yangtze River Delta, the integration of the Yangtze River Delta Economic Zone is elevated to a national strategy, emphasizing the importance of realizing sustainable urban development by means of territorial space optimization [26,27]. As an important central city in the Yangtze River Delta region, Suzhou is a key node for promoting the integrated ecological and green development of the Yangtze River Delta [28]. Therefore, this study focused on the complex impacts of PLE space changes on urban carbon metabolism, taking Suzhou City as an example. We constructed a carbon metabolic network model based on the classification of PLE space, accounted for the carbon emissions and sequestration of each node in the network, and analyzed the changes in horizontal carbon flow among the nodes in the network with the land use transition matrix. Then we evaluated the influences of the change in urban PLE space on the carbon metabolic network by ENA. The results of the study can provide a scientific reference for optimizing urban spatial layout and promoting low-carbon urban land use.

2. Materials and Methods

2.1. Study Area

Suzhou City (119°55′–121°20′ E, 30°47′–32°02′ N) is located in the southeast of Jiangsu Province and on the east bank of Taihu Lake (Figure 1), with a total area of 8657.32 km^2. The topography in Suzhou is flat and low, with plains accounting for 54.8% of the total area, numerous lakes, and dense rivers. As of 2018, Suzhou had a resident population of 10.72 million people, an urbanization rate of 76.05%, and a GDP of CNY 1.86 trillion (USD 0.27 trillion), accounting for 20.08% of the Jiangsu Province.

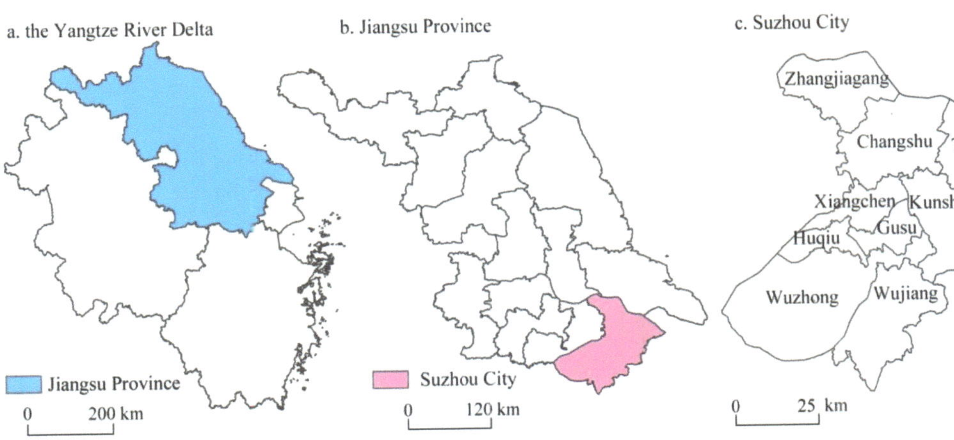

Figure 1. Location of Suzhou City.

With the rapid development of Suzhou's economy and the gradual expansion of its industrial scale, the total carbon emissions from socio-economic activities have been increasing, and the ecological spaces with carbon sink function such as forest and water have been compressed due to the continuous promotion of the urbanization process, which has posed a great challenge for Suzhou in achieving the "carbon peak and carbon neutral" strategic goal. How to optimize the spatial layout of the city and explore long-term and profound emissions reduction paths while ensuring scientific development has become the key to its low-carbon transformation.

2.2. Data Sources

The socio-economic data used in this study, including agriculture, industry, populatior energy consumption data and so on, were derived from the Suzhou Statistical Yearbook Jiangsu Statistical Yearbook, China Transport Statistical Yearbook, China Energy Statistica Yearbook, and Suzhou National Economic and Social Development Statistical Bulletin fo the corresponding years, etc. (http://tj.jiangsu.gov.cn/, accessed on 22 April 2022). Lanc use data with a resolution of 30 m in 2000, 2005, 2010, and 2018, including 6 primary land categories such as cultivated land and forest land and 25 secondary land categories, were provided by the Chinese Academy of Sciences Resource and Environmental Science Data Center (http://www.resdc.cn, accessed on 22 April 2022). Based on the PLE dominant function of land use, and referring to the research of Yang et at. [29], we merged the basic land use types taking their dominant function into account and established a classification system of the PLE space (Table 1).

Table 1. Classification system of the PLE space and carbon metabolism function.

PLE Space Types Based on PLE Dominant Function			**Secondary Classification of Land Use System**	**Carbon Metabolism Function**
First Category	**Code**	**Secondary Category**		
Production space	CU	Agricultural production space	Paddy field, dry farm	carbon source, carbon sink
	IN	Industrial production space	Industrial, mining, and transportation construction land	carbon source
Living space	UR	Urban living space	Urban land	carbon source
	RU	Rural living space	Rural settlements	carbon source
Ecological space	G	Grassland ecological space	Grassland with low, medium, and high cover	carbon sink
	F	Forest ecological space	Open woodland, shrubland, wooded land, other woodland	carbon sink
	W	Water ecological space	Rivers, lakes	carbon sink
	B	Other ecological space	Bare rocky gravel land, bare land, saline land	-

2.3. Methods

2.3.1. The Carbon Metabolic Network Model

The carbon metabolic network model in urban PLE space constructed in this study took into account the carbon exchange of network nodes between the biosphere and the atmosphere, as well as the potential carbon flow in the horizontal direction resulting from the transitions between the PLE space types (Figure 2). The manner of carbon exchange between each network node and the atmosphere determines its carbon metabolism function (Table 1). In this study, IN, UR, and RU were the socio-economic components of the urban system, which were the main sources of urban carbon emissions and possessed the carbon source function; G, F, and W were the natural compartments of the urban system, which were responsible for the carbon sequestration process in urban carbon metabolism and possessed the carbon sink function; CU had dual attributes and assumed both functions. B was hardly used as the carrier of socio-economic activities, so it assumed no carbon exchange with the atmosphere.

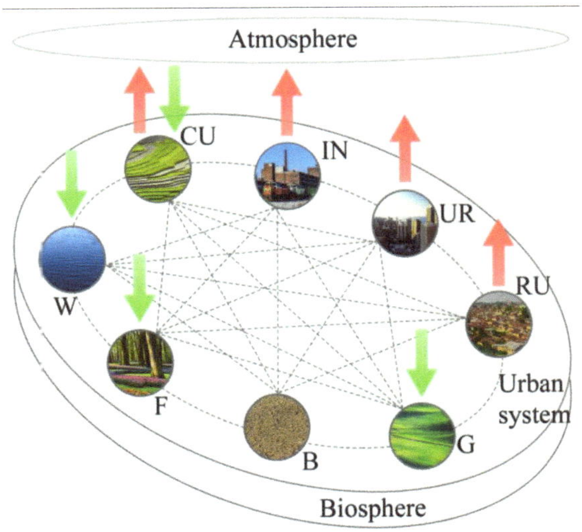

Figure 2. The conceptual diagram of the carbon metabolic network model in urban PLE space. Note: IN, industrial production space; UR, urban living space; RU, rural living space; G, grassland ecological space; F, forest ecological space; W, water ecological space; CU, agricultural production space; B, other ecological space; Red arrow, carbon source function; Green arrow, carbon sink function.

1. Carbon Emissions Accounting

Carbon emissions from CU were mainly from agricultural production activities, livestock respiration, and crop cultivation [30]. The agricultural production activities mainly included crop sowing, agricultural machinery activities, irrigation process, and fertilizer application; the livestock types mainly included pigs, cattle, and sheep according to the specific characteristics of the study area; the agricultural crop in this study was mainly rice. Its total carbon emissions ECU was specifically calculated as followed:

$$E_{CU} = E_a + E_b + E_c \tag{1}$$

$$E_a = K_1 f_1 + K_2 f_2 + K_3 f_3 + K_4 f_4 \tag{2}$$

$$E_b = K_5 f_5 + K_6 f_6 + K_7 f_7 \tag{3}$$

$$E_c = K_8 f_8 \tag{4}$$

where E_{CU} was the total carbon emissions from CU; E_a was the carbon emissions from agricultural production activities; E_b was the carbon emissions from livestock respiration; E_c was the carbon emissions from crop respiration; K_1 was the total sown area of crops; K_2 was the total power of agricultural machinery; K_3 was the irrigated area; K_4 was the fertilizer application; K_5, K_6, and K_7 were the annual pig, cattle, and sheep rearing in the study area, respectively; K_8 was the rice cultivation area; f_i was the carbon emissions conversion coefficient corresponding to each agricultural activity (Supplementary Table S1).

IN included industrial, mining, and transportation construction land, so its carbon emissions were also divided into industrial energy consumption carbon emissions and transportation carbon emissions. Its total carbon emissions E_{IN} were specifically calculated as follows:

$$E_{IN} = E_d + E_e \tag{5}$$

$$E_d = \sum_{i=1}^{n} C_i \times d_i \tag{6}$$

$$E_e = K_9 f_9 + K_{10} f_{10} + K_{11} f_{11} + K_{12} f_{12} + K_{13} f_{13} + K_{14} f_{14} \tag{7}$$

where E_{IN} was the total carbon emissions from IN; E_d was the carbon emissions from industrial energy consumption; E_e was the carbon emissions from transportation; C_i was the consumption of the ith energy source; d_i was the carbon emissions coefficient of the ith energy source; K_9, K_{10}, K_{11}, and K_{12} were the annual running kilometers of buses, cabs, private cars, and motorcycles, respectively; K_{13} and K_{14} were the annual passenger and freight volume of roads and railroads, respectively; f_i was the carbon emissions coefficient of each transportation type (Supplementary Table S1).

Carbon emissions from UR were mainly the carbon emissions of urban residents' respiration and the carbon generated by urban residents' electricity consumption. Similarly, the carbon emissions from RU mainly came from the carbon emissions of rural residents' respiration and the carbon generated by rural residents' electricity consumption. Their total carbon emissions E_{UR} and E_{RU} were specifically calculated as followed:

$$E_{UR}= K_{15}f_{15}+K_{16}f_{16} \quad (8)$$

$$E_{RU}= K_{17}f_{17}+K_{18}f_{18} \quad (9)$$

where E_{UR} was total carbon emissions from UR; E_{RU} was total carbon emissions from RU; K_{15} was urban living electricity consumption; K_{16} was urban resident population; K_{17} was rural living electricity consumption; K_{18} was rural resident population; f_i was carbon emissions coefficient of each index (Supplementary Table S1).

2. Carbon Sink Accounting

CU, G, F, and W can absorb carbon from the atmosphere and sequester it. In this study, the carbon sink coefficient was used to calculate the amount of carbon sink by combining the area of each space that possessed the carbon sink function. The specific calculation formula was as follows:

$$V_i = \sum_{i=1}^{n} S_i \times k_i \quad (10)$$

where V_i was the carbon sequestration of the ith type of PLE space; S_i was the area of the ith type of PLE space; k_i was the carbon sink coefficient of the ith type of PLE space (Supplementary Table S2). In order to improve the accuracy of calculating the carbon sequestration of each space, all secondary land types were considered in this study.

3. Horizontal Carbon Flow Accounting

In this study, the net carbon emissions or sequestration per unit area (i.e., carbon metabolism density) of each PLE space type was used to characterize the carbon metabolism capacity of the PLE space type. The change of PLE space type led to the change in its carbon metabolism density, and the horizontal carbon flow among network nodes can be quantified according to the change of area of each PLE space type, which was reflected by the land use transfer matrix [31]. The specific calculation formula was as follows:

$$f_{ij} = \Delta W \times \Delta S \quad (11)$$

$$\Delta W = W_i - W_j = \frac{V_i}{S_i} - \frac{V_j}{S_j} \quad (12)$$

where f_{ij} was the horizontal carbon flow from PLE space type j to i; ΔW was the difference in carbon metabolism density between PLE space type i and j; ΔS was the area transferred from PLE space type j to i; S_i and S_j were the areas of PLE space type i and j, respectively; V_i and V_j were the net carbon exchange between PLE space type i/j and the atmosphere. Because carbon metabolism capacity included carbon emissions and sequestration, there was positive and negative horizontal carbon flow among the PLE spaces in the network. If $\Delta W > 0$, it indicated a positive carbon flow, which represented a decrease in carbon emissions or an increase in carbon sequestration, and helped to alleviate the carbon metabolism imbalance of the urban system; if $\Delta W < 0$, it indicated a negative carbon flow, which represented an increase in carbon emissions or a decrease in carbon sequestration, aggravating the carbon metabolism imbalance in the urban system. Based on the above equation, the carbon flow among PLE spaces was calculated to constitute the direct carbon flow matrix F in the horizontal direction.

2.3.2. Ecological Network Analysis (ENA) Method

The ENA method originated from the analysis of currency circulation in economic problems. In 1973, Hannon first applied the input–output model in economics to analyze the energy flow in an ecosystem and the hierarchical structure of the entire ecosystem, forming the prototype of the ENA method [32]. As research progressed, the ENA method was widely used to analyze the structure of ecosystems and various ecological attributes, and the quantitative models used were also enriched [33–35]. In this study, we used the throughflow, utility, and structure analysis of the ENA method to quantitatively assess the carbon throughflow of each node, the ecological relationships between each PLE space type, and the hierarchical structure of the network, to reflect the impact of urban PLE space changes on urban carbon metabolism.

1. Throughflow analysis

Each PLE space type had the potential for absorbing carbon or releasing carbon from the environment. In this study, the state variable x_k was defined to reflect the change in the carbon stock of PLE space type k to balance the total outflow and inflow of carbon in the network [36]. If $x_k > 0$, then the carbon throughflow T_k of the space was equal to the sum of all carbon flow out of the space plus the state variable; if $x_k < 0$, then the carbon throughflow T_k of the space was equal to the sum of all carbon flow into the space minus the state variable [19,37]. The specific calculation of throughflow $\mathrm{T_k}$ was as follows:

$$T_k = \sum_{i=1}^{n} f_{ik} + y_k + (x_k)_+ = \sum_{j=1}^{n} f_{kj} + z_k - (x_k)_- \quad (13)$$

where f_{ik} was the horizontal carbon flow from PLE space type k to i; y_k was the carbon flow released from PLE space type k to the environment; $(x_k)_+$ represented the increase of carbon stock; f_{kj} was the horizontal carbon flow from PLE space type j to k; z_k was the carbon flow absorbed from the environment by PLE space type k; $(x_k)_-$ represented the decrease of carbon stock; n was the number of network nodes, and n was 8 in this study.

2. Utility analysis

Based on the direct carbon flow matrix F calculated above and the throughflow analysis results, with reference to the results obtained in previous research [38,39], the utility analysis method was used to determine the ecological relationships between each PLE space type. The calculation steps were as follows: ① Based on the carbon throughflow of each PLE space type, the direct carbon flow matrix F was normalized to compose the direct utility matrix D, which reflected the direct utility of the horizontal carbon flow between each node. ② Based on the direct utility matrix D, the overall utility matrix U was calculated to fully reflect the direct or indirect utility relationships of the horizontal carbon flow. ③ Based on the properties of positive and negative of the elements in the overall utility matrix U, the ecological relationships in the network were determined; ④ the mutualism index MI and the synergy index SI were defined to quantify the comprehensive impact of the PLE space changes on the urban carbon metabolic network. The specific calculation was as follows:

$$d_{ij} = (f_{ij} - f_{ji})/T_i \quad (14)$$

$$U = (u_{ij}) = D^0 + D^1 + D^2 + D^3 + \ldots + D^m = (I - D)^{-1} \quad (15)$$

$$MI = \frac{N_+}{N_-} \quad (16)$$

$$SI = \sum_{j=1}^{n} \sum_{i=1}^{n} u_{ij} \quad (17)$$

where d_{ij} was the net utility of carbon flow from PLE space type j to i; u_{ij} represented the elements in the overall utility matrix U; I represented the unit matrix; the superscript m of the direct utility matrix D represented the m paths that the carbon flow had to go through to reach the final PLE space type; $N_{+/-}$ represented the number of positive/negative elements in the overall utility matrix U. If MI > 1, it indicated that the PLE space changes had a positive effect on the urban carbon metabolic network, and the larger the *MI*, the stronger the positive effect. If MI < 1, it indicated that the PLE space changes had a negative effect on the urban carbon metabolic network, and the larger the *MI*, the stronger the negative effect. *SI* was the sum of the elements in the overall utility matrix U. A larger SI meant that the positive effect of horizontal carbon flow was more obvious, and the urban carbon metabolic network was more coordinated.

Based on the properties of positive and negative of elements in the overall utility matrix U, the ecological relationships between each PLE space type in the network can be determined (Table 2). Although there were theoretically nine ecological relationships, four of them were more common (mutualism, competition, exploitation, and control) [40,41]. The control and exploitation relationships were similar in nature, so we combined them into a single category (exploitation relationship), both indicating that one PLE space type obtained more carbon storage in the process of carbon transfer, while the other space type lost carbon storage. The mutualism relationship indicated that both PLE space types obtained more carbon storage as a result of horizontal carbon flow in the network, while the competition relationship indicated the contrary. Both were the outcome of indirect effects of carbon flow on each PLE space type in the network.

Table 2. Ecological relationships between PLE space types in the network.

Signs of Elements in Matrix U	Positive (+)	Neutral (0)	Negative (−)
Positive (+)	(+, +) mutualism	(+, 0) commensalism	(+, −) exploitation
Neutral (0)	(0, +) commensalism host	(0, 0) neutralism	(0, −) amensalism
Negative (−)	(−, +) control	(−, 0) amensal host	(−, −) competition

3. Structure Analysis

The utility analysis can reflect the ecological relationships in the network, but it was difficult to reflect the overall structure of the urban carbon metabolic network, and the level and specific role of each node in the network. Therefore, the biological concepts of producer, consumer, and decomposer were introduced [42], referring to previous research [43,44], to analyze the hierarchical structure of the carbon metabolic network model in urban PLE space (Figure 3), and quantify the capacity of each node to absorb and export horizontal carbon flow to reflect its importance in the network.

Grassland (G), forest (F), and water (W) ecological space, as producers in the network structure, were located at the bottom of network, providing natural resources to the upper layer, and maintaining the carbon balance of the whole carbon metabolic network through carbon sinks. Agricultural production space (CU) and rural living space (RU), as primary consumers, were at the junction of natural and socio-economic compartments, and used the natural resources provided by producers for agricultural production and primary agricultural products processing in order to provide products to more advanced consumers, and their carbon metabolic capacity was also relatively low. Industrial production space (IN), as secondary consumer, carried out secondary processing of agricultural products and transported them to senior consumers, which was an important node in the generation of horizontal carbon flow. Urban living space (UR) acted as a senior consumer and used these products and natural resources to improve human living standards and develop the economy. Other ecological space (B) acted as a decomposer and was highly susceptible to human activities, often serving as a "transfer station" for PLE space type change, e.g., bare space may be formed by the conversion of abandoned buildings in living and production space, but it can be transformed into ecological space through a series of treatment measures. Thus, its ability to release or absorb carbon flow was very low.

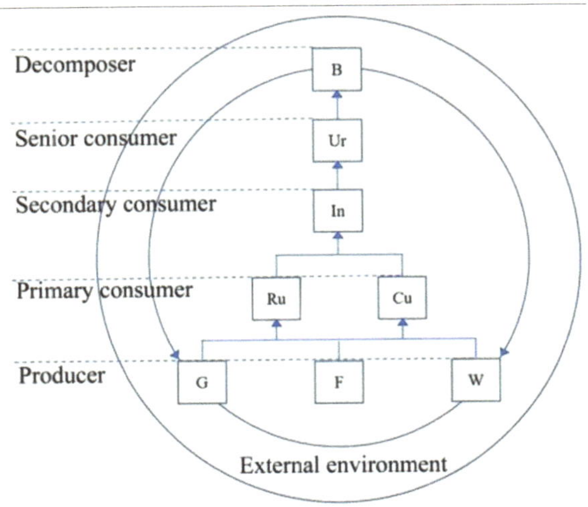

Figure 3. The hierarchical structure of the carbon metabolic network in urban PLE space. Note: IN, industrial production space; UR, urban living space; RU, rural living space; G, grassland ecological space; F, forest ecological space; W, water ecological space; CU, agricultural production space; B, other ecological space; Red arrow, carbon source function; Green arrow, carbon sink function.

Based on the results of the throughflow analysis above, we can obtain the structural utility matrix Y. By calculating the elements in Y, we can derive the driving weight and pull weight of each PLE space type in the network. The driving weight indicated the ability of the space type to influence other space types by exporting horizontal carbon flow, while the pull weight represented the ability of the space type to receive carbon flow [45]. The specific calculation formula was as follows [46]:

$$Y = diag(T)(I - G)^{-1} \tag{18}$$

$$W_j = \sum_{i=1}^{n} y_{ij} \sum_{i=1}^{n}\sum_{j=1}^{n} y_{ij} \tag{19}$$

$$W_i = \sum_{j=1}^{n} y_{ij} \sum_{i=1}^{n}\sum_{j=1}^{n} y_{ij} \tag{20}$$

where I was the unit matrix; $diag(T)$ was the diagonal matrix composed of throughflow T_i; $G = [g_{ij}] = [f_{ij}/T_j]$; f_{ij} was the horizontal carbon flow from PLE space type j to i; T_j was the carbon throughflow of PLE space type j; W_j was the driving weight of PLE space type j; W_i was the pull weight of PLE space type i.

3. Results

3.1. Carbon Emissions and Carbon Sequestration Changes

The results of carbon emissions and sequestration accounting in Suzhou from 2000 to 2018 are shown in Table 3. The total carbon sequestration in Suzhou showed a fluctuating change from rising to falling during the 18 years. From 2000 to 2005, the total carbon sequestration increased from 12.40×10^4 t C to 12.51×10^4 t C. The carbon sequestration in CU decreased by 9.17%, while the carbon sequestration in W increased by 12%, accounting for 90.05% of the total. From 2005 to 2010, the total carbon sequestration rapidly declined, with an average annual decrease of 4.23%, and the carbon sequestration of the PLE space which possessed the carbon sink function experienced different degrees of decline. From 2010 to 2018, the decline in total carbon sequestration tended to level off, with a rate of 4.93%. The carbon sequestration in CU and F had not changed much, while the carbon sequestration of W decreased 6.43%, which is the most important space type for carbon sink.

Table 3. Accounting of carbon emissions and sequestration of each PLE space type/(t C/a).

PLE Space Type	2000		2005		2010		2018	
	Carbon Emissions	Carbon Sequestration	Carbon Emissions	Carbon Sequestration	Carbon Emissions	Carbon Sequestration	Carbon Emissions	Carbon Sequestration
CU	447,016.27	3012.05	299,002.39	2735.98	259,636.20	2063.94	143,945.71	1976.50
IN	930,984.47	0	2,098,900.08	0	3,758,572.52	0	11,880,749.79	0
UR	341,810.20	0	480,438.64	0	944,859.50	0	1,385,291.59	0
RU	297,185.88	0	336,873.62	0	591,065.42	0	793,878.20	0
G	0	312.00	0	310.99	0	275.41	0	1172.36
F	0	9010.98	0	9034.67	0	8783.14	0	8737.29
W	0	111,668.43	0	113,008.93	0	87,491.62	0	81,862.22
O	0	0	0	0	0	0	0	0

Note: IN, industrial production space; UR, urban living space; RU, rural living space; G, grassland ecological spac F, forest ecological space; W, water ecological space; CU, agricultural production space; B, other ecological space.

The total carbon emissions had been increasing in the past 18 years, resulting in a more serious carbon imbalance in the city. From 2000 to 2005, the total carbon emissions rose, with a rate of 59.41%, and the carbon emission from IN increased from 46.16% to 65.28%, while the carbon emissions from CU decreased from 22.16% to 9.30%. From 2005 to 2010, the total carbon emissions continued to rise to 5.55×10^6 t C, the carbon emissions from IN increased to 3.76×10^6 t C, and the carbon emissions from UR increased by 96.67% accounting for 17.01% of the total. From 2010 to 2018, the total carbon emissions showed an exponential increase, and the annual average increase of 27.01% from IN was an important reason for this.

3.2. Horizontal Carbon Flow Changes

The results of the horizontal carbon flow accounting are shown in Supplementary Tables S3–S5. The total amount of negative carbon flow in Suzhou shows an increase over the three study periods. From 2000 to 2005, the total negative carbon flow was 1.22×10^6 t C, mainly from the transition of CU→IN and CU→UR, accounting for 78.22% and 10.07% of the total, respectively. From 2005 to 2010, the total negative carbon flow was 3.26×10^6 t C, which was 2.68 times of the previous study period, and the transition of CU→IN and W→IN contributed 59.25% and 13.76% of the total, respectively. From 2010 to 2018, the total negative carbon flow rose further, to 7.26×10^6 t C, mainly from CU→IN, W→IN, and G→IN, which accounted for 45.35%, 40.34%, and 6.90% of the total, respectively.

The spatial distribution of negative carbon flow is shown in Figure 4. In order to reveal the multifaceted characteristics of the spatial changes of carbon flow more comprehensively, Standard deviational ellipse (SDE), a spatial statistical method widely used in many fields such as sociology and geology was introduced (Supplementary Figure S1). From 2000 to 2005, the high-value regions of negative carbon flow were spatially concentrated in the middle of the study area, represented by Gusu and Kunshan. From 2005 to 2010, the area of negative carbon flow increased significantly, with larger negative carbon flow appearing in a block-like pattern in northeastern Gusu and central Changshu, and more scattered distribution in Kunshan and Taicang. From 2010 to 2018, the area of negative carbon flow was reduced, the frequency of low-value negative carbon flow increased, and the high-value negative carbon flow shifted to a scattered distribution pattern. The SDE analysis showed that the negative carbon flow spread along the north–south direction, mainly in the central and northern part of Suzhou; the short axis showed a trend of first narrowing and then increasing, and the azimuth angle was from $-10° \rightarrow -17° \rightarrow 7°$, indicating that the spatial distribution of negative carbon flow was more concentrated and dense during 2005–2010 compared with the previous period, and Kunshan, which was the core area of industrialization development of Suzhou, produced a greater pulling effect. During 2010–2018, the spatial distribution of negative carbon flow became more discrete. The less-developed Taicang raised the speed of economic growth and became the main driving force of the SDE azimuth deflection.

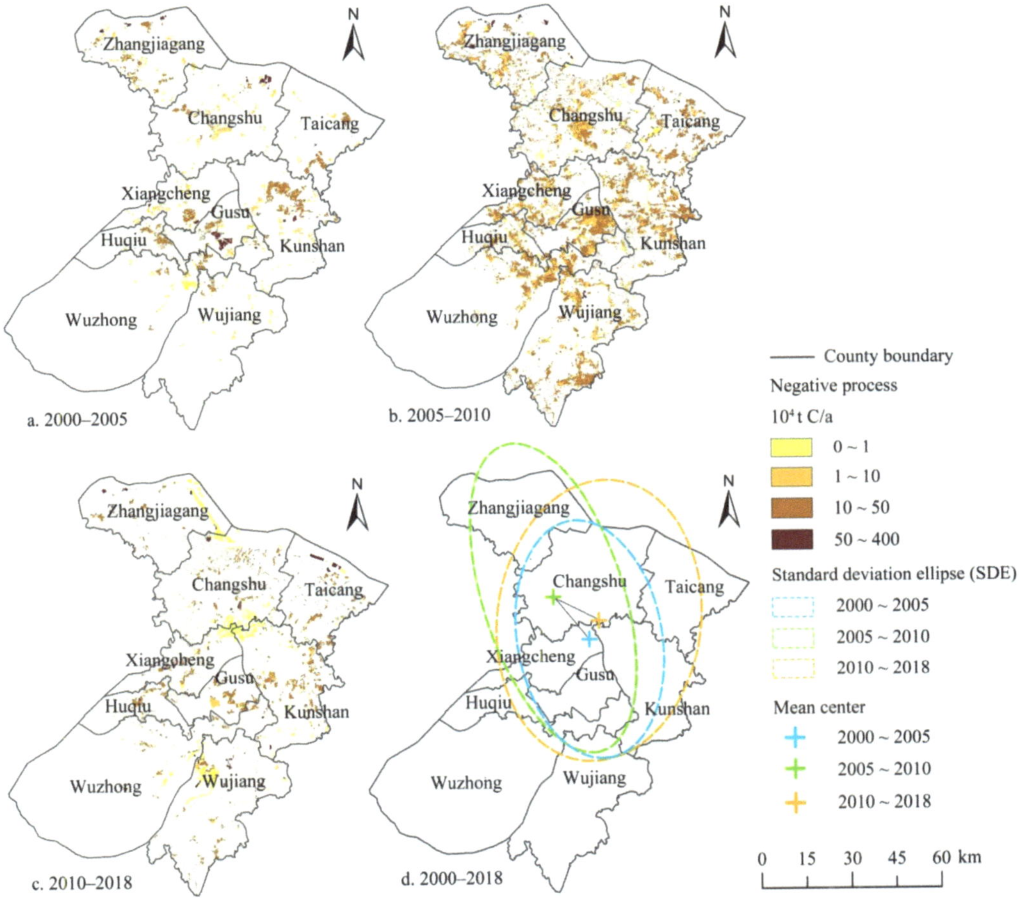

Figure 4. Spatial distribution of negative carbon flow.

The total positive carbon flow showed an upward and then downward trend during three study periods; it was difficult to offset the huge negative carbon flow in general, and thus the urban carbon balance remained under great pressure. From 2000 to 2005, the total positive carbon flow was 9.88×10^4 t C, which was only 8.12% of the total negative carbon flow, mainly from the transition of RU→CU, IN→UR, and IN→CU, accounting for 21.52%, 18.13%, and 12.50% of the total, respectively. From 2005 to 2010, the total positive carbon flow showed a huge increase, which can offset 58.30% of the negative carbon flow, mainly from the transition of IN→UR, which accounted for 83.05% of the total. From 2010 to 2018, the total positive carbon flow was 7.09×10^5 t C, accounting for 9.76% of the total negative carbon flow, mainly from the transition of IN→G, which contributed 36.47% of the total. Meanwhile, the transition of RU→CU and IN→W also contributed 8.13% and 4.87% of the total, respectively.

The spatial distribution of positive carbon flow changes is shown in Figure 5. From 2000 to 2005, the positive carbon flow had a scattered distribution pattern in all counties and districts with relatively low value. From 2005 to 2010, the positive carbon flow appeared in a large area and high-value regions mainly concentrated in the central part of Gusu and Kunshan, resulting from the transition of IN→UR. The area of positive carbon flow decreased from 2010 to 2018, and the high-value regions were mainly distributed in the central Gusu, Taicang, and Wujiang, the first from the transition of IN to G, and the latter two mainly from the transition of artificial compartments to natural compartments. The SDE analysis showed that the positive carbon flow appeared along the north–south direction as a whole, similar to the negative flow, but the distribution range and the mean center shifted more southward than the negative. The short axis of the SDE did not change much in each study period, the area of SDE showed a fluctuation trend of first becoming smaller and then larger, and the azimuth angle changed more from 21°→26°→−9°. It showed that the transition of IN in Kunshan and Gusu had a large traction effect from 2005 to 2010. During the following eight years, the areas of positive carbon flow decreased, and the distribution pattern tended to be more fragmented. Huqiu and Wuzhong had played a positive role in the environmental management of Taihu Lake, which contributed to the reverse deflection of the azimuth of SDE.

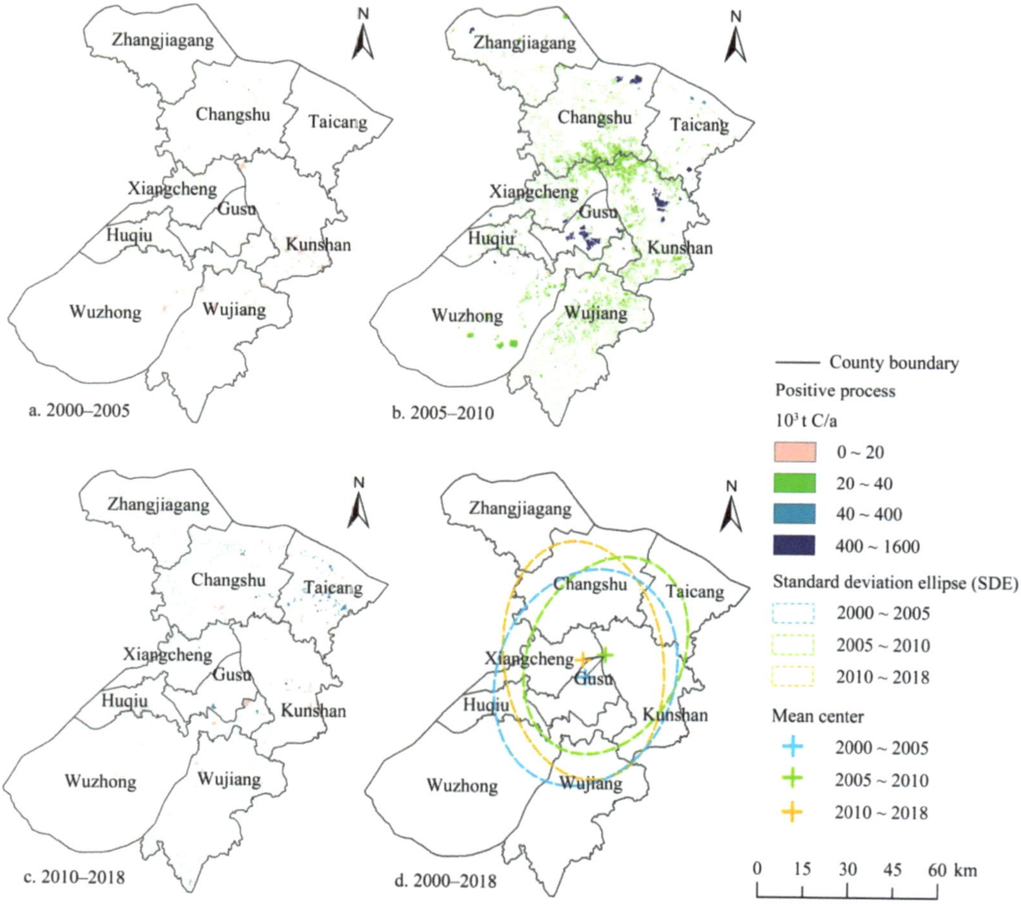

Figure 5. Spatial distribution of positive carbon flow.

3.3. The Ecological Relationships between PLE Spaces

From 2000 to 2018, the proportion of each ecological relationship between PLE spaces in Suzhou had been changing (see Figure 6a, Supplementary Tables S6–S8). From 2000 to 2005, the exploitation relationship was absolutely dominant and accounted for 75% of the total, and each PLE space type was fiercely plundering each other's carbon stock. Although that had unexpectedly improved the overall utility of carbon flow in the network, the development mode in which one side gained and one side lost was not conducive to the long-term carbon health of the city. The competition relationship accounted for 17.86% of the total, mainly existing in such artificial compartments as IN, RU, and UR. The proportion of the mutualism relationship was only 7.14%, all of which were related to B, reflecting their neutralizing role as decomposer in the urban ecological process. From 2010 to 2018, the proportion of the exploitation relationship increased slightly to 53.57%, the proportion of the competition relationship increased to 32.14%, while the proportion of the mutualism relationship decreased to 14.29%.

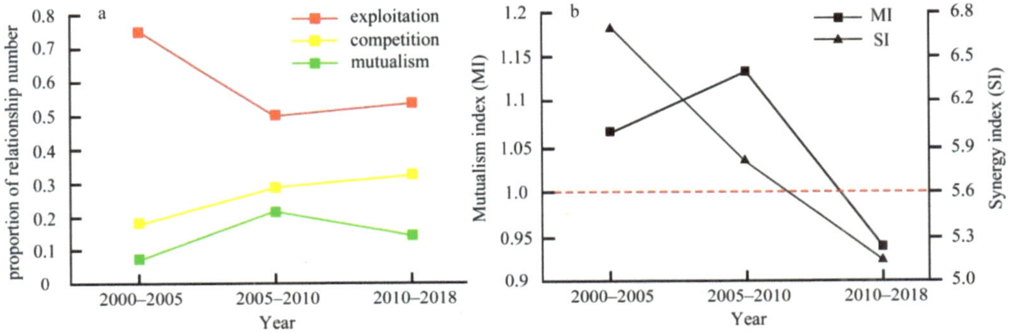

Figure 6. (**a**) Quantity proportion of ecological relationships between PLE spaces, (**b**) The changes in the Mutualism index (MI) and Synergy index (SI) from 2000 to 2018.

During 2000–2018, the spatial distribution of ecological relationships between PLE spaces in Suzhou is shown in Figure 7. From 2000 to 2005, exploitation relationship was distributed in all districts and counties, being was the highest in Xiangcheng, Gusu, and Kunshan, accounting for 12.68%, 11.02%, and 11.02% of their regional areas, respectively. The spatial distribution of the competition relationship was more dispersed, with Kunshan and Xiangcheng still dominating, accounting for 4.15% and 3.18% of their regional areas. The distribution of the mutualism relationship was relatively less dispersed, mainly concentrated in Wuzhong. From 2005 to 2010, the changes in PLE space intensified and the spatial distribution of each ecological relationship became intersecting. The main distribution regions of the exploitation relationship had not changed. The competition relationship increased significantly in all districts and counties, especially in Xiangcheng, accounting for 18.22% of its regional area, mainly from IN and UR. The mutualism relationship was mainly distributed in central Suzhou and along Taihu Lake, occurring mainly from UR and W. From 2010 to 2018, the distribution area of each ecological relationship decreased. The exploitation relationship increased in Taicang, where the ratio of the area increased to 9.53%. The competition relationship was more apparent in Zhangjiagang, Wujiang, and Changshu. The spatial distribution of the mutualism relationship was becoming less scattered and mainly concentrated in Taicang.

The overall utility matrix U was used to calculate the mutualism index MI of the carbon metabolic network in urban PLE space in Suzhou for each study period (see Figure 6b). The MI for 2000–2005 and 2005–2010 were greater than 1, which indicated that the spatial transition of PLE space had a positive impact on the network. From 2005 to 2010, MI attained the highest value of 1.13, which was mainly due to the greater prominence of the mutualism relationship between UR, G, F, and B. From 2010 to 2018, MI decreased to 0.94 and was less than 1, indicating that the spatial transition of PLE space had a negative impact on the network. The synergy index SI was 6.69→5.81→5.17 in the three study periods, showing a decreasing trend, mainly due to the increase in the competition relationship in the network, which generated intense competition for carbon storage. The self-feedback of each PLE space type for horizontal carbon flow was reduced, leading to a certain negative effect on the network.

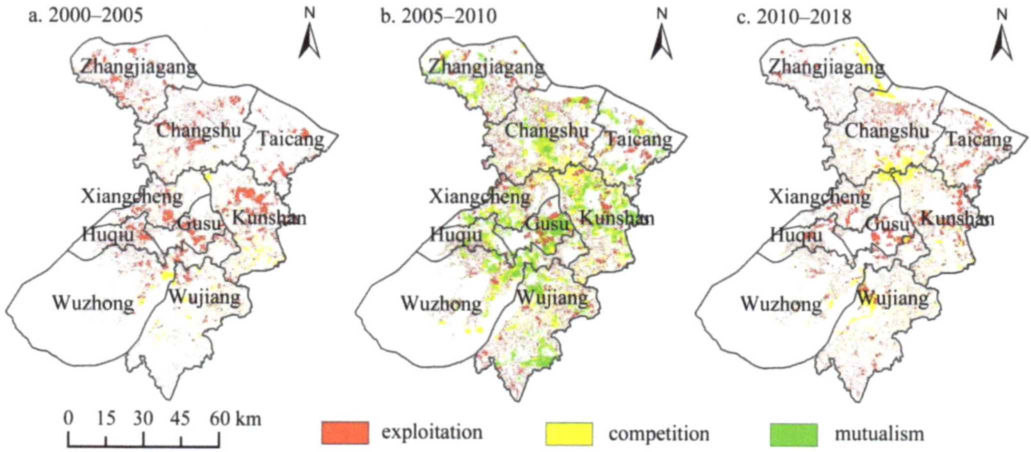

Figure 7. Spatial distribution of ecological relationships between PLE spaces.

3.4. Hierarchical Structure of Urban Carbon Metabolism Network

The driving and pull weights of each PLE space type in the network during 2000–2018 are shown in Figure 8. At the driving end, the network hierarchy from 2000 to 2005 showed an irregular state, with CU as primary consumer taking the largest driving weight (47.65%) followed by IN as secondary consumer (41.79%), while F, G, and W as producers only contributed 1.93%. The network experienced a distinct production shortage, with yield struggling to meet the demand for urban carbon metabolism. From 2005 to 2010, this shortage was alleviated, and the driving weight of producers increased to 8.25%. As primary consumer, the driving weight of CU decreased to 31.41%. In order to maintain the stability of the whole network, primary consumers assumed part of the ecological functions of producers. From 2010 to 2018, the network hierarchy gradually evolved into a pyramidal structure, with the driving weight of producers further increasing, especially the driving weight of W, rising to 35.71%. The weight of the secondary consumer (IN) decreased to 10.74%, and the weight of the senior consumer (UR) decreased to 1.73%.

At the pull end, from 2000–2005 the network hierarchy showed an unhealthy structure similar to an ellipse. The secondary consumer (IN) contributed the largest pull weight (61.91%) and absorbed a large amount of carbon flow for development, followed by CU as the primary consumer (24.22%). The producers and decomposers bore very low pull weight during this period, which was not conducive to the realization of the urban carbon balance This phenomenon changed during 2005–2010, with a decrease in the pull weight of CU (24.22%→14.54%), an increase in the weight of the senior consumer (UR) (8.27%→21.46%), and the ability of the secondary consumer (IN) to absorb carbon flow decreased slightly (−7.04%). The problem of smaller carbon flow absorption by producers was slightly alleviated, and the producers' pull weight increased to 4.53%, but it was still difficult to maintain the carbon balance of the whole network. From 2010 to 2018, the pull weight of the secondary consumer (IN) and the senior consumer (UR) further decreased, and the producers' weight increased further (+19.80%).

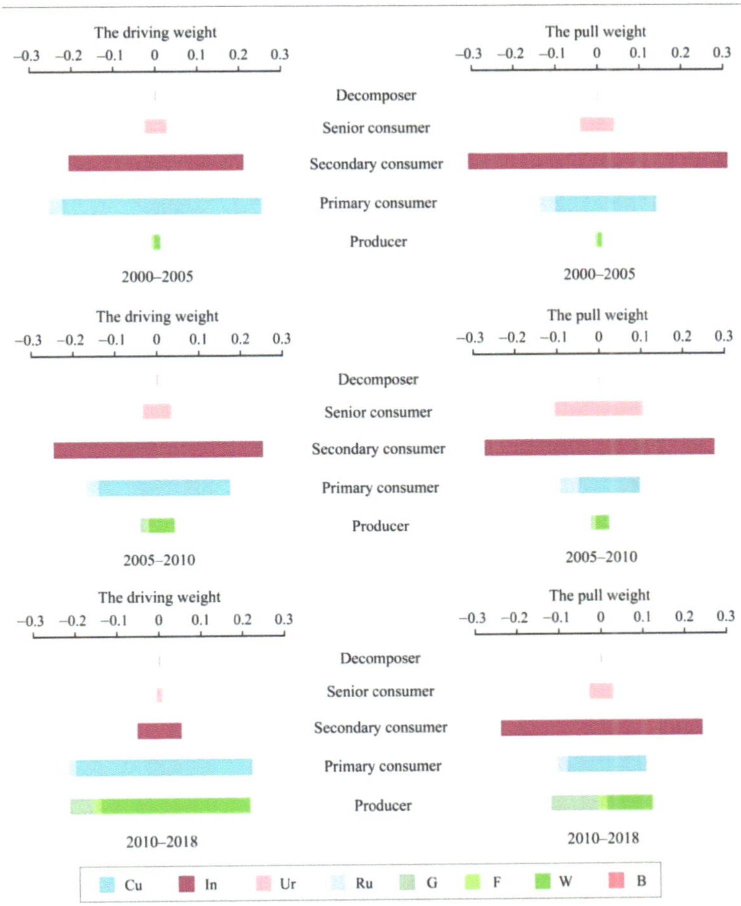

Figure 8. The hierarchical structure of carbon metabolic network in urban PLE space.

4. Discussion

4.1. The Changes in PLE Space Types and Carbon Flow

Urban space is a complex multi-layered spatial system consisting of various elements such as geographical features, human resources, economic conditions, and policies and management [47]. The transition of PLE space reflects the changes in the relationship between the natural environment and socio-economic factors from the perspective of urban land use, which is the main external expression of urban spatial evolution [48]. The transformation of PLE space brought about the transition of PLE dominant functions of land. It is a process of orderly allocation and spatial reconfiguration of limited urban land resources among different dominant functions [49], which greatly affects the carbon cycle of the urban space and the carbon metabolism efficiency of the regional land system. Yang et al. [29] studied the eco-environmental effects and changes in terrestrial carbon stocks caused by land transition in the Beijing–Tianjin–Hebei (BTH) metropolitan region based on the classification of PLE dominant functions, further confirming the practicality of this classification method. Therefore, we adopt a similar classification method based on PLE dominant functions to identify the nodes in the urban carbon metabolic network, considering the multi-functionality of urban land with regard to socio-economic aspects. It was more appropriate to the evolution law of the urban land system and will help to formulate more low-carbon land management policies adapted to each stage of urban socio-economic development.

Suzhou is a representative region of China's economic transformation and development, reflecting the change process of the regional development path since China's reform and opening up [28]. It had undergone phased urbanization and industrialization from 2000 to 2018 [50], and the layout of PLE space had changed dramatically. In this study, we measured carbon emissions and sequestration at four time points of each PLE space type based on statistical data, and calculated the horizontal carbon flow in the carbon metabolic network in three study periods. The negative carbon flow in Suzhou City mainly came from the transition from CU to IN, accounting for more than 40% of the total in each study period. This was similar to the research on Nanjing City by Zhao et al. [51]. And it was also appropriate to the characteristics of Suzhou's industrial structure adjustment, which showed a shift from agriculture to industry and services as the main driving force of the city's economic development. However, through further research, we found that the proportion of negative carbon flow generated from the transition from CU to IN was decreasing over the three study periods, and the spatial pattern of negative carbon flow changed from a large concentration of high-value regions in 2005–2010 to a fragmented distribution in 2010–2018. This was mainly because Suzhou's economic development had entered a new stage, the expansion of UR in core urban regions such as Kunshan and Gusu had gradually stabilized, the transformation of large-scale agricultural to non-agricultural land had come to an end, and the restrictions on the expansion of IN had gradually strengthened. The positive carbon flow was mainly generated from the transition within artificial compartments in the first two study periods, especially from 2005 to 2010 when the conversion of IN to UR generated a large amount of positive carbon flow; this reflected the extremely high carbon emissions density of IN on the one hand, and confirmed the key role of industrial land restrictions in promoting local carbon balance on the other [52]. During 2010–2018, due to the vigorous promotion of policies such as the Suzhou Low Carbon Development Plan (2010–2020), the transition of production and living space to ecological space dominated the positive carbon flow. Therefore, scientific spatial planning was indispensable in low-carbon land management [4].

4.2. The ecological Relationships between PLE Space and Urban Expansion

The utility analysis in the ENA method had revealed the ecological relationships formed in the process of carbon transfer in urban PLE space. From 2000 to 2005, the exploitation relationship was absolutely dominant, mainly reflected in the deprivation of ecological space and CU by artificial compartments, especially in Kunshan and Gusu, which caused the rapid increase in carbon emissions in the central part of Suzhou. The reason for this deprivation was the development orientation at the early stage of Suzhou's urbanization, which was dominated by small towns [53]. This orientation led to the expansion of towns in a chaotic way, which reduced the efficiency of spatial utilization and formed an inefficient and disordered spatial pattern of urbanization, which was not conducive to the effective control of urban carbon emissions. Similar conclusions had been reached in some studies from Switzerland [54] and Mexico [55], highlighting the important influence of urban spatial form optimization on carbon emissions reduction.

From 2005 to 2010, Suzhou experienced the most rapid transition of PLE space, gradually getting rid of the "Southern Jiangsu Pattern", which was dominated by the expansion of small towns and industrial development in the countryside [56], and presented a larger scale and more concentrated spatial characteristics of urbanization. Given these characteristics, on the one hand, the orderly expansion of cities further compressed the areas of agricultural and ecological space and forced them to compete with each other in a smaller space, coupling with China's unique policy of protecting farmland, i.e., compensating for the loss of farmland due to urban expansion by occupying other space types to maintain the dynamic balance of farmland quantity [57]. This policy further exacerbated the appropriation of agricultural land for ecological space, and the increase in carbon emissions from this occupation was noteworthy due to the large differences in carbon metabolic capacity. A study from the United States showed that slash-and-burn cultivation resulted in a loss of 436.8 TgC/year of carbon sequestration in woodlands [58]. Similar phenomena had occurred in Hangzhou [46], which shared many similar urban characteristics with Suzhou. On the other hand, it was also the government-led expansion of cities that enabled some strong environmental protection measures to be effectively implemented from the top down, resulting in some transient appearance of the mutualism relationship, not only within ecological space but also between UR and ecological space. This confirmed the possibility of promoting a healthy carbon cycle by planning urban green landscapes and protecting key ecological space [59].

The rate of urban spatial and functional expansion in Suzhou had slowed down from 2010 to 2018, and the peri-urban areas around the core urban regions had developed into fully urbanized regions with an enhanced stability of the ecological relationship. A noteworthy development was the disappearance of the mutualism relationship between key nodes in network. Although many ecological red line protection plans had been implemented in Suzhou during this time period [60], and the rate of PLE space conversion had slowed down, a certain number of peri-urban or rural areas still existed in certain districts, which were far away from main city areas [61]. How to intensively and efficiently turn these areas into UR by means of spatial planning and moderately increased urban population density, in order to improve energy consumption efficiency and reduce carbon emissions, was the main concern in Suzhou's future low-carbon development.

In our study, the impacts of urban PLE space transition on the carbon metabolic network were quantified by the mutualism index MI and the synergy index SI. The MI reached a maximum of 1.13 from 2005 to 2010, indicating that the transition of PLE space had a positive impact on the urban carbon metabolic network during this period. But it was still far below the 1.91 found by Chen and Chen in the carbon metabolic system of Vienna [39]. This difference may be the result of the difference in the maturity of the two cities: Vienna is an old and relatively mature city with stable development [62], while Suzhou has undergone an unparalleled rapid urbanization process during the past 18 years, and the city lacked stability in this phase of rapid change. Moreover, the two indices showed different trends in each study period. The main reason was that MI reflected primarily the degree of influence of the ecological relationships between each PLE space type on the carbon metabolic network, and the calculation was based on the positive and negative properties of each element in the overall utility matrix [39]; thus, the competition relationship and the mutualism relationship were crucial to MI, and the number of different network nodes also had a greater influence on it. Zhang et al. [18] and Xia et al. [19] used the same method to analyze the mutualism index of urban carbon metabolism in Beijing with 8 and 18 network nodes, and the mean difference was as high as 0.94. However, SI, originating from the sum of the values of elements in the overall utility matrix, could quantify the utility of carbon flow in the network more finely and reduce the errors due to the different ways of node classification. The brief emergence of the partial mutualism relationship in Suzhou from 2005 to 2010 improved the MI of the network, but in the subsequent study period, the mutualism relationship shifted back to exploitation relationship, indicating the fragility of this brief mutualism relationship. However, SI captured this difference well and showed a decreasing trend in the three study periods. Therefore, when analyzing the overall utility of carbon flow in the network, the results of both indices should be combined to reflect the ecological utility of carbon flow more comprehensively.

4.3. Policy Recommendations for Low-Carbon Urban Development

The low-carbon city has been a hot concept in China in recent years [63]. The National New Urbanization Plan (2014–2020) released by the Chinese government in 2014 emphasized that urban development should be people-oriented and environment-friendly [64]. Although people have reached a consensus on the concept of low-carbon, its realization path still needs to be explored in practice because of the different PLE space patterns and socio-economic development stages of Chinese cities. As a representative highly urbanized region in China, the rapid urbanization in Suzhou over the past 18 years had caused a major increase in carbon emissions from the land type with carbon source function, further negative ecological relationships in the carbon metabolic network, and the imbalance of the hierarchical structure. The essential reason was that the economy, society, and ecological system interacting with each other in the course of the rapid development [65] resulted in the intense conflict between PLE functions of land, and the formation of a less compatible spatial pattern of PLE types [66]. The effective combination of the ENA method and the carbon metabolic network in urban PLE space analysis framework can provide a more special and multifaceted perspective. By exploring the comprehensive impact of the transition of PLE space on the ecological relationship and hierarchical structure of the network, more targeted low-carbon land management policies can be proposed: (1) For fully urbanized regions such as Kunshan and Gusu, in the central part of the study area, where the senior consumer (UR) was the core, on the one hand, it is necessary to restrain the scale of urban space, promote the integrated development of land and transportation, and protect key ecological regions within the city to increase the amount of urban carbon sink; on the other hand, it is necessary to promote the low-carbon transformation of human activities at the level of social life [67] through the optimization of habitable space and the reasonable layout of public facilities, achieving carbon emissions reduction. (2) For industrial parks such as Kunshan Industrial Park, etc., where the secondary consumer (IN) was the core, the production function was its dominant function. Under the premise of ensuring sufficient employment opportunities and meeting production targets, measures such as optimizing the energy structure, introducing clean renewable energy, and improving resource utilization efficiency should be taken to reduce the energy consumption, carbon emissions density, and the negative carbon flow in the network. (3) For peri-urban and rural areas e.g., Changshu and Zhangjiagang in the northern part of the study area, where CU and RU as primary consumers were the core, the production function was its dominant function, so the main way to achieve carbon emissions reduction is to improve agricultural production efficiency and land intensification. Measures such as promoting large-scale operations, introducing the smart agricultural production model [68], and promoting high-efficiency, low carbon cycle agriculture, should be taken. With the improvement of production efficiency the conflict resulting from the supplementing of farmland at the cost of occupying the ecological space will be eased, thus promoting the mutual symbiosis between CU and ecological space to achieve carbon balance in this region. (4) For ecological reserves, e.g., Taihu Lake, Yangcheng Lake, etc., where producers (F, G, and W) were the core, the ecological function was its dominant function. Priority should be given to maintaining the scale of ecological space, improving the regional habitat quality, and strengthening ecological restoration and treatment to ensure the stability of the natural carbon sink function. In addition, it is necessary to control the disorderly spread of urban land and strictly adhere to the ecological protection red line, so that the pull weight of consumers in this region could remain low, thus realizing a healthy pyramid structure at the pull end of the carbon metabolic network, and promoting a positive cycle of carbon metabolism.

4.4. Limitations and Future Directions of This Study

This present study was limited by data acquisition and quantitative assessment methods, and there was room for further exploration in some aspects. This study classified the land use types in the network into eight categories based on PLE dominant functions, and the different classification ways had a great impact on the complexity and practicality of the network structure; thus, we should try to classify them in a more precise way, and determine the optimal classification scheme for different characteristics of study areas in the future. Moreover, on this basis, it would be interesting to apply the method used in this study to compare several cities or regions with different characteristics as a way to test the validity of the model. In addition, this study mainly analyzed the changes in urban carbon metabolism from the perspective of land use, and lacked the in-depth investigation of the driving mechanism of the carbon metabolism network changes. The influence of more diverse factors, such as the scale and quality of urbanization [69,70], and the recessive functions of land, on the urban carbon metabolic network should be further considered in future studies.

5. Conclusions

This study constructed a carbon metabolic network model in urban PLE space, accounted for the carbon emissions and sequestration of each node in the network, analyzed the changes in horizontal carbon flow among nodes in the network, in combination with the land use transfer matrix, and evaluated the comprehensive effects of the PLE space changes on the network by using the ENA method.

The results showed that the total carbon sequestration in Suzhou from 2000 to 2018 showed a fluctuating change of rising and then declining, while the total carbon emissions grew rapidly. CU and IN were the key nodes for the generation of negative carbon flow in the network, and the spatial distribution of negative carbon flow showed a discrete–concentrated–discrete change trend. The generation of positive carbon flow mainly came from the mutual transformation between production and living space, and Gusu was the main region for its generation. The exploitation relationship was the dominant ecological relationship in the network, which mainly occurred between production and ecological space, and gradually shifted from Gusu and Kunshan to Wujiang, etc., in spatial distribution. The mutualism relationship was abundant in 2005–2010 and gradually decreased in 2010–2018. The frequency of competition relationship appeared gradually increased from 17.86% in 2000–2005 to 32.14% in 2010–2018. The mutualism index MI showed a trend of first increasing and then decreasing, while the synergy index SI continued to decrease, and the negative effect brought by horizontal carbon flow in the network was relatively significant. The ecological network hierarchy at the driving end evolved from an irregular shape dominated by primary consumers (especially CU) in 2000–2005, to a pyramidal-like shape dominated by producers (F, G, and W); at the pull end, although it presented an elliptical structure dominated by secondary consumers (IN), its pull weight kept shrinking from 61.91% in 2000–2005 to 48.21% in 2010–2018, and the weight of producers (F, G, and W) rose from 1.72% to 24.33%.

The contribution of this study was to provide a new idea on how to identify the nodes in an urban carbon metabolic network. The classification of network nodes based on PLE space can better measure the impact of socio-economic development on the urban carbon cycle. The urban carbon metabolic network model constructed in this study was able to quantify and spatialize the horizontal carbon flow due to land use transition, providing an example and methodology guidance for further investigation of the land–carbon relationship. Through the ENA method, we evaluated the integrated effect of PLE space changes on the urban carbon metabolic network and proposed more targeted low-carbon land management policies. The results of this study can provide a scientific reference for optimizing the spatial layout of Suzhou City to achieve low-carbon urban land use on the one hand, and have potential applications as guidance for spatial analysis of carbon transfer at other scales and regions, as well as for low-carbon adjustment of land use patterns, on the other hand.

Supplementary Materials: The following supporting information can be downloaded at: https://www.mdpi.com/article/10.3390/land11091445/s1, Table S1: Carbon emissions coefficients of each item of carbon sources; Table S2: Carbon sequestration coefficients of each secondary land type; Table S3: Direct carbon flow matrix F from 2000 to 2005/(t C/a); Table S4: Direct carbon flow matrix F from 2005 to 2010/(t C/a); Table S5: Direct carbon flow matrix F from 2010 to 2018/(t C/a); Table S6: Ecological relationships between PLE space types in the network from 2000 to 2005; Table S7: Ecological relationships between PLE space types in the network from 2005 to 2010; Table S8: Ecological relationships between PLE space types in the network from 2010 to 2018; Figure S1: Diagrammatic sketch of SDE.

Author Contributions: Conceptualization, Y.L.; methodology and software, C.X.; validation and formal analysis, X.F.; investigation, E.Y.; resources, X.F.; data curation, X.F.; writing—original draft preparation, X.F.; writing—review and editing, X.F. and Y.L.; visualization, J.Y. and L.Z.; supervision,

Y.L.; project administration, Y.L.; funding acquisition, Y.L. All authors have read and agreed to the published version of the manuscript.

Funding: This research was funded by the Humanities and Social Science Foundation of the Ministry of Education of China (22A10335004); the National Natural Science Foundation of China (Grant No. 41771244; 72004014); the Key Program of Zhejiang Provincial Natural Science Foundation of China (LZ21D010002).

Data Availability Statement: Not applicable.

Acknowledgments: The authors are grateful to the editors and reviewers for their valuable suggestions.

Conflicts of Interest: The authors declare that they have no conflict of interest.

References

1. Grimm, N.B.; Faeth, S.H.; Golubiewski, N.E.; Redman, C.L.; Wu, J.; Bai, X.; Briggs, J.M. Global change and the ecology of cities. *Science* **2008**, *319*, 756–760. [CrossRef]
2. Zhao, R.Q.; Huang, X.J.; Peng, B.Z. Research on carbon cycle and carbon balance of Nanjing urban system. *Acta Geogr. Sin.* **2012**, *67*, 758–770. (In Chinese)
3. Xu, Q.; Yang, R.; Dong, Y.X.; Liu, Y.X.; Qiu, L.R. The influence of rapid urbanization and land use changes on terrestrial carbon sources/sinks in Guangzhou, China. *Ecol. Indic.* **2016**, *70*, 304–316. [CrossRef]
4. Lin, Q.; Zhang, L.; Qiu, B.; Zhao, Y.; Wei, C. Spatiotemporal Analysis of Land Use Patterns on Carbon Emissions in China. *Land* **2021**, *10*, 141. [CrossRef]
5. Xia, C.; Li, Y.; Xu, T.; Chen, Q.; Ye, Y.; Shi, Z.; Liu, J.; Ding, Q.; Li, X. Analyzing spatial patterns of urban carbon metabolism and its response to change of urban size: A Case of the Yangtze River Delta, China. *Ecol. Indic.* **2019**, *104*, 615–625. [CrossRef]
6. Ergas, C.; Clement, M.; McGee, J. Urban density and the metabolic reach of metropolitan areas: A panel analysis of per capita transportation emissions at the county-level. *Soc. Sci. Res.* **2016**, *58*, 243–253. [CrossRef] [PubMed]
7. Xia, L.L.; Zhang, Y.; Li, M.J. A review of urban carbon metabolism. *Acta Ecol. Sin.* **2017**, *37*, 4268–4277. (In Chinese) [CrossRef]
8. Thomson, G.; Newman, P. Urban fabrics and urban metabolism—From sustainable to regenerative cities. *Resour. Conserv. Recycl.* **2018**, *132*, 218–229. [CrossRef]
9. Zhang, Y.; Yang, Z.; Fath, B.D. Ecological network analysis of an urban water metabolic system: Model development, and a case study for Beijing. *Sci. Total Environ.* **2010**, *408*, 4702–4711. [CrossRef]
10. Forkes, J. Nitrogen balance for the urban food metabolism of Toronto, Canada. *Resour. Conserv. Recycl.* **2007**, *52*, 74–94. [CrossRef]
11. Pincetl, S.; Bunje, P.; Holmes, T. An expanded urban metabolism method: Toward a systems approach for assessing urban energy processes and causes. *Landsc. Urban Plan.* **2012**, *107*, 193–202. [CrossRef]
12. Cortés-Borda, D.; Ruiz-Hernández, A.; Guillén-Gosálbez, G.; Llop, M.; Guimerà, R.; Sales-Pardo, M. Identifying strategies for mitigating the global warming impact of the EU-25 economy using a multi-objective input–output approach. *Energy Policy* **2015**, *77*, 21–30. [CrossRef]
13. Ye, H.; Wang, K.; Zhao, X.; Chen, F.; Li, X.; Pan, L. Relationship between construction characteristics and carbon emissions from urban household operational energy usage. *Energy Build.* **2011**, *43*, 147–152. [CrossRef]
14. Kennedy, C.; Steinberger, J.; Gasson, B.; Hansen, Y.; Hillman, T.; Havránek, M.; Pataki, D.; Phdungsilp, A.; Ramaswami, A.; Mendez, G.V. Methodology for inventorying greenhouse gas emissions from global cities. *Energy Policy* **2010**, *38*, 4828–4837. [CrossRef]
15. Li, J.; Huang, G.; Liu, L. Ecological network analysis for urban metabolism and carbon emissions based on input-output tables: A case study of Guangdong Province. *Ecol. Model.* **2018**, *383*, 118–126. [CrossRef]
16. Li, S.; Zhang, Y.; Yang, Z.; Liu, H.; Zhang, J. Ecological relationship analysis of the urban metabolic system of Beijing, China. *Environ. Pollut.* **2012**, *170*, 169–176. [CrossRef] [PubMed]
17. Zhang, Y.; Zheng, H.; Fath, B.D. Analysis of the energy metabolism of urban socioeconomic sectors and the associated carbon footprints: Model development and a case study for Beijing. *Energy Policy* **2014**, *73*, 540–551. [CrossRef]
18. Zhang, Y.; Xia, L.; Fath, B.D.; Yang, Z.; Yin, X.; Su, M.; Liu, G.; Li, Y. Development of a spatially explicit network model of urban metabolism and analysis of the distribution of ecological relationships: Case study of Beijing, China. *J. Clean. Prod.* **2016**, *112*, 4304–4317. [CrossRef]
19. Xia, L.; Fath, B.D.; Scharler, U.M.; Zhang, Y. Spatial variation in the ecological relationships among the components of Beijing's carbon metabolic system. *Sci. Total Environ.* **2016**, *544*, 103–113. [CrossRef]
20. Purushothaman, S.; Patil, S.; Francis, I. Assessing the impact of policy-driven agricultural practices in Karnataka, India. *Sustain. Sci.* **2013**, *8*, 173–185. [CrossRef]
21. Zhou, D.; Xu, J.; Lin, Z. Conflict or coordination? Assessing land use multi-functionalization using production-living-ecology analysis. *Sci. Total Environ.* **2017**, *577*, 136–147. [CrossRef] [PubMed]
22. Duan, Y.; Wang, H.; Huang, A.; Xu, Y.; Lu, L.; Ji, Z. Identification and spatial-temporal evolution of rural "production-living-ecological" space from the perspective of villagers' behavior—A case study of Ertai Town, Zhangjiakou City. *Land Use Policy* **2021**, *106*, 105457. [CrossRef]

23. Zhao, Y.; Cheng, J.; Zhu, Y.; Zhao, Y. Spatiotemporal Evolution and Regional Differences in the Production-Living-Ecological Space of the Urban Agglomeration in the Middle Reaches of the Yangtze River. *Int. J. Environ. Res. Public Health* **2021**, *18*, 12497. [CrossRef]
24. Li, G.D.; Fang, C.L. Quantitative function identification and analysis of urban ecological-production-living spaces. *Acta Geogr. Sin.* **2016**, *71*, 49–65. (In Chinese) [CrossRef]
25. Tu, X.; Fu, C.; Huang, A.; Chen, H.; Ding, X. DBSCAN Spatial Clustering Analysis of Urban "Production–Living–Ecological" Space Based on POI Data: A Case Study of Central Urban Wuhan, China. *Int. J. Environ. Res. Public Health* **2022**, *19*, 5153. [CrossRef]
26. Guo, S.; Han, M.; Yang, Y.; Di, H. Embodied energy flows in China's Economic Zones: Jing-Jin-Ji, Yangtze-River-Delta and Pearl-River-Delta. *J. Clean. Prod.* **2020**, *268*, 121710. [CrossRef]
27. Du, Q.; Guo, X.; Bao, T.; Huang, Y.; Han, X. CO_2 flows in the inter-regional and inter-sectoral network of the Yangtze River Economic Zone. Environ. *Sci. Pollut. Res.* **2020**, *27*, 16293–16316. [CrossRef]
28. Long, H.; Liu, Y.; Wu, X.; Dong, G. Spatio-temporal dynamic patterns of farmland and rural settlements in Su–Xi–Chang Region: Implications for building a new countryside in Coastal China. *Land Use Policy* **2009**, *26*, 322–333. [CrossRef]
29. Yang, Y.; Bao, W.; Li, Y.; Wang, Y.; Chen, Z. Land Use Transition and Its Eco-Environmental Effects in the Beijing–Tianjin–Hebei Urban Agglomeration: A Production–Living–Ecological Perspective. *Land* **2020**, *9*, 285. [CrossRef]
30. West, T.O.; Marland, G. A synthesis of carbon sequestration, carbon emissions, and net carbon flux in agriculture: Comparing tillage practices in the United States. *Agric. Ecosyst. Environ.* **2002**, *91*, 217–232. [CrossRef]
31. Xia, C.; Li, Y.; Xu, T.; Ye, Y.; Shi, Z.; Peng, Y.; Liu, J. Quantifying the spatial patterns of urban carbon metabolism: A case study of Hangzhou, China. *Ecol. Indic.* **2018**, *95*, 474–484. [CrossRef]
32. Hannon, B. The structure of ecosystems. *J. Theor. Biol.* **1973**, *41*, 535–546. [CrossRef]
33. Fath, B.D. Network mutualism: Positive community-level relations in ecosystems. *Ecol. Model.* **2007**, *208*, 56–67. [CrossRef]
34. Hines, D.; Lisa, J.; Song, B.; Tobias, C.; Borrett, S. Estimating the effects of seawater intrusion on an estuarine nitrogen cycle by comparative network analysis. *Mar. Ecol. Prog. Ser.* **2015**, *524*, 137–154. [CrossRef]
35. Dame, J.K.; Christian, R.R. Uncertainty and the use of network analysis for ecosystem-based fishery management. *Fisheries* **2006**, *31*, 331–341. [CrossRef]
36. Finn, J.T. Flow analysis of models of the hubbard brook ecosystem. *Ecology* **1980**, *61*, 562–571. [CrossRef]
37. Chen, S.; Xu, B.; Chen, B. Unfolding the interplay between carbon flows and socioeconomic development in a city: What can network analysis offer? *Appl. Energy* **2018**, *211*, 403–412. [CrossRef]
38. Fath, B.D.; Patten, B.C. Quantifying resource homogenization using network flow analysis. *Ecol. Model.* **1999**, *123*, 193–205. [CrossRef]
39. Chen, S.; Chen, B. Network environ perspective for urban metabolism and carbon emissions: A case study of Vienna, Austria. *Environ. Sci. Technol.* **2012**, *46*, 4498–4506. [CrossRef]
40. Borrett, S.R. Bibliometric review of ecological network analysis: 2010–2016. *Ecol. Model.* **2018**, *382*, 63–82. [CrossRef]
41. Zhang, Y.; Liu, H.; Li, Y.; Yang, Z.; Li, S.; Yang, N. Ecological network analysis of China's societal metabolism. *J. Environ. Manag.* **2012**, *93*, 254–263. [CrossRef] [PubMed]
42. Piezer, K.; Petit-Boix, A.; Sanjuan-Delmás, D.; Briese, E.; Celik, I.; Rieradevall, J.; Gabarrell, X.; Josa, A.; Apul, D. Ecological network analysis of growing tomatoes in an urban rooftop greenhouse. *Sci. Total Environ.* **2019**, *651*, 1495–1504. [CrossRef] [PubMed]
43. Xia, L.; Zhang, Y.; Yu, X.; Li, Y.; Zhang, X. Hierarchical structure analysis of urban carbon metabolism: A case study of Beijing, China. *Ecol. Indic.* **2019**, *107*, 105602. [CrossRef]
44. Zhang, Y.; Zheng, H.; Yang, Z.; Su, M.; Liu, G.; Li, Y. Multi-regional input–output model and ecological network analysis for regional embodied energy accounting in China. *Energy Policy* **2015**, *86*, 651–663. [CrossRef]
45. Schramski, J.R.; Gattie, D.K.; Patten, B.C.; Borrett, S.R.; Fath, B.D.; Thomas, C.R.; Whipple, S.J. Indirect effects and distributed control in ecosystems: Distributed control in the environ networks of a seven-compartment model of nitrogen flow in the Neuse River Estuary, USA—Steady-state analysis. *Ecol. Model.* **2006**, *194*, 189–201. [CrossRef]
46. Xia, C.; Chen, B. Urban land-carbon nexus based on ecological network analysis. *Appl. Energy* **2020**, *276*, 115465. [CrossRef]
47. Liu, Y. Introduction to land use and rural sustainability in China. *Land Use Policy* **2018**, *74*, 1–4. [CrossRef]
48. Deng, Y.; Yang, R. Influence Mechanism of Production-Living-Ecological Space Changes in the Urbanization Process of Guangdong Province, China. *Land* **2021**, *10*, 1357. [CrossRef]
49. De Groot, R. Function-analysis and valuation as a tool to assess land use conflicts in planning for sustainable, multi-functional landscapes. *Landsc. Urban Plan.* **2006**, *75*, 175–186. [CrossRef]
50. Liang, X.; Jin, X.; Ren, J.; Gu, Z.; Zhou, Y. A research framework of land use transition in Suzhou City coupled with land use structure and landscape multifunctionality. *Sci. Total Environ.* **2020**, *737*, 139932. [CrossRef]
51. Zhao, R.; Huang, X.; Zhong, T.; Liu, Y.; Chuai, X. Carbon Flow of urban system and its policy implications: The case of Nanjing. *Renew. Sustain. Energy Rev.* **2014**, *33*, 589–601. [CrossRef]
52. Chuai, X.; Huang, X.; Wang, W.; Zhao, R.; Zhang, M.; Wu, C. Land use, total carbon emissions change and low carbon land management in Coastal Jiangsu, China. *J. Clean. Prod.* **2015**, *103*, 77–86. [CrossRef]

53. Fan, L.Y. Analysis of process and characteristics of Sunan rural urbanization from the perspective of social space: A case study o Suzhou. *Urban Plan. Forum* **2015**, *4*, 27–35. (In Chinese) [CrossRef]
54. Mohajeri, N.; Perera, A.T.D.; Coccolo, S.; Mosca, L.; Le Guen, M.; Scartezzini, J.L. Integrating urban form and distribute energy systems: Assessment of sustainable development scenarios for a Swiss Village to 2050. *Renew. Energy* **2019**, *143*, 810–82 [CrossRef]
55. Carpio, A.; Ponce-Lopez, R.; Lozano-Garcia, D.F. Urban Form, Land use, and cover change and their impact on carbon emission in the Monterrey Metropolitan Area, Mexico. *Urban Clim.* **2021**, *39*, 100947. [CrossRef]
56. Yuan, F.; Wei, Y.D.; Chen, W. Economic transition, industrial location and corporate networks: Remaking the Sunan Model i Wuxi City, China. *Habitat Int.* **2014**, *42*, 58–68. [CrossRef]
57. Wu, Y.; Shan, L.; Guo, Z.; Peng, Y. Cultivated land protection policies in China facing 2030: Dynamic balance system versus basi farmland zoning. *Habitat Int.* **2017**, *69*, 126–138. [CrossRef]
58. Liu, J.; Sleeter, B.M.; Zhu, Z.; Loveland, T.R.; Sohl, T.; Howard, S.M.; Key, C.H.; Hawbaker, T.; Liu, S.; Reed, B.; et al. Critical lan change information enhances the understanding of carbon balance in the United States. *Glob. Chang. Biol.* **2020**, *26*, 3920–392 [CrossRef]
59. Zhang, D.; Huang, Q.; He, C.; Yin, D.; Liu, Z. Planning urban landscape to maintain key ecosystem services in a rapidly urbanizin area: A scenario analysis in the Beijing-Tianjin-Hebei Urban Agglomeration, China. *Ecol. Indic.* **2019**, *96*, 559–571. [CrossRef]
60. Jia, Y.Y.; Li, H.; Wang, Z.Q.; Du, X.L.; Yang, Y.; Li, J.L. Quantitative evaluation of eco-environment quality of the terrestria ecological red line area in Wuzhong District, Suzhou City. *J. Ecol. Rural Environ.* **2017**, *33*, 525–532. (In Chinese) [CrossRef]
61. Liao, X.; Shu, T.H.; Shen, L.Y.; Ren, Y.T.; Yang, N. Identification and evolution of peri-urban areas in the context of urban-rura integrated development: A case study of Suzhou City. *Prog. Geogr.* **2021**, *40*, 1847–1860. (In Chinese) [CrossRef]
62. Hendriks, C.; Obernosterer, R.; Müller, D.; Kytzia, S.; Baccini, P.; Brunner, P.H. Material Flow Analysis: A tool to suppor environmental policy decision making. Case-studies on the city of Vienna and the Swiss Lowlands. *Local Environ.* **2000**, *5*, 311–32 [CrossRef]
63. Yu, L. Low carbon eco-city: New approach for Chinese urbanisation. *Habitat Int.* **2014**, *44*, 102–110. [CrossRef]
64. Liu, W.; Qin, B. Low-Carbon City Initiatives in China: A review from the policy paradigm perspective. *Cities* **2016**, *51*, 131–13 [CrossRef]
65. Bernués, A.; Rodríguez-Ortega, T.; Alfnes, F.; Clemetsen, M.; Eik, L.O. Quantifying the multifunctionality of fjord and mountair agriculture by means of sociocultural and economic valuation of ecosystem services. *Land Use Policy* **2015**, *48*, 170–178. [CrossRef
66. Wei, L.; Zhang, Y.; Wang, L.; Mi, X.; Wu, X.; Cheng, Z. Spatiotemporal Evolution Patterns of "Production-Living-Ecological Spaces and the Coordination Level and Optimization of the Functions in Jilin Province. *Sustainability* **2021**, *13*, 13192. [CrossRef
67. Walker, G.; Karvonen, A.; Guy, S. Zero carbon homes and zero carbon living: Sociomaterial interdependencies in carbor governance. *Trans. Inst. Br. Geogr.* **2015**, *40*, 494–506. [CrossRef]
68. Xin, Y.; Tao, F. Have the agricultural production systems in the North China Plain changed towards to climate smart agricultur since 2000? *J. Clean. Prod.* **2021**, *299*, 126940. [CrossRef]
69. Li, Y.; Shen, J.; Xia, C.; Xiang, M.; Cao, Y.; Yang, J. The impact of urban scale on carbon metabolism—A case study o Hangzhou, China. *J. Clean. Prod.* **2021**, *292*, 126055. [CrossRef]
70. Xia, C.; Xiang, M.; Fang, K.; Li, Y.; Ye, Y.; Shi, Z.; Liu, J. Spatial-Temporal Distribution of Carbon Emissions by Daily Travel and Its Response to Urban Form: A Case Study of Hangzhou, China. *J. Clean. Prod.* **2020**, *257*, 120797. [CrossRef]

land

MDPI

Article

Analysis of Spatial Carbon Metabolism by ENA: A Case Study of Tongzhou District, Beijing

Yongchao Qu [1,2], Jian Zhang [1], Chongyuan Xu [1], Yichao Gao [1], Shanwen Zheng [1,*] and Meiling Xia [3]

1 Faculty of Architecture, Civil and Transportation Engineering, Beijing University of Technology, Beijing 100124, China
2 Beijing Municipal Institute of City Planning & Design, Beijing 100045, China
3 Beijing Tsinghua TongHeng Urban Planning & Design Institute, Beijing 100085, China
* Correspondence: zhengshanwen@bjut.edu.cn; Tel.: +86-10-67391645

Abstract: Carbon metabolism research has attracted worldwide attention as an important way to cope with climate change, promote carbon emission reduction, increase carbon sequestration, and support low-carbon city construction. Ecological network analysis (ENA) plays an important role in network analysis and simulation of carbon metabolism. However, current studies largely focus on single elements or local processes while rarely analyzing the spatial coupling between land use and carbon metabolism. Therefore, taking Tongzhou District as an example, based on the data of land use change and energy consumption, this study constructed an analysis framework based on ENA to explore the comprehensive impact of land use changes on carbon metabolism. The results show the following: (1) From 2014 to 2020, the total carbon emissions increased year by year. Carbon emissions of other construction land (OCL) were dominant, while the carbon sequestration capacity of forest land (FL) increased by 236%. The positive carbon metabolic density remained relatively stable, while the negative carbon metabolic density decreased year by year. (2) The negative carbon flow was concentrated in the transfer of other land to OCL, accounting for 40.2% of the total negative "carbon flow." The positive carbon flow was primarily from the transfer of other land to FL. (3) From 2014 to 2016, the spatial ecological relationships of carbon flow were dominated by exploitation and control. From 2016 to 2018, competition relationships intensified due to the expansion of the field; from 2016 to 2018, exploitation and control relationships, competition relationships, and mutualism relationships increased significantly and were evenly distributed. This study provides decision-making guidance for the subsequent formulation of government carbon emission reduction policies.

Keywords: carbon metabolism; carbon flow; ENA; spatial and temporal characteristics; Tongzhou District

Citation: Qu, Y.; Zhang, J.; Xu, C.; Gao, Y.; Zheng, S.; Xia, M. Analysis of Spatial Carbon Metabolism by ENA: A Case Study of Tongzhou District, Beijing. *Land* **2022**, *11*, 1573. https://doi.org/10.3390/land11091573

Academic Editor: Yan Li

Received: 16 August 2022
Accepted: 7 September 2022
Published: 15 September 2022

1. Introduction

Carbon emission reduction and carbon sequestration enhancement are key factors for the global response to climate change [1–3]. Carbon metabolism can effectively track the inflow and outflow of carbon in cities and can help to find the key nodes for carbon emission reduction and carbon sequestration enhancement [4,5]. Looking at the carbon element of the whole city from the perspective of metabolism is helpful for clarifying the intensity and efficiency of carbon circulation among various components of the urban ecosystem and for comprehensively understanding the mechanism and operation laws of the carbon element cycle. Land use and land cover change are key factors affecting carbon emissions and carbon sequestration [6–8]. It was estimated that over the past 150 years, the contribution of land use change to carbon emissions has accounted for about one third of human activity since the industrial revolution [9]. Therefore, it is of great significance to study the driving mechanisms of land use changes on carbon metabolism in order to explore the development path of low-carbon cities.

In recent years, the research on land use change and carbon metabolism has primarily focused on the impact of land use changes on the carbon cycle [10–18], the multi-scale measure of carbon emission intensity and efficiency [19–23], the impact of spatial policy and urban planning on carbon metabolism [24–32], and the utility analysis and coupling relationship between carbon emissions and carbon sequestration [33–39]. Such studies could be seen as preliminary investigations of land use changes and carbon metabolism attempting to clarify the impact of land use changes on carbon emissions and sequestration from a single element or a local perspective. For example, Zhu et al. examined the impact of land use changes on carbon emissions in Zhejiang Province over the past four decades [11]. Tang et al. performed a spatial–temporal assessment of carbon emissions and carbon sequestration from land use changes in the Mekong River Basin [15]. Kellett et al. developed a systematic approach to the study of carbon cycles and emissions at the urban community scale [19]. Zheng et al. analyzed the effect of sustainable development policies of resource-based cities on carbon emission reduction [25]. Ecological network analysis (ENA) is widely used in the study of urban metabolism [34,35,38], because it has the advantage of quantitatively determining the ecological flow direction and intensity in complex networks, which is helpful to determine the path of structural optimization. For example, Xia and Du calculated the respective carbon fluxes in Hangzhou and Zhaotong using ENA [8,35]; the research results provided effective guidance and a new perspective for the formulation of urban low-carbon spatial development planning. Such studies simply focused on quantitative relationships while ignoring the spatial analysis of carbon metabolic processes and coarsely studied the natural subjects involved in carbon metabolic processes. Therefore, the combined effects of land use change on carbon metabolism need to be investigated.

To fill these gaps, this study employed Tongzhou District as an example, where we selected three time periods with high temporal accuracy (2014–2016, 2016–2018, and 2018–2020) to establish a land use–carbon metabolism analysis model based on ENA. We discuss the impact of land use changes on spatial carbon metabolism. In particular, we sought to (1) explore the impact of land use changes on carbon flow; (2) examine the ecological relationships and spatial distribution of carbon metabolism in the territorial space, and (3) provide guidance for urban carbon emission reduction and carbon sequestration enhancement.

2. Materials and Methods

2.1. Study Area

The study area was located in Tongzhou District, Beijing, with an area of 906 km^2 (Figure 1). This area is a key node in the geographical pattern of the plains in the east of the capital from mountain to sea. In recent years, Tongzhou District's economy has developed rapidly, and the government has invested hundreds of billions of yuan every year to support and cultivate new economic growth points, with the GDP growing from CNY 54.89 billion in 2014 to CNY 110.3 billion in 2020. Economic development has brought about drastic changes in land use. Moreover, the government has actively promoted green and low-carbon transition and has formulated policies in finance, planning, management, and other fields to improve carbon metabolism and actively create carbon-neutral examples in China. The economic development and policy measures of Tongzhou District are typical for the study of the impact of land use change on carbon metabolism. Therefore, it is of great practical value to deeply discuss the eco-environmental effects in the process of the spatial evolution of rapidly developing areas.

2.2. Data Sources and Preparation

Land use data for 2014–2020 were obtained from the interpretation results of thematic mapper remote sensing image supervision classification. To improve the accuracy of the data, each remote sensing image was preprocessed with survey correction and geometric correction.

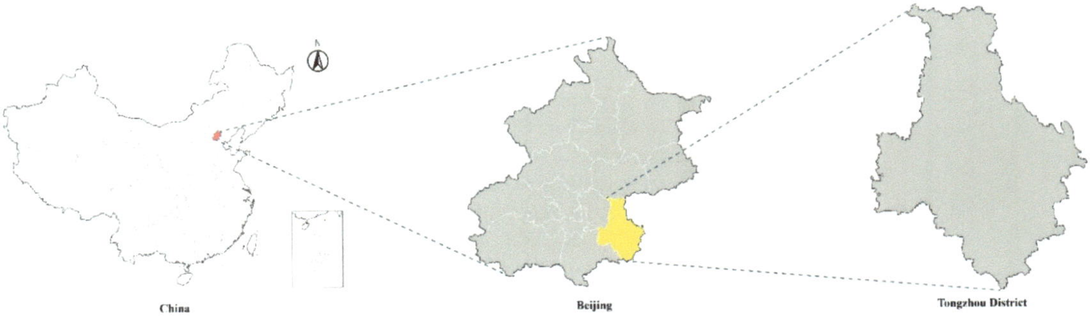

Figure 1. Location map of Tongzhou District.

The data for energy, resident population, household car ownership, motorcycle ownership, bicycle self-ownership, taxi ownership, suburban passenger car ownership, highway cargo turnover, fertilizer application amount, and pig, cattle, and sheep breeding were obtained from the Statistical Yearbook of Beijing in the corresponding year. The carbon emission generated by the energy consumption of industrial land and other construction land in 2018 and 2020 was calculated based on the total equivalent amounts of standard coal. The data on total power and irrigation area of agricultural machinery were converted to the ratio of the sown area to the area of the city in the same year.

2.3. Methods

2.3.1. Calculating the Carbon Transitions Caused by Land Use Changes

Carbon metabolism includes carbon emission and carbon sequestration. According to Xia, Zhang, Du, and Liu et al.'s research on carbon metabolism [7,8,33,35,37], the territorial space can be divided into seven compartments, i.e., forest land (FL), cultivated land (CL), grassland (GL), water and wetland (WL), industrial land (IL), road and traffic land (R&TL), and other construction land (OCL). These categories can be combined with the actual situation of land use in Tongzhou District. IL, R&TL, and OCL are the main contributors to carbon emissions. FL, GL, and WL are important carbon sequestration zones. CL includes the activities of agricultural production and animal husbandry and has the dual functions of carbon emission and carbon sequestration (Figure 2). Due to the complexity of obtaining indirect carbon emission data in the processes of production and living, only direct carbon emissions caused by vertical land use were calculated in this study.

The calculation formulas for carbon emissions are as follows [40]:

$$V_N = V_I + V_T + V_U + V_C \quad (1)$$

$$V_I = \sum E_i F_i \quad (2)$$

$$V_T = V_p + V_F \quad (3)$$

$$V_p = K_1 M_h + K_2 M_t + K_3 M_m + K_4 M_z + K_5 M_b \quad (4)$$

$$V_F = \sum K_6 M_F \quad (5)$$

$$V_U = \sum E_u F_i + K_7 P \quad (6)$$

$$V_C = V_A + V_L \quad (7)$$

$$V_A = K_8 M + K_9 S_i + K_{10} D \quad (8)$$

$$V_L = K_{11}(P_i + G_i) + K_{12}\, C_i \quad (9)$$

where V_N is the spatial total carbon emission; V_I is the carbon emission of IL; V_T is the R&TL use carbon emission; V_U is the OCL carbon emission; V_C is the carbon emissions of CL; E_i is the industrial energy consumption amount (of standard coal); F_i is the energy carbon emission coefficient; V_p is the passenger transport carbon emission; V_F is the freigh transport carbon emission; K_i is a carbon emission coefficient; and M_h, M_t, M_m, M_z an M_b are, respectively, the mileages of family cars, suburban passenger buses, motorcycle mopeds, and taxis. According to relevant research, the average annual mileage per car i China is about 15,000 km, and the average annual mileage per motorcycle and moped i 4000 km. The mileage of taxis was selected as 250 km per vehicle per day, and the mileage o suburban passenger vehicles was calculated according to the numbers of suburban passer ger transport vehicles, passenger routes, and the total mileage of passenger transport in th Statistical Yearbook of Beijing 2014, 2016, 2018, and 2020. M_F is the highway cargo turnove Considering the small amount of railway cargo transportation in Tongzhou District, thi study did not consider the carbon emissions generated by freight transportation. E_u is th energy consumption (standard coal) of OCL in the territorial space, and P is the size of th resident population. V_A is the carbon emission from agricultural activities; V_L is the carbo emission from livestock metabolism; M is the total driving force of agricultural machiner S_i is the irrigated area of the entire region; and D is the scale of fertilizer application. P G_i, and C_i are pig, sheep, and cattle feeds, respectively. All K and F values are shown i Table 1.

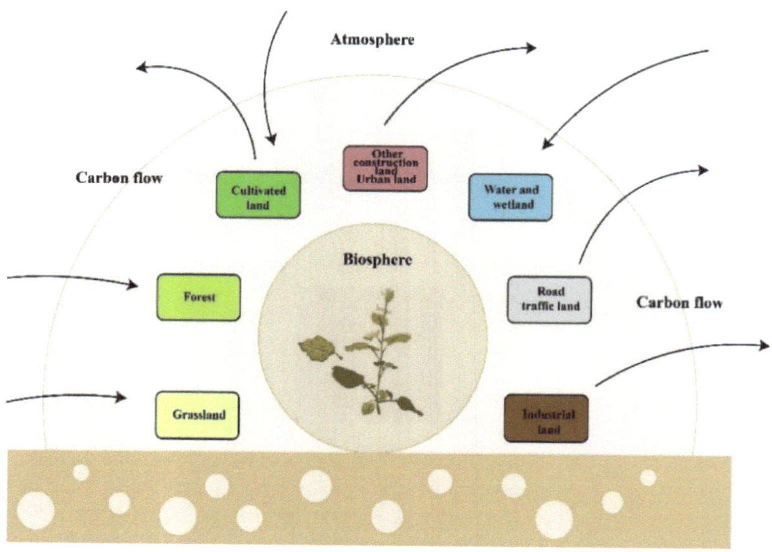

Figure 2. Spatial network model of carbon metabolism.

Table 1. Carbon emission coefficient of land use in the territorial space.

Item	Carbon Emission Coefficient	Unit	Code
Raw coal	0.7559	kg/kgce	F_1
Coke	0.8550	kg/kgce	F_2
Fuel oil	0.6185	kg/kgce	F_3
Gasoline	0.5538	kg/kgce	F_4
Diesel oil	0.5921	kg/kgce	F_5
Liquified petroleum gas	0.5042	kg/kgce	F_6
Natural gas	0.4483	kg/kgce	F_7
Family car	22.3	kg/100 km	K_1
Suburban passenger car	88.1	kg/100 km	K_2
Motorbike	6.7	kg/100 km	K_3
Moped	6.7	kg/100 km	K_4
Taxi	28.3	kg/100 km	K_5
Passenger transport and freight transport of highway	0.0556	$Kg{\cdot}t^{-1}{\cdot}km^{-1}$	K_6
Human breathing	79	$kg{\cdot}a^{-1}{\cdot}per^{-1}$	K_7
Agricultural machinery	0.18	kg/kw	K_8
Irrigation	6.7	kg/hm^2	K_9
Fertilizer	0.858	kg/kg	K_{10}
Pigs and sheep breathing	82	$kg{\cdot}a^{-1}{\cdot}h^{-1}$	K_{11}
Cow breathing	796	$kg{\cdot}a^{-1}{\cdot}h^{-1}$	K_{12}

Note: the carbon emission coefficients of energy, road transportation, agricultural production, and animal and human metabolism were derived from the relevant research results of Xia, Sun, Zhang, Xie, Min, and Solomon S. et al. [8,41–45].

The calculation formula for carbon sequestration is as follows:

$$V_s = \sum K_n S_n \tag{10}$$

where V_s is the sum of carbon sequestration in the territorial space, S_n is the area of land types with carbon sequestration in various territorial spaces, and K_n is the carbon sequestration coefficient of land sub-types in various territorial spaces referring to the studies of Gui, Fang, Liu, Duan, and Walsh et al. in Table 2 [46–52].

Table 2. Carbon sequestration coefficient of each land use compartment in the territorial space.

Land Type	Coefficient of Carbon Sequestration	Unit
Cultivated land (CL)	0.0007	$kgC/(m^2{\cdot}a)$
Forest land (FL)	0.0657	$kgC/(m^2{\cdot}a)$
Water and wetland (WL)	0.0402	$kgC/(m^2{\cdot}a)$
Grassland (GL)	0.0206	$kgC/(m^2{\cdot}a)$

Based on the definitions of carbon emission and carbon sequestration capacity, carbon metabolic density ΔW can be defined as follows:

$$\Delta W = W_i - W_j = V_i/S_i - V_j/S_j \tag{11}$$

$$F_{ij} = \Delta W \Delta S \tag{12}$$

where i and j represent two types of land subdivision; F_{ij} represents the "carbon flow" generated after the land type changes from j to i; W_i represents the net "carbon flow" density of subdivision i; W_j represents the net "carbon flow" density of subdivision j; and ΔS represents the land use transfer area. If $\Delta W > 0$, this indicates that urban carbon metabolism is positive, and urban carbon sequestration is increasing or carbon emissions are gradually decreasing. The opposite, in contrast, suggests that the city's carbon metabolism is out of balance, with the city's carbon sequestration decreasing or carbon emissions increasing.

2.3.2. Determining Ecological Relationships between Different Land Uses

ENA is an analytical method used to delineate the interactions between ecosystem components and to identify the internal attributes of the system as a whole [53]. The method is widely used in the analysis of economic and urban metabolic systems [54,55], in industrial ecological models [56,57], in landscape ecology [58–60], and in other fields [61,62]. For example, Ukidwe et al. used network analysis to study the ecological relationships of the chemical industry [56]. Chen et al. studied the structural complexity of ecological industrial systems [57]. Ecological networks quantitatively describe the material and energy flow between different compartments of an ecosystem by simulating the compartments and paths of biological networks [53]. ENA was used to quantitatively analyze the interaction and intensity of the two compartments in the carbon metabolism network under the action of the overall network. Furthermore, we referred to the dynamic analysis of Finn's equilibrium variable [63] and the conclusions of Zhang et al.'s study on the changes in carbon storage in Beijing's urban carbon cycle [33], i.e., all compartments may obtain and store carbon from the environment and can also emit carbon into the environment, but the overall inflow is equal to the overall outflow.

The effective utilization matrix (D) can reflect the direct effect of each "carbon flow"; matrix elements d_{ij} represent the effective utilization rate of "carbon flow" between compartments i and j. According to D, the dimensionless overall utility matrix U can be obtained, and the specific calculations are as follows:

$$d_{ij} = (f_{ij} - f_{ji})/T_i \tag{13}$$

$$D_{ij} = \begin{pmatrix} \frac{f_{jj}-f_{jj}}{T_j} & \frac{f_{ji}-f_{ij}}{T_j} \\ \frac{f_{ij}-f_{ji}}{T_i} & \frac{f_{ii}-f_{ii}}{T_i} \end{pmatrix} \tag{14}$$

$$U = (U_{ij}) = \sum_{m=0}^{n} D^m = (I-D)^{-1} \tag{15}$$

where U_{ij} is an element of matrix U, and n is the type of land used in the territorial space; I is the identity matrix that represents the self-feedback effect of various land uses in the process of carbon flow exchange; F_{ij} is the horizontal carbon flow from land j to land i; and T_i is the carbon flux of land type i. The ecological relationship between the two land uses can be judged according to the complete utility matrix U. There are four common relationships in ecological networks: plunder, restriction, mutualism, and competition.

The interaction mode between compartments in the ecological network can be obtained through the overall utility matrix U. According to relevant studies, there are nine ecological relationships in theory (listed in Table 3), among which the exploitation and control relationship indicates that one compartment uses another compartment. The competitive relationship indicates that the competition between two compartments leads to utility loss, and the mutualistic symbiosis relationship indicates that the two compartments increase utility in the interaction.

Table 3. Relationships between components of the network.

Matrix Notation	Positive +	Neutral 0	Negative −
Positive +	(+,+) mutualism	(+,0) symbiotic	(+,−) snatch
Neutral 0	(0,+) symbiotic	(0,0) neutrality	(0,−) amensalism
Negative −	(−,+) limitation	(−,0) control	(−,−) competition

Note: "+" means positive; "−" means negative; "0" stands for neutral.

The positive and negative elements in the complete utility matrix U can reflect whether the interactions between nodes in the ecological network are positive or negative. The comprehensive impact of carbon balance is defined as M, and the calculation formula is as follows:

$$M = S+/S- \quad (16)$$

where $S+$ and $S-$, respectively, represent the number of positive and negative elements in the matrix U. $M > 1$ indicates that the comprehensive effect of land use change on carbon metabolism is positive, and the reciprocal synergy of the whole ecological synergy is greater than the negative competition. In contrast, when $M < 1$, the comprehensive effect is negative. The smaller the value of M, the stronger the negative effect.

3. Results

3.1. Spatial Carbon Metabolism

3.1.1. Carbon Emission

Figure 3 illustrates the carbon emissions caused by land use changes from 2014 to 2020. The total carbon emission of Tongzhou District in 2020 was 413.5 × 10^4 TC, an increase of 10.94% compared with 2014, with an average annual growth of about 1.82%. The growth rate was higher in 2016–2018, with an increase of about 17.5 × 10^4 TC. With the increasing demands of Tongzhou District's economic development and population density, carbon emissions, especially those of R&TL and OCL, increased rapidly. In 2020, carbon emissions from OCL dominated, with 286.2 × 10^4 TC contributing 69% of the total carbon emissions. The proportion of CL carbon emissions in total carbon emissions was small, being 2.36%, 2%, 0.55%, and 0.31% in 2014, 2016, 2018, and 2020, respectively.

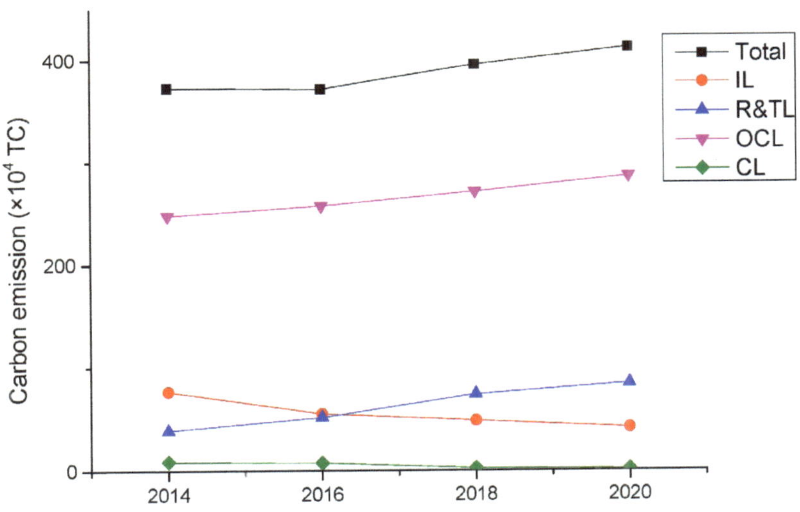

Figure 3. Total carbon emissions.

3.1.2. Carbon Sequestration

The total carbon sequestration of Tongzhou District in 2020 was 38.3 × 10^4 TC, an increase of 230% compared with 11.7 × 10^4 TC in 2014 (Figure 4). Comparing carbon emissions, the increase was significant. FL was the most important functional land for carbon sequestration, contributing 69% of the total carbon sequestration. In general, the total spatial carbon emission was greater than the carbon absorption in all three periods, and the net carbon flux in the vertical direction was negative. Due to the promotion of low-carbon development policies, the proportion of carbon emissions offset by carbon sequestration increased from 3.1% in 2014 to 9.3% in 2020. The carbon balance has improved, but it is still dominated by carbon emissions. Ecological problems need to be paid attention to. The control of carbon emission sources and intensity is the key to realizing the balance of carbon metabolism in the territorial space.

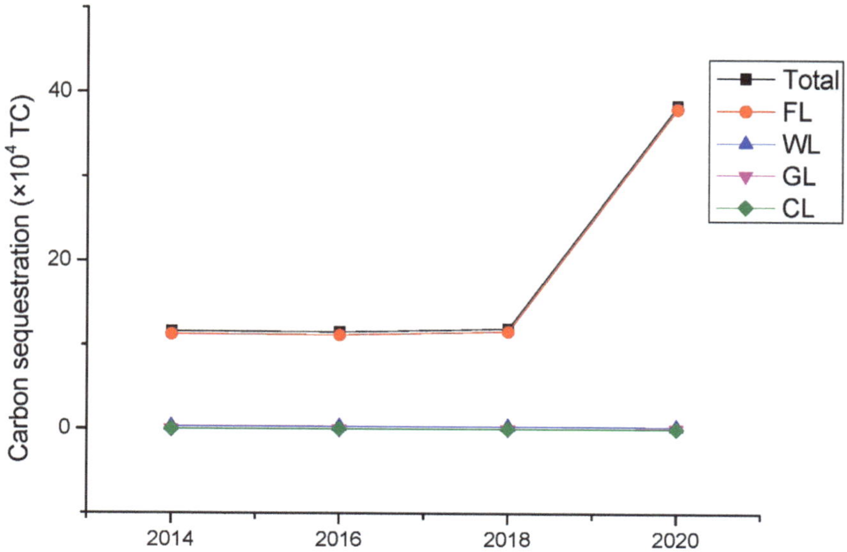

Figure 4. Total carbon sequestration.

3.1.3. Carbon Flow from 2014 to 2020

Based on the calculations of carbon flux in the vertical direction of Tongzhou District and the data of land use change, the carbon metabolic densities of each land use from 2014 to 2020 were obtained (Figure 5). From 2014 to 2020, FL, GL, and WL were the positive carbon metabolism compartments in the entire space of Tongzhou District, and the carbon sequestration density remained relatively stable. Therefore, the change in carbon sequestration was largely related to the change in land area use. R&TL, IL, OCL, and CL all fell within the range of negative carbon metabolism. The net carbon emission of the negative carbon metabolism compartment, which occupied less than 47% of the total area of the sub-center, was about 11–32 times the net carbon sequestration of the positive carbon metabolism compartment. Due to the large amount of carbon produced by CL, the carbon metabolic density was negative, although CL had the function of carbon sequestration. OCL dominated the negative carbon metabolism, and the carbon metabolism density decreased by 33.7% during the period of 2014–2020. The carbon metabolism density of R&TL was reduced by 61.6%. The sub-compartment of IL is an important factor for negative carbon metabolism. With the industrial land reduction in Tongzhou District, the carbon metabolism density in 2020 was reduced to 71% of that in 2014.

According to the distribution of "carbon flow" caused by land use changes in different periods, the net "carbon flow" driven by low-carbon city construction continued to increase, with an increase of 13.3% in 2016–2018 compared with 2014–2016 (Table 4). Although the net "carbon flow" in 2014–2018 remained negative, and the positive "carbon flow" was less than the negative "carbon flow," the carbon metabolism improved each year, and the net "carbon flow" in 2018–2020 increased by 897% compared with that in 2016–2018. The net "carbon flow" in the region changed from negative to positive, with a significant increase. The main reason was that R&TL, CL, and IL were converted to FL.

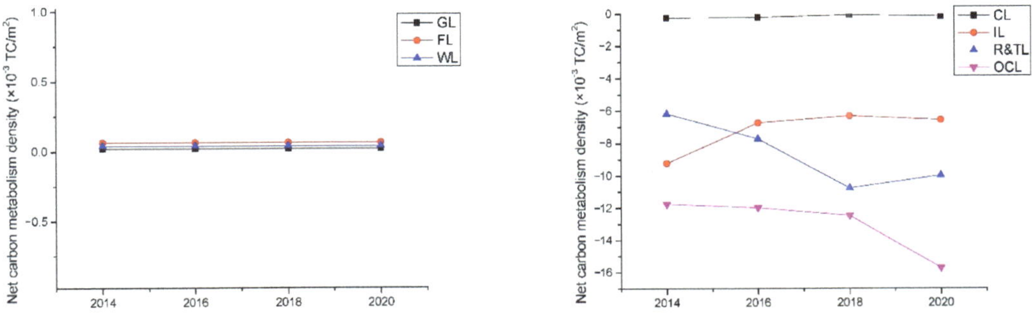

(a) (b)

Figure 5. Carbon metabolism density. (**a**) Net carbon metabolism density of GL, FL, and WL; (**b**) net carbon metabolism density of CL, IL, R&TL, and OCL.

In the three research periods, the negative "carbon flow" mainly came from the conversion of IL to OCL (2014–2016), CL to OCL (2016–2018), and CL to OCL (2018–2020), accounting for 40.2%, 32.3%, and 23.6% of the total negative "carbon flow", respectively. From 2018 to 2020, the contributions of the conversion of CL to R&TL and FL to OCL were not negligible, accounting for 17.4% and 12.2% of the total negative carbon flow, respectively. This indicates that the high carbon emissions caused by R&TL and OCL occupation of CL and FL in the early stages of low-carbon city construction have become important factors affecting the balance of carbon metabolism in regional territorial spaces. In addition, in order to solve the problem of work–housing balance caused by population redistribution, the supply of land for housing, offices, and scientific and technological innovation centers has increased, exerting pressure on the balance of carbon metabolism. The positive "carbon flow" was primarily from the conversion of OCL to IL (2014–2016), OCL to FL (2016–2018), and OCL to FL (2018–2020), accounting for 54%, 40.2%, and 31.8% of the total positive "carbon flow", respectively. FL, as an important source of carbon sequestration, improved the capacity of carbon metabolism in the territorial space.

3.2. *Spatial Ecological Relationships of Carbon Flow*

3.2.1. Ecological Relationships of Carbon Flow

The complete utility matrix of land use carbon flow was calculated using the ENA method, and then the ecological relationships among land use types from 2014 to 2020 were obtained, including competition, control, exploitation, and mutualism. Since exploitation and control relationships have the same essence, they were combined as exploitation and control relationships for statistical analysis. The white areas in Table 5 and Figure 6 represent no land use changes.

From 2014 to 2016, the ecological relationships of Tongzhou District land in the carbon metabolism system were dominated by exploitation and control relationships, accounting for 54.3% of all ecological relationships from the calculation in Table 6. These exploitation and control relationships existed in OCL, R&TL, and IL, accounting for 57% of these relationships. In terms of space, this was mainly distributed in cultural tourism areas, administrative office areas, and new urban areas, indicating that the conversion of OCL, R&TL, and IL in these areas was large, and ecological conflicts were significant. There was strong competition for carbon storage among different areas. Mutualism and competition were sparsely distributed in the region.

Table 4. Exchange of carbon flow.

Exchange Value ($\times 10^6$ kgc/a)	2014–2016		2016–2018		2018–2020	
	Transitions	Direction	Transitions	Direction	Transitions	Direction
Beneficial carbon flow	113.86	+	35.13	+	833.62	+
Harmful carbon flow	−168.48	−	−82.48	−	−456.14	−
Net carbon flow	−54.62	−	−47.35	−	377.48	+
B1 (GL,FL)	—	—	0.01	GL→FL	0.04	GL→FL
B2 (R&TL,GL)	—	—	—	—	22.39	R&TL→GL
B3 (R&TL,CL)	0.08	R&TL→CL	0.9	R&TL→CL	11.89	R&TL→CL
B4 (R&TL,FL)	—	—	0.5	R&TL→FL	66.34	R&TL→FL
B5 (R&TL,WL)	0.01	R&TL→WL	0.8	R&TL→WL	19.71	R&TL→WL
B6 (CL,GL)	—	—	—	—	0.27	CL→GL
B7 (CL,FL)	0.15	CL→FL	0.85	CL→FL	29.74	CL→FL
B8 (CL,WL)	—	—	0.03	CL→WL	0.63	CL→WL
B9 (IL,GL)	—	—	—	—	16.58	IL→GL
B10 (IL,R&TL)	4.3	IL→R&TL	—	—	—	—
B11 (IL,CL)	—	—	2.36	IL→CL	4.16	IL→CL
B12 (IL,FL)	0.14	IL→FL	9.72	IL→FL	58.66	IL→FL
B13 (IL,WL)	—	—	0.15	IL→WL	1.16	IL→WL
B14 (OCL,GL)	—	—	0.03	OCL→GL	210.92	OCL→GL
B15 (OCL,R&TL)	33.43	OCL→R&TL	0.21	OCL→R&TL	29.25	OCL→R&TL
B16 (OCL,CL)	11.2	OCL→CL	4.13	OCL→CL	59.69	OCL→CL
B17 (OCL,IL)	61.47	OCL→IL	0.69	OCL→IL	1.56	OCL→IL
B18 (OCL,FL)	2.41	OCL→FL	14.13	OCL→FL	265.44	OCL→FL
B19 (OCL,WL)	0.67	OCL→WL	0.55	OCL→WL	28.96	OCL→WL
B20 (WL,FL)	—	—	0.02	WL→FL	0.35	WL→FL
B21 (R&TL,IL)	—	—	0.05	R&TL→IL	5.88	R&TL→IL
H1 (GL,R&TL)	—	—	—	—	−0.2	GL→R&TL
H2 (GL,IL)	−0.13	GL→R&TL	−0.04	GL→IL	−0.04	GL→IL
H3 (GL,OCL)	−0.28	GL→OCL	−0.13	GL→OCL	−0.34	GL→OCL
H4 (R&TL,IL)	−1.7	R&TL→IL	—	—	—	—
H5 (R&TL,OCL)	−18.21	R&TL→OCL	−0.9	R&TL→OCL	−7.42	R&TL→OCL
H6 (CL,R&TL)	−0.31	CL→R&TL	−12.36	CL→R&TL	−79.41	CL→R&TL
H7 (CL,IL)	−10.98	CL→IL	−5.33	CL→IL	−25.98	CL→IL
H8 (CL,OCL)	−38.75	CL→OCL	−26.63	CL→OCL	−107.57	CL→OCL
H9 (IL,OCL)	−67.72	IL→OCL	−3.53	IL→OCL	−46.85	IL→OCL
H10 (FL,GL)	—	—	—	—	−0.23	—
H11 (FL,CL)	−0.08	FL→CL	−0.03	FL→CL	−0.66	FL→CL
H12 (FL,R&TL)	−0.2	FL→R&TL	−4.23	FL→R&TL	−38.38	FL→R&TL
H13 (FL,IL)	−2.94	FL→IL	−1.67	FL→IL	−6.31	FL→IL
H14 (FL,WL)	—	—	—	—	−0.06	FL→WL
H15 (FL,OCL)	−15.46	FL→OCL	−17.85	FL→OCL	−55.78	FL→OCL
H16 (WL,GL)	—	—	—	—	−0.07	WL→GL
H17 (WL,CL)	−0.03	WL→CL	−0.04	WL→CL	−0.1	WL→CL
H18 (WL,IL)	−2.72	WL→IL	−0.62	WL→IL	−6.67	WL→IL
H19 (WL,R&TL)	−0.21	WL→R&TL	−1.76	WL→R&TL	−29.7	WL→R&TL
H20 (WL,OCL)	−8.76	WL→OCL	−7.35	WL→OCL	−43.19	WL→OCL
H21 (IL,R&TL)	—	—	−0.01	IL→R&TL	−7.18	IL→R&TL

Note: "Bn" represents the nth positive carbon flow; "Hn" represents the nth negative carbon flow; "+" represents positive carbon flow direction; "−" represents negative carbon flow; "—" indicates that there is no exchange between two components; GL: grassland; FL: forest land; R&TL: road and traffic land; CL: cultivated land; WL: water and wetland; IL: industrial land; OCL: other construction land.

Table 5. Changes in ecological relationships of Tongzhou District.

2014–2016	CL	FL	GL	WL	R&TL	IL	OCL
CL		−	−	−	+	−	+
FL	−		−	−	+	−	+
GL	−	−		−	+	+	+
WL	−	−	−		+	−	+
R&TL	+	−	+	−		−	−
IL	−	−	−	−	+		+
OCL	−	−	−	−	+	−	
2016–2018	**CL**	**FL**	**GL**	**WL**	**R&TL**	**IL**	**OCL**
CL		−	−	−	+	+	+
FL	−		−	−	+	−	+
GL	−	−		−	−	+	+
WL	−	−	−		+	−	+
R&TL	−	−	+	−		−	−
IL	−	−	−	−	−		+
OCL	−	−	−	−	−	−	
2018–2020	**CL**	**FL**	**GL**	**WL**	**R&TL**	**IL**	**OCL**
CL		−	+	−	+	+	+
FL	−		+	−	+	−	+
GL	−	−		+	+	−	+
WL	−	−	+		+	−	+
R&TL	−	−	+	−		+	−
IL	−	−	+	−	−		+
OCL	−	−	+	−	+	−	

Note: "+/−" represents the positive and negative elements in the complete utility matrix U, respectively; yellow—competition; blue—control; red—exploitation; green—mutualism.

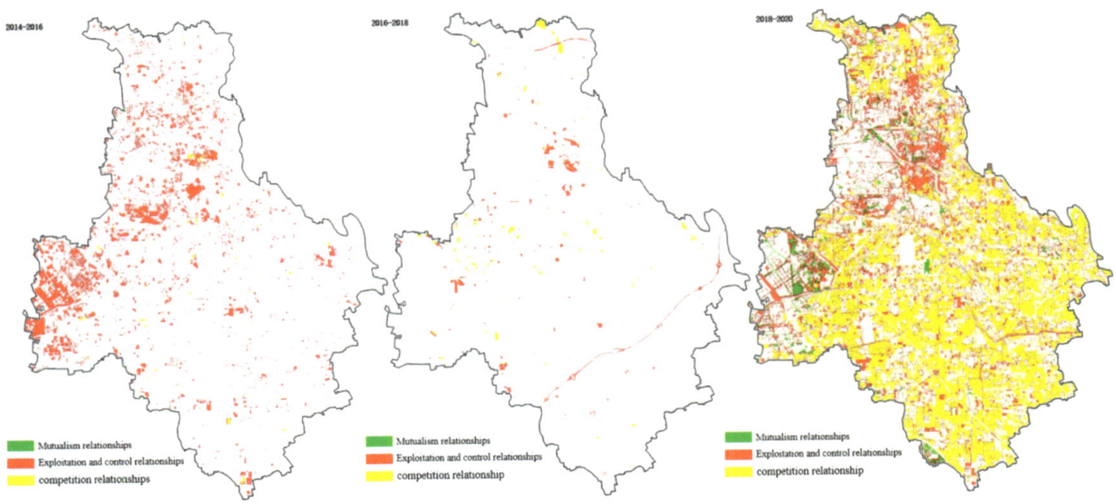

(a) (b) (c)

Figure 6. Spatial distribution of ecological relationships: (**a**) spatial distribution of ecological relationships from 2014 to 2016; (**b**) spatial distribution of ecological relationships from 2016 to 2018; (**c**) spatial distribution of ecological relationships from 2018 to 2020.

Table 6. Distribution of ecological relationships.

Land Use Type	CL	FL	GL	WL	R&TL	IL	OCL
Competition relationships	9	11	6	11	2	8	1
Exploitation and control relationships	9	7	8	6	13	10	16
Mutualism relationships	1	0	4	1	3	0	1

Note: since the nature of the relationship between exploitation and control is the same, they were combined into the relationship between exploitation and control for statistical analysis.

From 2016 to 2018, the spatial distribution of exploitation and control relationships decreased significantly, primarily due to a large number of construction projects being stalled and hence less spatial conversion of land use. Only the built-up areas had exploitation and control relationships. At the same time, due to the large conversion areas of FL, WL, and CL in this region, the competition relationship was strengthened compared with the previous period, being primarily distributed in the peripheral areas. Competition relationships were still scattered and relatively few.

From 2018 to 2020, the spatial distribution of exploitation and control relationships increased slightly. This was mainly due to the speed of urbanization in the region driven by related urban renewal and municipal infrastructure projects. The spatial distribution of competition increased significantly, covering almost the entire township area evenly, largely because the government accelerated the construction of national forest areas and carried out large-scale plain afforestation work, resulting in a significant increase in the area of FL and hence significant spatial conversion between FL and other land uses. These competition relationships also increased significantly compared with the previous period, being primarily distributed in the blue and green land.

3.2.2. Overall Ecological Utility Function

According to the results of the complete utility matrix, the value of the overall ecological utility function M was 0.54, less than 1, indicating the utility of the horizontal flow of carbon metabolism in the study area. The overall regional comprehensive effect of land use change on carbon metabolism balance in the study area from 2014 to 2020 was negative.

4. Discussion

4.1. Impact of Land Use Changes on Carbon Flow

4.1.1. Impact of Negative Carbon Flow

The carbon flow associated with land use changes was determined by both land use change area and carbon metabolism density. The negative carbon flow in the three study periods largely came from the transfer of other land to OCL, unlike other cities in China. For example, the negative carbon flow in Hangzhou [6,7], Zhaotong [35], and other cities [36] was in large part due to the transfer of other land to IL, and this was determined by the development stage and development policies of the cities. In the early 20th century, the development of Hangzhou and other cities attracted the investment of enterprises through the sale of cheap IL, often resulting in the transformation of CL into IL [6,7,35]. However, the low-carbon city development route has been established in the early stage of the construction of Tongzhou District [64,65], being converted to business offices, cultural tourism, administration, and other functions. IL was used for high-tech enterprises bearing low carbon emissions, activities that were not the main source of negative carbon flow. Therefore, the industrial development path selection of the central and local governments is crucial to regional low-carbon development. Industrial policies should not only emphasize industrial development and the improvement of economic efficiency but should also pay attention to the support and encouragement of green industries, reduce the carbon emission intensity of the industry, and promote the sustainable development of the region.

4.1.2. Impact of Positive Carbon Flow

The positive carbon flow was primarily due to the transfer of other land to FL. From 2016 to 2018 and from 2018 to 2020, the proportions of carbon flow to FL were 72% and 51%, respectively. Afforestation and forest management as well as preventing deforestation and forest degradation were also important measures used to increase carbon storage in ecosystems in other cities around the world [66]. Since 2015, Beijing has been committed to green development and has made all-out efforts to promote a new round of afforestation projects covering an area of 1 million mu. Eight super-large parks have been built in Tongzhou District, and the forest coverage rate of the entire area has reached 35.24%. The contribution of GL increases to the positive carbon flow cannot be ignored, as this accounted for 25.3% of the total positive carbon flow. Therefore, optimizing the spatial layout of woodland and GL, planting vegetation reasonably, and promoting the healthy growth of plants in spatial planning have strong development potential for improving the carbon sequestration capacity of green spaces.

4.2. Effects of Land Use Changes on Ecological Relationships

4.2.1. Exploitation and Control Relationships

Analyzing the ecological relationship of carbon metabolism under the influence of land use change is helpful in formulating policies for optimizing land use structure [10,11]. The results of the utility analysis show that these exploitation and control relationships were the main ecological relationships involved in carbon metabolism in Tongzhou District, and these were spatially distributed in the concentrated development and construction areas. Beijing implements a strict policy of intensive land conservation [67], resulting in very limited land available for development and construction. This intensifies the land competition among OCL, R&TL, and IL in the process of urbanization, especially in high-value areas. The government re-develops land through urban renewal and other means to explore new economic growth points, but this process will generate massive population and economic activities, leading to increases in energy consumption and carbon emissions. Therefore, in the process of urbanization, low-carbon development requires a clear and reasonable scale of land use and optimization of urban spatial structure in order to reduce urban carbon emissions and promote the balanced development of regional carbon metabolism.

4.2.2. Competition Relationships

The competition relationship accounted for 37.8% of the overall ecological relationship, being primarily in the positive carbon metabolism compartments and manifested as the competition between FL, CL, WL, and GL for space. With the gradual establishment of territorial space planning systems, the role of natural ecological space in carbon metabolism analysis is becoming increasingly important. In the traditional management mode, this type of land belongs to the management of different industry departments, especially FL, CL, and WL, which have certain rigid control requirements. When the index is difficult to meet, deforestation or occupation of cultivated land often occurs, and this can significantly change the vegetation coverage of the surface in a short time. For example, from 2018 to 2020, the government vigorously adopted economic means to promote plain afforestation projects, resulting in shrinkage of the cultivated land area and a substantial increase in the woodland area; this had a significant impact on the regional carbon metabolism balance. Therefore, in the development of low-carbon cities, it is necessary to respect ecological laws and natural processes, combine the characteristics of natural endowments in various regions, reasonably determine the scale and proportion of natural elements such as woodland, farmland, and water areas, guide rational layouts, alleviate the competition between land types, and promote the balanced development of regional carbon metabolism.

4.2.3. Mutualism Relationships

During the study period, mutualism existed in both positive and negative carbon metabolism compartments, but the overall proportion was relatively small and scattered in the area, occurring largely in GL and T&RL, indicating that there was no significant land use change in these two compartments. Mutualism emphasizes the harmonious symbiosis, mutual benefit, and mutual use between social and economic development and the external natural environment. During the study period, mutualism was mainly distributed in the natural ecological space within the urban construction area of Tongzhou District, which highlighted the urban characteristics of mixed blue and green land, and a large number of green spaces inside the city have been reserved to provide high-quality and convenient leisure and recreation space for citizens since 2014. Such spaces are also areas that conform to planning guidance, increasing the ecological utility of the area in the process of interaction. Therefore, reasonable urban space policy is of great value to the improvement of urban comprehensive ecological utility.

5. Conclusions and Suggestions

5.1. Conclusions

In this study, an analysis framework between land use change and carbon metabolism was constructed based on ENA to explore the comprehensive impact on carbon metabolism. Taking Tongzhou District as an example, we analyzed the spatial characteristics of carbon emissions and carbon sequestration in three time periods from 2014 to 2020 and depicted the carbon flow and spatial distribution related to land use. The following conclusions were drawn.

From 2014 to 2020, the total carbon emission of Tongzhou District increased year by year. The carbon emission of OCL was dominant, and the carbon emission of IL decreased significantly. The carbon sequestration capacity of FL increased by 236%. The positive carbon metabolic density remained relatively stable, while the negative carbon metabolic density decreased year by year.

From 2014 to 2020, the negative carbon flow in Tongzhou District was concentrated in the transfer IL to OCL (2014–2016), CL to OCL (2016–2018), and CL to OCL (2018–2020) accounting for 40.2% of the total negative "carbon flow". The positive carbon flow mainly came from the conversion of OCL to IL (2014–2016), OCL to FL (2016–2018), and OCL to FL (2018–2020). Since the negative "carbon flow" originated from the transfer of other land to OCL, reducing the carbon emission density of OCL and promoting the centralized and intensive development of urban construction land are key to realizing low-carbon development in the process of urban development and construction.

From 2014 to 2016, the spatial ecological relationships of carbon flow in Tongzhou District were dominated by exploitation and control relationships. In 2016–2018, competition intensified, largely in expanded areas; in 2016–2018, exploitation and control relationships, competition relationships, and mutualism relationships increased significantly and were evenly distributed. However, the overall ecological utility value was 0.54, indicating that the regional carbon metabolism was unbalanced. In terms of urban industrial development, the government should pay attention to supporting and encouraging green industries, optimize the spatial structure of urban construction land, and reduce urban carbon emissions. Moreover, we should respect ecological laws, optimize the scale and distribution of ecological factors, and improve the carbon sequestration capacity of green spaces.

5.2. Suggestions

Based on the above conclusions, suggestions for the construction of low-carbon cities in Tongzhou District are proposed as follows.

In terms of social life, education should be developed to increase residents' awareness of low carbon and promote lifestyle transformation. The lifestyle of residents in OCL has a direct impact on carbon emissions. The adoption of a green, healthy, and environmentally friendly lifestyle should be encouraged, thereby reducing the emission of carbon dioxide and other greenhouse gases. It should be noted that OCL is mainly concentrated in urban construction areas. Therefore, raising the awareness of a low-carbon lifestyle and encouraging the use of clean energy are important aspects of urban construction and low-carbon development in construction areas.

In terms of public management, spatial planning guidance and natural ecological environment protection should be strengthened. In fact, natural ecological space (especially FL) is the main source of positive carbon flow. It is necessary to strengthen the protection of green space and carry out the construction of laws and regulations for the development and management of low-carbon ecological cities. Important green spaces should be reserved in low-carbon urban planning while paying attention to strengthening afforestation and forest management to prevent deforestation and forest degradation.

In terms of economic development, the mode of economic development needs to be transformed to encourage green industries and gradually promote the decoupling of economic growth and carbon emissions. The development of a low-carbon economy needs to adopt an economic development mode to adapt to the global changes, improve the energy utilization rate, and choose a green industrial development path, in order to achieve an economic growth rate higher than the carbon emission rate without slowing down economic development and lowering living standards. As a result, the industrial development path selection of central and local governments is crucial to regional low-carbon development.

5.3. Limitations and Future Insights

The present study also has certain limitations, e.g., this study did not explore the driving mechanisms of ecological relationship change and lacked an analysis of the impact of construction policies and management on carbon metabolism. In addition, we also did not perform a quantitative analysis of the differences in vegetation types and ignored the difference in carbon emissions between urban and rural houses in OCL. Given these limitations, the analysis needs to be further strengthened.

Author Contributions: Conceptualization, Y.Q., J.Z. and S.Z.; methodology, Y.Q.; formal analysis, Y.Q. and M.X.; writing—original draft preparation, Y.Q. and M.X.; writing—review and editing, Y.Q., J.Z., C.X., Y.G. and S.Z.; supervision, Y.Q., J.Z. and S.Z.; funding acquisition, J.Z. and S.Z. All authors have read and agreed to the published version of the manuscript.

Funding: This work was supported by grants from the National Natural Science Foundation of China (No. 52078008, No. 51908004) and the Beijing Municipal Natural Science Foundation (No. 8202004).

Institutional Review Board Statement: Not applicable.

Data Availability Statement: Not applicable.

Acknowledgments: We thank Chuyu Xia for her constructive comments and suggestions to improve the quality of this article.

Conflicts of Interest: The authors declare no conflict of interest.

References

1. Intergovernmental Panel on Climate Change (IPCC). *Special Report on Global Warming of 1.5 °C[R]*; Cambridge University Press: Cambridge, UK, 2018.
2. United Nations Environment Programme (UNEP). *Emissions Gap Report 2020[R]*; United Nations Environment Programme and UNEP DTU Partnership: Nairobi, Kenya, 2020.
3. He, J.; Yang, Y.; Liao, Z.; Xu, A.; Fang, K. Linking SDG 7 to assess the renewable energy footprint of nations by 2030. *Appl. Energy* **2022**, *317*, 119167. [CrossRef]
4. Wolman, A. The metabolism of cities. *Sci. Am.* **1965**, *213*, 179–190. [CrossRef] [PubMed]
5. Xia, L.L.; Zhang, Y.; Li, M.J. A review of urban carbon metabolism. *Acta Ecol. Sin.* **2017**, *37*, 4268–4277.
6. Xia, C.; Li, Y.; Xu, T.; Chen, Q.; Ye, Y.; Shi, Z.; Liu, J.; Ding, Q.; Li, X. Analyzing spatial patterns of urban carbon metabolism and its response to change of urban size: A case of the Yangtze River Delta, China. *Ecol. Indic.* **2019**, *104*, 615–625. [CrossRef]
7. Xia, C.; Xiang, M.; Fang, K.; Li, Y.; Ye, Y.; Shi, Z.; Liu, J. Spatial-temporal distribution of carbon emissions by daily travel and its response to urban form: A case study of Hangzhou, China. *J. Clean. Prod.* **2020**, *257*, 120797. [CrossRef]

8. Li, Y.; Shen, J.; Xia, C.; Xiang, M.; Cao, Y.; Yang, J. The impact of urban scale on carbon metabolism –a case study of Hangzho China. *J. Clean. Prod.* **2021**, *292*, 126055. [CrossRef]
9. Houghton, R.A. The annual net flux of carbon to the atmosphere from changes in land use 1850–1990. *Tellus B Chem. Phy. Meteorol.* **1999**, *51*, 298–313. [CrossRef]
10. Xia, C.; Chen, B. Urban land-carbon nexus based on ecological network analysis. *Appl. Energy* **2020**, *276*, 115465. [CrossRef]
11. Zhu, E.; Deng, J.; Zhou, M.; Gan, M.; Jiang, R.; Wang, K.; AmirReza, S. Carbon emissions induced by land-use and land-cove change from 1970 to 2010 in Zhejiang, China. *Sci. Total Environ.* **2019**, *646*, 930–939. [CrossRef]
12. Houghton, R.A.; House, J.I.; Pongratz, J.; Van Der Werf, G.R.; Defries, R.S.; Hansen, M.C.; Le Quéré, C.; Ramankutty, N. Carbo emissions from land use and land-cover change. *Biogeosciences* **2012**, *9*, 5125–5142. [CrossRef]
13. Findell, K.L.; Berg, A.; Gentine, P.; Krasting, J.P.; Lintner, B.R.; Malyshev, S.; Santanello, J.A.; Shevliakova, E. The impact o anthropogenic land use and land cover change on regional climate extremes. *Nat. Commun.* **2017**, *8*, 989. [CrossRef] [PubMed]
14. Kay, S.; Rega, C.; Moreno, G.; den Herder, M.; Palma, J.H.N.; Borek, R.; Herzog, F. Agroforestry creates carbon sinks whils enhancing the environment in agricultural landscapes in Europe. *Land Use Policy* **2019**, *83*, 581–593. [CrossRef]
15. Tang, X.; Woodcock, C.E.; Olofsson, P.; Hutyra, L.R. Spatiotemporal assessment of land use/land cover change and associate carbon emissions and uptake in the Mekong River Basin. *Remote Sens. Environ.* **2021**, *256*, 112336. [CrossRef]
16. Xia, C. Multi-Scale Studies on Urban Carbon Metabolism from the Perspective of Land Use and Scenario Analysis of Emissio Reduction, China. Ph.D. Thesis, Zhejiang University, Zhejiang, China, 2019. [CrossRef]
17. Yu, Z.; Chen, L.; Tong, H.; Chen, L.; Zhang, T.; Li, L.; Yuan, L.; Xiao, J.; Wu, R.; Bai, L.; et al. Spatial correlations of land-use carbo emissions in the Yangtze River Delta region: A perspective from social network analysis. *Ecol. Indic.* **2022**, *142*, 109147. [CrossRef
18. Zhao, J.; Shao, Z.; Xia, C.; Fang, K.; Chen, R.; Zhou, J. Ecosystem services assessment based on land use simulation: A case stud in the Heihe River Basin, China. *Ecol. Indic.* **2022**, *143*, 109402. [CrossRef]
19. Kellett, R.; Christen, A.; Coops, N.C.; van der Laan, M.; Crawford, B.; Tooke, T.R.; Olchovski, I. A systems approach to carbo cycling and emissions modeling at an urban neighborhood scale. *Landsc. Urban Plan.* **2013**, *110*, 48–58. [CrossRef]
20. Long, Y.; Ayyoob, S.; Huang, L.; Chen, J. Urban carbon accounting: An overview. *Urban Clim.* **2022**, *44*, 101195.
21. Chen, X.; Meng, Q.; Shi, J.; Liu, Y.; Sun, J.; Shen, W. Regional Differences and Convergence of Carbon Emissions Intensity in Citie along the Yellow River Basin in China. *Land* **2022**, *11*, 1042. [CrossRef]
22. Sun, Y.; Jia, J.; Ju, M.; Chen, C. Spatiotemporal Dynamics of Direct Carbon Emission and Policy Implication of Energy Transitio for China's Residential Consumption Sector by the Methods of Social Network Analysis and Geographically Weighted Regressio *Land* **2022**, *11*, 1039. [CrossRef]
23. Chen, S.-Q.; Chen, B. Network environ perspective for urban metabolism and carbon emissions: A case study of Vienna, Austria *Environ. Sci. Technol.* **2012**, *46*, 4498–4506. [CrossRef]
24. Chen, H.; Meng, C.; Cao, Q. Measurement and Influencing Factors of Low Carbon Urban Land Use Efficiency—Based o Non-Radial Directional Distance Function. *Land* **2022**, *11*, 1052. [CrossRef]
25. Zheng, H.; Ge, L. Carbon emissions reduction effects of sustainable development policy in resource-based cities from the perspective of resource dependence: Theory and Chinese experience. *Resour. Policy* **2022**, *78*, 102799. [CrossRef]
26. Vaccari, F.P.; Gioli, B.; Toscano, P.; Perrone, C. Carbon dioxide balance assessment of the city of Florence (Italy), and implications for urban planning. *Landsc. Urban Plan.* **2013**, *120*, 138–146. [CrossRef]
27. Wang, S.-H.; Huang, S.-L.; Huang, P.-J. Can spatial planning really mitigate carbon dioxide emissions in urban areas? A case study in Taipei, Taiwan. *Landsc. Urban Plan.* **2018**, *169*, 22–36. [CrossRef]
28. Fang, K. Moving away from sustainability. *Nat. Sustain.* **2022**, *5*, 5–6. [CrossRef]
29. Ye, R.; Gao, Z.; Fang, K.; Liu, K.; Chen, J. Moving from subsidy stimulation to endogenous development: A system dynamics analysis of China's NEVs in the post-subsidy era. *Technol. Forecast. Soc. Chang.* **2021**, *168*, 120757. [CrossRef]
30. Zhang, Q.; Fang, K.; Chen, J.; Liu, H.; Liu, P. The role of sectoral coverage in emission abatement costs: Evidence from China's industrial sectors. *Environ. Res. Lett.* **2022**, *17*, 045002. [CrossRef]
31. Huang, H.; Yang, H.; Chen, Y.; Chen, T.; Bai, L.; Peng, Z.R. Urban Green Space Optimization Based on a Climate Health Risk Appraisal—A case study of Beijing city, China Science Direct. *Urban For. Urban Green.* **2021**, *62*, 127154. [CrossRef]
32. Huang, H.; Chen, K.; Zhang, H.; Ren, L. Planning and coordinated response mechanism of economic and ecological services in urban expansion. *Econ. Res.-Ekon. Istraživanja* **2022**, *35*, 1–21. [CrossRef]
33. Zhang, Y.; Xia, L.-L.; Fath, B.-D.; Yang, Z.-F.; Yin, X.-A.; Su, M.-R.; Liu, G.-Y.; Li, Y.-X. Development of a spatially explicit network model of urban metabolism and analysis of the distribution of ecological relationships: Case study of Beijing, China. *J. Clean. Prod.* **2016**, *112*, 4306–4317. [CrossRef]
34. Chen, S.-Q.; Chen, B. Determining carbon metabolism in urban areas though network environ theory. *Procedia Environ. Sci.* **2012**, *13*, 2246–2255. [CrossRef]
35. Du, J.-S.; Fu, J.-Y.; Hao, M.-M. Analyzing the carbon metabolism of "Production-LivingEcological" space based on ecological network utility in Zhaotong, China. *J. Nat. Resour.* **2021**, *36*, 1208–1223.
36. Lin, G.; Jiang, D.; Fu, J.; Cao, C.; Li, X.; Li, Z.; Huang, S. Carbon flow analysis for production-living-ecological space: A case study of Tangshan, China. *Sci. Technol. Rev.* **2020**, *38*, 107–114. [CrossRef]
37. Liu, C.; Sun, W.; Li, X. Characteristics of spatiotemporal variations in coupling coordination between integrated carbon emission and sequestration index: A case study of the Yangtze River Delta, China. *Ecol. Indic.* **2022**, *135*, 108520. [CrossRef]

38. Xia, C.Y.; Li, Y.; Ye, Y.M.; Shi, Z.; Liu, J.M.; Li, X.S. Analyzing urban carbon metabolism based on ecological network utility: A case study of Hangzhou City. *Acta Ecol. Sin.* **2018**, *38*, 73–85.
39. Xia, L.L.; Fath, B.D.; Scharler, U.M.; Zhang, Y. Spatial variation in the ecological relationships among the components of Beijing's carbon metabolic system. *Sci. Total Environ.* **2016**, *544*, 101–113. [CrossRef]
40. Xia, C.; Li, Y.; Xu, T.; Ye, Y.; Shi, Z.; Peng, Y.; Liu, J. Quantifying the spatial patterns of urban carbon metabolism: A case study of Hangzhou, China. *Ecol. Indic.* **2018**, *95*, 474–484. [CrossRef]
41. Sun, H.; Liang, H.M.; Chang, X.-L.; Cui, Q.-C.; Tao, Y. Land Use Patterns on Carbon Emission and Spatial Association in China. *Econ. Geogr.* **2015**, *35*, 154–162.
42. Zhang, Q.; Tao, X.; Yang, P. Research On Carbon Emissions From Metropolis Urban Passenger Transport and Countermeasures, China Population. *Resour. Environ.* **2012**, *22*, 35–42.
43. Xie, T.; Wang, J. Comparative study on carbon emissions of transportation industry. *Compr. Transp.* **2011**, *8*, 20–24.
44. Min, J. A Research of the Impacts of Agricultural Products in International Market on Greenhouse Gas Emissions in the Chinese Agricultural Production. Ph.D. Dissertation, Chinese Doctoral Dissertations Full-Text Database, Nanjing Agricultural University, Nanjing, China, 2012.
45. Solomon, S. ; *Intergovernmental Panel on Climate Change. Climate Change the Physical Science Basis*; American Geophysical Union: Boulder, CO, USA, 2007.
46. Yaoqiu, K.; Ting-ping, O.; Yi, Z.; Yu, L.; Chao, L.; De-hui, W. Present Situation of Carbon Source and Sink and Potential for Increase of Carbon Sink in Guangdong Province, China Population. *Resour. Environ.* **2010**, *20*, 56–61.
47. Panqin, C.; Xiaoke, W.; Limao, W. *Carbon Budget and Sink Enhancement Strategies of Terrestrial Ecosystems in China*; Science Press: Beijing, China, 2008.
48. Gui, X.; Wang, J. *X000047 China's Terrestrial Ecosystem Carbon Budget and Countermeasures for Increasing Sinks*; Book Almanac of China, Ed.; Hubei People's Publishing House: Hubei, China, 2009; Volume 760.
49. Fang, J.; Zhao, G.; Shi, P.; Chen, A. Estimation of China's terrestrial vegetation carbon sink from 1981 to 2000. *Sci. China Ser. D Earth Sci.* **2007**, *6*, 804–812.
50. Liu, G.; Fu, B.; Fang, J. Carbon dynamics of Chinese forests and its contribution to global carbon balance. *Acta Ecol. Sin.* **2000**, *5*, 733–740.
51. Duan, X.-N.; Han, B.; Fei, X.K. Carbon sequestration and its potential by wetland ecosystems in China. *Acta Ecol.* **2008**, *2*, 463–469.
52. Walsh, J.J. Importance of continental margins in the marine biogeochemical cycling of carbon and nitrogen. *Nature* **1991**, *350*, 53–55. [CrossRef]
53. Li, Z.C.; Xu, J.Y.; Wu, C.Y.; Zhang, Y. A review of studies using ecological network analysis. *Acta Ecol. Sin.* **2011**, *31*, 5396–5405.
54. Ayres, R.U.; Kneese, A.V. Production, consumption and externalities. *Am. Econ. Rev.* **1969**, *59*, 282–297.
55. Yang, S.; Fath, B.; Chen, B. Ecological network analysis of embodied particulate matter 2.5—A case study of Beijing. *Appl. Energy* **2016**, *184*, 882–888. [CrossRef]
56. Ukidwe, N.R.; Bakshi, B.R. Resource intensities of chemical industry sectors in the United States via input-output network models. *Comput. Chem. Eng.* **2008**, *32*, 2050–2064. [CrossRef]
57. Chen, L.; Wang, R.S.; Yang, J.X.; Shi, Y.L. Structural complexity analysis for industrial ecosystems: A case study on LuBei industrial ecosystem in China. *Ecol. Complex.* **2010**, *7*, 179–187. [CrossRef]
58. Lu, Y.; Chen, B.; Hayat, T.; Alsaedi, A. Communal carbon metabolism: Methodology and case study. *J. Clean. Prod.* **2015**, *163*, S315–S321. [CrossRef]
59. Bolin, B. Changes of land biota and their importance for the carbon cycle. *Science* **1977**, *196*, 613–615. [CrossRef] [PubMed]
60. Zhang, Y.; Yang, Z.F.; Fath, B.D. Ecological network analysis of an urban water metabolic system: Model development and a case study for Beijing. *Sci. Total Environ.* **2010**, *408*, 4702–4711. [CrossRef]
61. Fang, D.; Chen, B. Ecological network analysis for a virtual water network. *Environ. Sci. Technol.* **2015**, *49*, 6722–6730. [CrossRef] [PubMed]
62. Shaikh, F.; Ji, Q.; Fan, Y. Evaluating China's natural gas supply security based on ecological network analysis. *J. Clean. Prod.* **2016**, *139*, 1196–1206. [CrossRef]
63. Finn, J.T. Measures of ecosystem structure and function derived from analysis of flows. *J. Theor. Biol.* **1976**, *56*, 363–380. [CrossRef]
64. Wu, Z.-Q.; Li, X. Design Thinking Innovation in the Planning of Beijing City Sub-Center. *Urban Plan. Forum.* **2019**, *S1*, 138–141. [CrossRef]
65. Li, Y.-X.; Wang, K.-Y. Experience and Implications of Domestic and Foreign Secondary City Center Construction for Beijing Tongzhou. *China Anc. City.* **2018**, *6*, 10–18. [CrossRef]
66. Finn, J.T. Flow analysis of models of the hubbard brook ecosystem. *Ecology* **1980**, *61*, 562–571. [CrossRef]
67. Zhong, H.-X.; Li, F.-X. Land Policy Selection Based on the Principle of Social Utility Maximization in Beijing. *China Land Sci.* **2009**, *23*, 67–71.

land

MDPI

Article

Spatiotemporal Evolution Trends of Urban Total Factor Carbon Efficiency under the Dual-Carbon Background

Haiyan Luo * and Xiaoe Qu

School of Economics and Finance, Xi'an Jiaotong University, Xi'an 710061, China
* Correspondence: lhy418761793@stu.xjtu.edu.cn; Tel.: +86-18629696335

Abstract: In order to grasp the development laws of carbon efficiency and help China to achieve the dual-carbon goal as soon as possible, we used the super-efficiency EBM model to calculate the total factor carbon efficiency of cities, based on panel data for 283 prefecture-level cities in China from 2004 to 2019. The spatiotemporal variations in carbon efficiency were analyzed using the Dagum Gini coefficient, LISA aggregation and standard deviational ellipse. The dynamic evolution trends were analyzed using kernel density estimation and the spatial Markov chain. The results showed that the overall difference in total factor carbon efficiency between the 283 cities decreased, and the difference within the three regions was greater than that between regions. The total factor carbon efficiency of the cities had a significant spatial correlation, and the spatial distribution patterns showed a centripetal aggregation from northeast to southwest. The dynamic evolution characteristics of the carbon emission intensity in different regions were quite different and the polarization of eastern and central cities was more obvious. There was a significant path dependence effect on the transition probability of total factor carbon efficiency between cities, and the carbon efficiency level of neighboring cities could affect the transition probability of the carbon efficiency of the cities. Based on the above conclusions, we also put forward relevant policy recommendations for technological changes on the energy supply side, innovative development patterns and the governance of regional policies.

Keywords: total factor carbon efficiency; spatiotemporal differentiation; dynamic evolution

Citation: Luo, H.; Qu, X. Spatiotemporal Evolution Trends of Urban Total Factor Carbon Efficiency under the Dual-Carbon Background. *Land* **2023**, *12*, 69. https://doi.org/10.3390/land12010069

Academic Editor: Yan Li

Received: 25 October 2022
Revised: 17 December 2022
Accepted: 21 December 2022
Published: 26 December 2022

1. Introduction

Across the world, in recent years, extreme climate events have occurred frequently, and global warming has been increasing. Ecological and environmental issues have become severe challenges to be addressed by all countries and countries are facing the increasingly urgent task of emission reduction. According to the Global Risks Report released by the Global Economic Forum in Davos in 2022, the top three risks are environmental issues. Of these, "poor climate action" was identified as the biggest risk facing the world over the next 10 years. In recent years, China has faced a number of development problems, such as insufficient energy and resource supplies, low energy use efficiency and the continuous deterioration of the ecological environment. The insufficient resource and environmental carrying capacity has become a bottleneck restricting China's high-quality economic development. As the current largest energy consumer and greenhouse gas emitter, China has adopted the attitude of a responsible major country in global climate negotiations and has promised to implement more favorable policies and measures for environmental governance, strive to peak carbon emissions by 2030 and endeavor to achieve carbon neutrality by 2060. Nevertheless, even as the second largest economy in the world, China is still the largest developing country. Therefore, at this stage, the primary task is still to develop the economy. Coordinating the relationship between economic development and emission reduction has become a dilemma that every large developing country must face [1]. Improving carbon emission efficiency to minimize resource and environmental constraints, reducing energy dependence on fossil fuels and high-carbon industries and promoting green and low-carbon transformations are important strategies for achieving the dual-carbon goal. These strategies are also key to boosting high-quality economic development. China has a vast territory and there are huge differences in factor endowments, energy structures and technology levels between the regions, which leads to large differences in total factor carbon efficiency (TFCE) between the regions. In this context, the scientific and objective evaluation of TFCE and the analysis of its spatiotemporal differentiation and dynamic evolution trends could not only provide a systematic and comprehensive understanding of the development of China's carbon efficiency level but also enable the central and local governments to grasp the development trends of TFCE in various regions in a timely manner. In turn, this could facilitate the formulation of various policies and measures related to carbon reduction and the reasonable allocation of energy-saving tasks so as to achieve the coordinated and overall development of regional carbon emission efficiency and lay solid foundations for the realization of the dual-carbon goal.

This work is arranged as follows: the second part is the literature review, sorting out the existing related literature, and then concluding the gaps in the existing literature and the innovation of this paper; the third part discusses the measurement method and variable selection of urban TFCE; the fourth part analyzes the spatiotemporal differentiation characteristics of TFCE; the fifth part studies the dynamic evolution trend of urban TFCE; and the last section summarizes the research conclusions and provides relevant policy suggestions accordingly.

2. Literature Review

2.1. Carbon Efficiency and Its Measurement Method

Driven by the dual-carbon goal, research on carbon emission efficiency has become an increasingly hot topic in academic circles. The origins of carbon emission efficiency can be traced back to the concept of carbon productivity, which was proposed by Kaya and Yokobori [2]. Nevertheless, carbon productivity does not consider the interactions between production input factors and cannot reflect the multidimensional characteristics of carbon emission efficiency in terms of economic production. The essence of carbon emission rates is to consider the technical production efficiency of carbon emissions [3]. therefore, carbon emission rates can be understood as obtaining more outputs with the same or fewer carbon emissions by considering the interactions between various input factors, such as capital and labor [4]. As a key part of China's carbon peak and neutrality goals, the scientific and comprehensive measurement of carbon efficiency plays a crucial role in subsequent related research.

In terms of the measurement and evaluation of carbon efficiency indicators, existing studies have mainly involved single factor efficiency and total factor efficiency, among which the measurement of single factors refers to carbon emissions and the proportions of certain input factors; for example, the ratio of carbon emissions to GDP is used to measure carbon efficiency [5]. The measurement of total factor carbon efficiency usually involves the stochastic frontier analysis (SFA) method and the data envelopment analysis (DEA) method. Herrala et al. used the SFA method to measure the carbon efficiency of 170 pairs of countries in the world and found that there are huge differences in carbon efficiency levels and efficiency changes between the countries [6]. Du and Zou [7] and Chen and Huang [8] used the SFA method to calculate China's provincial carbon efficiency. Because the SFA method needs specific settings on functional forms and has very strict assumptions, Charnes et al. proposed a non-parametric analysis method, namely the DEA method, which can not only take into account various input and output indicators, such as the economy and the environment, but can also effectively avoid the problem of subjective parameter weighting without setting specific function forms [9]. The International Energy Agency (IEA) measures carbon emission efficiency in different countries at the industrial level using the DEA method [10]. Ramanathan used the DEA model to calculate carbon emission efficiency at the national level and explored the factors affecting carbon emissions [11]. Nonetheless, the early DEA model required the input and output variables to be reduced in the same proportion, which is inconsistent with actual economic development and th radial model cannot measure non-radial relaxation variables [12]. Based on the abov deficiencies, Tone introduced slack input and output variables into the objective functio and constructed the non-radial SBM (slack-based measure) model [12], which effectivel avoids the strict assumption that the factor inputs and outputs are reduced in the sam proportion and can also incorporate undesirable outputs into the efficiency measuremen process. Therefore, this method has become widely used in the measurement of carbo emission efficiency. Sun et al. used the SBM-DEA model to measure carbon efficienc levels [13]. Some scholars have also used the super-efficiency DEA model to measure an analyze inter-provincial carbon efficiency [14,15], industrial carbon emission efficiency [16 the carbon emission trading pilot market [17] and the carbon efficiency of the Yangtze an Yellow River basins in China [18].

2.2. *Regional Differences in Carbon Emissions and Carbon Efficiency*

Some Chinese scholars have analyzed the dynamic evolution characteristics of carbo efficiency and carbon intensity from the perspectives of regional differences and conve gence. Zhou and Song measured the industrial carbon efficiency of China and conclude that industrial carbon efficiency has been increasing and that there are significant regiona differences [19]. Lin et al. studied the spatiotemporal evolution characteristics of industria carbon efficiency in the Beijing–Tianjin–Hebei region using kernel density estimation. Thei empirical tests showed that there are large differences in carbon efficiency between indus tries and small differences between cities [16]. Jiang et al. analyzed the spatiotempora evolution characteristics of carbon efficiency in the Yangtze River and Yellow River basin and concluded that carbon efficiency in this region shows the distribution characteristic of high in the middle and low at both ends and that the regions with high values form clusters [18]. Wang et al. and Han et al. used Theil Index to analyze the regional difference of China's provincial carbon emissions and carbon productivity, respectively. The forme study concluded that the gap of China's provincial carbon emissions was expanding yea by year [20], while the latter concluded that the regional differences of provincial carbor productivity were large [21]. Zhang et al. analyzed China's carbon emission intensity using kernel density estimation and concluded that carbon intensity shows a distributior pattern of "low in the south and high in the north", with significant differences in dynamic evolution between different regions. Significant β convergence was observed in the carbor intensity of the whole country and major strategic regions [22].

2.3. *The Gaps in the Existing Literature and the Innovation of This Paper*

In general, the existing research on the connotations, extension and quantitative mea surement of carbon efficiency has achieved certain results, which have laid the foundations for subsequent related research. However, there are also some shortcomings. Firstly, the existing research on carbon efficiency has mainly focused on the macro level, such as na tional, provincial and industrial levels. Due to the limitations of data acquisition, relevant research on TFCE at the urban level in China is still lacking. Secondly, research on the spatial distribution of carbon efficiency is still in its infancy and the overall research is still relatively rough. There have been few studies on the spatiotemporal differentiation and dynamic evolution of urban TFCE, which is not conducive to the overall improvement of China's carbon efficiency and coordinated development between regions. Thirdly, in terms of the measurement and evaluation of carbon efficiency, most of the existing studies have used the super-efficiency SBM-DEA model for their calculations; however, the SBM model cannot provide the ratio information between the actual value of the input (output) index and the target value [23], which can lead to deviations in measurement results. Fourth, most of the existing studies have used Theil index and the kernel density estimation method to reveal the regional differences and dynamic evolution characteristics of the research objects, and few have used the Dagum Gini coefficient and standard deviational ellipse method to analyze carbon efficiency. The standard deviational ellipse and the center of gravity transfer analysis can intuitively reveal the various spatial characteristics of TFCE. They can also complement the Dagum Gini coefficient decomposition method and kernel density estimation method. Fifth, the existing research has paid attention to the spatial distribution of carbon emissions and carbon efficiency but has not analyzed the spatiotemporal transfer probability of carbon efficiency. Additionally, there have been even fewer relevant studies on urban TFCE.

The innovations of this paper are as follows. In terms of research object, this study focused on the city level, took panel data for 283 prefecture-level cities in China as samples and calculated urban carbon emissions by combining the energy consumption generated by major carbon emission sources (i.e., the consumption of electricity, liquefied gas, liquefied petroleum gas and thermal energy in the whole society) with their respective carbon emission coefficients. Then, the super-efficiency EBM model with radial mixing parameters was combined with the GML index to calculate TFCE at the city level. In terms of the research method, the Dagum Gini coefficient, Moran's I index, a LISA aggregation map and the standard deviational ellipse method were used to comprehensively analyze the regional differences in the distribution and spatiotemporal differentiation characteristics of TFCE. Then, kernel density estimation and the spatial Markov chain method were used to analyze the development status and spatiotemporal evolution characteristics of TFCE at the city level. The obtained conclusions are not only conducive to a comprehensive and systematic understanding of the development of TFCE at the city level in China but could also provide useful references for local governments to help them to formulate relevant strategies for the coordinated development of regional carbon efficiency.

3. Measurement of Urban TFCE

3.1. Method of Measurement

3.1.1. Super-Efficiency EBM Model

In order to effectively make up for the shortcomings in efficiency measurement using the CER and SBM models, Tone et al. constructed a model that includes both radial and SBM distance functions, namely the epsilon-based measure (EBM) model [24]. The super-efficiency EBM model relaxes the assumption regarding the proportion between the target value and the actual value of the input and can also calculate the difference between the target value and the actual value by solving the non-radial values of the input and output variables. Therefore, the efficiency value of the obtained decision-making unit (DMU) is more accurate [25]. In this study, we selected the super-efficiency EBM model to define the directional distance function (DDF), which was set as the global super-efficiency, non-orientation and variable returns to scale to measure TFCE at the city level in China [26]. The global super-efficiency EBM model was constructed as follows:

$$\gamma^* = \min \frac{\theta - \varepsilon_x \sum\limits_{i=1}^{m} \frac{\omega_i^- s_i^-}{x_{i0}}}{\eta + \varepsilon_y \sum\limits_{r=1}^{k} \frac{\omega_r^+ s_r^+}{y_{r0}} + \varepsilon_u \sum\limits_{p=1}^{h} \frac{\omega_p^{u-} s_p^{u-}}{u_{p0}}} \quad (1)$$

$$\begin{aligned} s.t.\quad & \sum_{j=1}^{n} x_{ij}\lambda_j + s_i^- = \theta X_{i0} \quad (i = 1,2,\ldots,m) \\ & \sum_{j=1}^{n} y_{rj}\lambda_j - s_r^+ = \eta y_{r0} \quad (r = 1,2,\ldots,k) \\ & \sum_{p=1}^{h} u_{pj}\lambda_j + s_p^{u-} = \eta u_{p0} \quad (p = 1,2,\ldots,h) \\ & \lambda_j \geq 0,\ S_i^-,\ S_r^+,\ S_p^{u-} \geq 0 \end{aligned} \quad (2)$$

In the above formula, γ^* is the TFCE value; n is the number of DMUs j; x is the input factor; y is the desirable output; u is the undesirable output; m, k and h represent the quantities of input factors, desirable outputs and undesired outputs, respectively; θ is the planning parameter of the radial part; η is the output expansion ratio; ε is the parameter referring to the importance of the non-radial part, where $0 \leq \varepsilon_x \leq 1$; ω_i^-, ω_r^+ and ω_p^{u-} are the weights of the i-th input, r-th expected output and p-th unexpected output, respectively and S_i^-, S_r^+ and S_p^{u-} are the relaxation variables of the i-th input, r-th desirable output and p-th undesirable output, respectively.

3.1.2. Global Malmquist Index

In 1953, the Swedish economist Malmquist proposed the Malmquist (ML) index [27] which can measure the dynamic rate of change of DMUs using the distance function. However, this index does not have circularity, nor can it analyze the rate of change of the efficiency of technical progress over set periods [28]. Additionally, the ML index may have the problem of no linear solutions in the calculation process. The Global Malmquist (GML) index calculates global production possibility sets based on data from all sample periods which can effectively solve these problems of non-transmission and no linear solutions. The GML index can be decomposed into the rate of change of technical progress (GETC) and the rate of change of technical efficiency (GEFC) [29]. Since the super-efficiency EBM model adopted in this study could only provide static efficiency values, it was necessary to measure and decompose China's city-level TFCE dynamically using the GML index. The expression of the GML index is as follows:

$$GML_t^{t+1}(x^t, y^t, u^t, x^{t+1}, y^{t+1}, u^{t+1}) = \frac{1 + D^G(x^t, y^t, u^t)}{1 + D^G(x^{t+1}, y^{t+1}, u^{t+1})} \quad (3)$$

In Formula (3), x^t, y^t and u^t and x^{t+1}, y^{t+1} and u^{t+1} represent the input, desirable output and undesirable output of DMUs in the t and t+1 periods, respectively, and $D^G(x^t, y^t, u^t)$ is the mixed distance function of the EBM model. The calculation method of this function is shown in Formula (1).

3.2. Index Selection and Data Processing

The input indicators of TFCE at the city level were defined as follows: human capital, measured by the number of employees in the selected three industries in each city; capital stock, for which the fixed capital investment flow data were deflated by the fixed asset price index after eliminating the price factor and subtracting the accumulated depreciation and the actual fixed capital stock was calculated using the perpetual inventory method; energy consumption, measured by the total energy consumption of each city. The output indicators of TFCE at the city level were defined as follows: the desirable output was measured by the actual GDP of each city; the undesirable output was the CO_2 emissions of each city. The measurement of carbon emissions at the city level in existing studies has mainly involved carbon emissions from electricity production, natural gas, liquefied petroleum gas (LPG), urban heating and transportation [30], among which the carbon emissions generated by power production account for nearly half of China's total carbon emissions [31] while urban heating, natural gas and LPG generate the next highest amounts of carbon emissions. Furthermore, the carbon emissions generated by transportation account for a small proportion of the total carbon emissions of each city [32] and the transportation carbon emission data at the city level are incomplete, so this indicator was removed from our measurement. Therefore, we referred to the method of Wu and Guo [33] for the measurement of carbon emissions in prefecture-level cities and selected four types of energy consumption for calculation: electricity consumption, liquefied gas consumption, LPG consumption and thermal energy consumption. The formula for the measurement of carbon emissions at the city level is as follows:

$$CO_2 = C_m + C_n + C_p + C_q = \gamma E_m + \lambda E_n + +\tau E_p + \varphi(\eta \times E_q) \quad (4)$$

In Formula (4), C_m, C_n and C_p represent the carbon emissions generated by liquefied gas, LPG and urban heating, respectively; E_m, E_n and E_p represent natural gas consumption, LPG consumption and urban heating consumption, respectively; γ, λ and τ are the carbon emission coefficients of the three energy types[1]; C_q represents the carbon emissions generated by urban electricity use, where $C_q = \varphi(\eta \times E_q)$, according to the research of Glaeser and Kahn [34]. The power grid in China is divided into six regions: South China, East China, Central China, North China, Northeast China and Northwest China. Each region has an emission factor and φ is the emission factor generated by electric energy consumption in each region, η represents the ratio of coal power generation to total power generation and E_q represents the electricity consumption of the whole society.

In this study, we selected panel data for 283 cities in China from 2004 to 2019 and used MAX-DEA software to calculate the TFCE. The data used for the variables came from the *China Statistical Yearbook*, *China Urban Statistical Yearbook*, *China Power Yearbook*, *China Labor Statistical Yearbook*, WIND database, CEIC database, etc.

4. The Spatiotemporal Differentiation Characteristics of Urban TFCE

4.1. Research Methods

4.1.1. Dagum Gini Coefficient and Decomposition Method

We used the Dagum Gini coefficient (G) and its decomposition method to explore the relative spatial differences in TFCE at the city level. Dagum decomposed the Gini coefficient into three components: within-group differential contribution (G_w), between-group differential contribution (G_{nb}) and hypervariable density contribution (G_t) [35]. The relationship between the Gini coefficient and its decomposition term is $G = G_w + G_{nb} + G_t$. In this paper, G_w represents the difference in TFCE value added between cities in different regions of China, G_{nb} represents the difference in TFCE value added between cities in the eastern, central and western regions of China and G_t is the remainder of the cross-influence of the Gini coefficient between the three regions. A smaller within-group Gini coefficient (between-group) indicates a smaller TFCE difference between within-group cities (between-group) and larger coefficients indicate larger differences. The Dagum Gini coefficient is calculated as follows:

$$G = 1/2\mu n^2\left(\sum_{j=1}^{k}\sum_{h=1}^{k}\sum_{i=1}^{n_j}\sum_{r=1}^{n_h}\left|y_{ji} - y_{hr}\right|\right) \tag{5}$$

In Formula (5), k represents the number of regions; j and h represent different regions; n represents the number of cities; i and r represent different cities; $n_j(n_h)$ represents the number of cities in region $j(h)$; y_{ji} (y_{hr}) represents the TFCE of city $i(r)$ in region $j(h)$; μ is the mean growth of TFCE in all cities in the four regions; G_{jj} is the Gini coefficient of region j; G_{jh} is the Gini coefficient between region j and h; G_w is the within-group difference contribution; G_{nb} is the between-group difference contribution; and G_t is the hypervariable density contribution after Dagum Gini coefficient decomposition. These terms are calculated as follows:

$$G_{jj} = 1/2\overline{y_j}n_j^2\sum_{i=1}^{n_j}\sum_{r=1}^{n_h}\left|y_{ji} - y_{hr}\right| \tag{6}$$

$$G_{jh} = \sum_{i=1}^{n_j}\sum_{r=1}^{n_h}\left|y_{ji} - y_{hr}\right|/n_j n_h(\overline{y_j} + \overline{y_h}) \tag{7}$$

$$G_w = \sum_{j=1}^{k} G_{jj}p_j q_j \tag{8}$$

$$G_{nb} = \sum_{j=2}^{k}\sum_{h=1}^{j-1} G_{jh}(p_j q_h + p_h q_j)D_{jh} \tag{9}$$

$$G_t = \sum_{j=2}^{k}\sum_{h=1}^{j-1} G_{jh}(p_j q_h + p_h q_j)\left(1 - D_{jh}\right) \quad (10)$$

$$d_{jh} = \int_0^{\infty} dF_j(y)\int_0^{y} (y-x)dF_h(x) \quad (11)$$

$$p_{jh} = \int_0^{\infty} dF_h(y)\int_0^{y} (y-x)dF_j(x) \quad (12)$$

$$D_{jh} = \left(d_{jh} - p_{jh}\right)/\left(d_{jh} + p_{jh}\right) \quad (13)$$

In the above formulae, $p_j = n_j/n$, $q_j = n_j\overline{y_j}/n\overline{y}$ and D_{jh} represents the interaction effect of TFCE between regions j and h. The calculation method for the difference is shown in Equation (13). Additionally, d_{jh} is the difference in TFCE between regions, representing the mathematical expectation of all $y_{ji} - y_{hr} > 0$ in regions j and h; p_{jh} is the hypervariable first-order matrix, representing the mathematical expectation of all $y_{hr} - y_{ji} > 0$ in regions and h; and $F_j(F_h)$ is the cumulative density distribution function of region j (h).

4.1.2. Global Spatial Correlation Test

In order to investigate whether there was spatial dependence in the data, it was necessary to conduct a spatial correlation test on the urban TFCE. The Moran's I index was used to test the global spatial correlation of TFCE at the city level and the selected spatial weight matrix was the geographical distance weight matrix. The formula for the global spatial autocorrelation is as follows:

$$Moran's\ I = (1/\sum_{i=1}^{n}\sum_{j=1}^{n} w_{ij}) \times \{\sum_{i=1}^{n}\sum_{j=1}^{n} w_i(Y_i - \overline{Y})(Y_j - \overline{Y})/\left[\sum_{i=1}^{n}\sum_{j=1}^{n}(Y_i - \overline{Y})^2/n\right]\} \quad (14)$$

In the above formula, i and j represent different cities, n represents the number of cities, Y represents the TFCE of the city, $\overline{Y} = \frac{1}{n}\sum_{i=1}^{n} Y_i$ represents the mean TFCE value of the city and W_{ij} represents the spatial weight matrix. A *Moran's* $I\in[-1, 1]$ value greater than 0 represents positive autocorrelation, a *Moran's* $I\in[-1, 1]$ value less than 0 represents negative autocorrelation and a *Moran's* $I\in[-1, 1]$ value equal to 0 represents no correlation.

4.1.3. Local Spatial Autocorrelation

The local spatial autocorrelation mainly uses the local Moran index to observe the type of spatial agglomeration to which the spatial unit belongs [36]. In this study, the local spatial correlation index (LISA aggregation) was used to measure the local spatial correlation of urban TFCE. LISA aggregation maps can visually show the significance of the local spatial aggregation of urban TFCE in the form of graphs, as well as reflecting the specific spatial correlation forms of research elements in the aggregation maps at the same time. Our LISA aggregate graphs were generated based on the local *Moran's I* index and the specific calculation formula was as follows:

$$Moran's\ I_{it} = [(Y_i - \overline{Y})\sum_{j=1}^{n} w_{ij}(Y_j - \overline{Y})]/S^2 \quad (15)$$

In Formula (15), *Moran's* I_{it} is the local Moran index and S^2 is the variance. The other parameters were set to be the same as those in Formula (14). Similar to the global Moran index, the local *Moran's* $I_{it} \in [-1, 1]$. The LISA aggregation maps were generated by combining the scatter maps obtained using the Moran index with the significance of the LISA aggregation.

4.1.4. Standard Deviational Ellipse

The standard deviational ellipse method is a classical method that is used to analyze spatial distribution characteristics [37]. A standard deviational ellipse is composed of the center of a circle, the azimuth angle, the major axis, the minor axis and the ellipse area [38]. The center of a standard deviational ellipse represents the center of gravity for the spatial distribution of the elements and the azimuth angle represents the main spatial distribution trends of the research elements. The major and minor axes reflect the dispersion degree of the spatial layout of the research elements in the primary and secondary directions, respectively, while the area of the standard deviational ellipse reflects the aggregation and dispersion degree of the spatial distribution forms of the research elements. The calculation formulae for the standard deviational ellipse and its components are as follows:

$$\overline{X}_w = \sqrt{\frac{\sum_{i=1}^{n} w_i x_i}{w_i^2}} \tag{16}$$

$$\overline{Y}_w = \sqrt{\frac{\sum_{i=1}^{n} w_i y_i}{w_i^2}} \tag{17}$$

$$\tan\theta = \frac{\left(\sum_{i=1}^{n} w_i^2 \widetilde{x}_i^2 - \sum_{i=1}^{n} w_i^2 \widetilde{y}_i^2\right) + \sqrt{\left(\sum_{i=1}^{n} w_i^2 \widetilde{x}_i^2 - \sum_{i=1}^{n} w_i^2 \widetilde{y}_i^2\right)^2 + 4\sum_{i=1}^{n} w_i^2 \widetilde{x}_i^2 \widetilde{y}_i^2}}{2\sum_{i=1}^{n} w_i^2 \widetilde{x}_i \widetilde{y}_i} \tag{18}$$

$$\sigma_x = \sqrt{\sum_{i=1}^{n} \left[(w_i \widetilde{x}_i \cos\theta - w_i \widetilde{y}_i \sin\theta)^2\right] / \sum_{i=1}^{n} w_i^2} \tag{19}$$

$$\sigma_x = \sqrt{\sum_{i=1}^{n} \left[(w_i \widetilde{x}_i \sin\theta + w_i \widetilde{y}_i \cos\theta)^2\right] / \sum_{i=1}^{n} w_i^2} \tag{20}$$

In Formulae (16)–(20), $\overline{X}$ and $\overline{Y}$ represent the weighted average center of the ellipse, x_i and y_i are the spatial coordinates of each feature point, $\widetilde{x}_i$ and $\widetilde{y}_i$ are the deviations of different feature points from the average center coordinates, w_i is the weight, θ is the azimuth angle, σ_x and σ_y represent the standard deviations along the x and y axes of the ellipse, respectively.

4.2. Empirical Analysis

4.2.1. Analysis of the Dagum Gini Coefficient Results

Table 1 presents the Dagum Gini coefficients of TFCE at the city level and the decomposition results for the three regions. The results showed that the mean of the overall Gini coefficient in China was 0.0537 and that the change in the overall difference had an inverted "N" shape, i.e., During the sample period, the overall difference in TFCE between the cities tended to decline in the fluctuation. Specifically, the overall relative difference in urban TFCE during 2007–2015 was lower than the average level and decreased continuously, while the difference in urban TFCE during 2016–2019 first decreased and then increased. From the perspective of differences in change trends, the Gini coefficient decreased from 0.1025 in 2005 to 0.0538 in 2019, with a decrease rate of 47.51%, which meant that the difference in TFCE between cities in China as a whole decreased.

Additionally, from the perspective of intra-regional differences, among the three selected regions, the mean urban TFCE Gini coefficients in the eastern and western regions were lower than the national average, while the Gini coefficient in the central region was higher than the national average, which indicated that the intra-regional differences in urban TFCE were very significant. Specifically, the Gini coefficient in the eastern region showed an inverted "N"-shaped feature and the TFCE variation trends in the central and western cities were basically the same. During 2005–2015, the TFCE variation trend had a "W"-shaped pattern, with alternating decreases and increases. The variation trend during 2016–2019 had an "N"-shaped pattern, but the Gini coefficient in the central region was higher than those in the eastern and western regions as a whole. This indicated that the TFCE changes in the central and western cities were unstable, while the Gini coefficients in the eastern region showed stable changes in the early stages and significant fluctuations in the later stages.

Table 1. The Dagum Gini coefficients of total factor carbon efficiency in China and the three selected regions.

Year	Gini Coefficient	Inner-Regional Difference			Inter-Regional Difference			Contribution Rate (%)			Inner-Regional Difference	Inter-Regional Difference	Hypervariable Density
		East	Central	West	East-Central	EastWest	Central-West	G_w	G_{nb}	G_t			
2005	0.1025	0.0886	0.1232	0.0888	0.1078	0.0908	0.1106	32.98	6.41	60.61	0.0338	0.0066	0.0621
2006	0.0624	0.0575	0.0896	0.0299	0.075	0.048	0.064	32.95	21.28	45.77	0.0206	0.0133	0.0286
2007	0.0489	0.0409	0.0563	0.0487	0.0492	0.0452	0.0529	33.29	8.57	58.14	0.0163	0.0042	0.0284
2008	0.0404	0.0308	0.0493	0.0405	0.0405	0.0362	0.0451	33.29	7.27	59.44	0.0135	0.0029	0.024
2009	0.047	0.0356	0.0556	0.0477	0.0468	0.0433	0.0525	32.91	12.5	54.6	0.0155	0.0059	0.0256
2010	0.0363	0.0338	0.0364	0.0377	0.0356	0.0368	0.0376	33.09	2.6	64.31	0.012	0.0009	0.0233
2011	0.0419	0.0362	0.0527	0.029	0.0452	0.0355	0.0462	32.61	20.27	47.12	0.0137	0.0085	0.0197
2012	0.0449	0.0343	0.0613	0.0351	0.0494	0.035	0.0499	33.39	20.17	46.44	0.015	0.0091	0.0209
2013	0.0444	0.0431	0.0354	0.0519	0.043	0.0487	0.045	32.17	3.27	64.56	0.0143	0.0015	0.0287
2014	0.0424	0.0387	0.0418	0.0453	0.0414	0.0424	0.0452	32.84	8.63	58.53	0.0139	0.0037	0.0248
2015	0.0367	0.0368	0.0284	0.0346	0.0343	0.0401	0.0414	30.38	14.56	55.06	0.0111	0.0053	0.0202
2016	0.0731	1003	0.0635	0.0413	0.0835	0.0784	0.0555	33.37	11.38	55.25	0.0244	0.0083	0.0404
2017	0.0799	0.0675	0.0856	0.0865	0.0774	0.0783	0.0863	33.17	3.39	63.44	0.0265	0.0027	0.0507
2018	0.0502	0.551	0.0507	0.0411	0.0538	0.0496	0.0471	33.3	2.3	64.4	0.0167	0.0012	0.0323
2019	0.0538	0.045	0.0621	0.0525	0.0547	0.0795	0.058	32.86	6.33	60.81	0.0177	0.0034	0.0327
Mean	0.0537	0.0496	0.0594	0.0474	0.0559	0.0505	0.0558	32.84	9.93	57.23	0.0177	0.0052	0.0308

Moreover, from the perspective of inter-regional differences, the variation trends in the relative differences between the three regions first decreased and then increased. The mean of the relative difference coefficient between the central, eastern and western regions was lower than the national average level, while the Gini coefficients between the eastern and central regions and the central and western regions were higher than the national average. Specifically, the relative difference between the eastern and central regions was the largest, followed by the difference between the central and western regions. The relative difference between the eastern and western regions was the smallest. In terms of stability trends, the relative difference coefficients between the three regions changed stably from 2007 to 2015 but fluctuated significantly after 2016.

From the perspective of contribution rates and the sources of differences, the contribution rate of regional differences in TFCE between 2005 and 2019 ranged from 30.38% to 33.39% and the contribution rate only changed slightly over time. The contribution rate of inter-regional differences had a large variation range, but its value was significantly smaller than the contribution rate within the regions, indicating that there were huge differences in total factor carbon efficiency between Chinese cities within the three regions and that the differences between regions were small. These differences led to significant differences in TFCE within the overall region of China, which means that in order to improve the overall TFCE level of Chinese cities, we should focus on the development gaps in each region. The contribution rate of hypervariable density showed a fluctuating trend from 2005 to 2019, during which time its value ranged between 45.77% and 64.56%. Under the influence of contribution rate, the intra-regional differences to the overall national difference ranged between 0.0111 and 0.0338 and decreased slightly over time. The inter-regional differences ranged between 0.0009 and 0.0133 and the change trend had an inverted "N" shape, with alternating decreases and increases. The interaction effect of intra-regional and inter-regional differences decreased slightly from 0.0621 at the beginning of the period to 0.0308 at the end of the period, with a decrease rate of 50.40%. In general, the intra-regional differences were higher than inter-regional differences and the hypervariable density was significantly higher than the intra-regional and inter-regional difference values, which meant that the interaction levels between intra-regional and inter-regional relative differences contributed the most to differences in urban TFCE, followed by intra-regional differences.

4.2.2. Global Spatial Correlation Test

The results of the urban TFCE global spatial correlation test are shown in Table 2. The Moran 'I index values for urban TFCE were all greater than 0 and were all significant at the level of at least 1%, which confirmed that there was a significant positive spatial correlation between the TFCE of Chinese cities. This meant that the TFCE of each city was not completely independent or randomly distributed in space. This is because there is a cluster effect in economic production activities, through which the flow of production, technology and other factors make economic activities in neighboring cities converge, resulting in the spatial spillover effect of TFCE between cities. In addition, carbon emissions are inherently spatially dispersed, especially in geographically concentrated areas, which leads to the existence of spatial aggregation in urban TFCE.

Table 2. The global Moran's I index of spatial correlation.

Year	I	E(I)	sd(I)	Z	P	Year	I	E(I)	sd(I)	Z	P
2005	0.1973	−0.0035	0.0005	9.3941	0.000	2013	0.0834	−0.0035	0.0005	4.0757	0.001
2006	0.2094	−0.0035	0.0005	9.9597	0.000	2014	0.1236	−0.0035	0.0005	5.9676	0.000
2007	0.1094	−0.0035	0.0005	5.2953	0.000	2015	0.0627	−0.0035	0.0004	3.1753	0.000
2008	0.0506	−0.0035	0.0005	2.5651	0.008	2016	0.1685	−0.0035	0.0005	8.0759	0.000
2009	0.0273	−0.0035	0.0002	2.0413	0.041	2017	0.0416	−0.0035	0.0002	3.0631	0.002
2010	0.1068	−0.0035	0.0005	5.2334	0.000	2018	0.1660	−0.0035	0.0005	7.9190	0.000
2011	0.1055	−0.0035	0.0004	5.2119	0.000	2019	0.2224	−0.0035	0.0005	10.6588	0.000
2012	0.0925	−0.0035	0.0004	4.6347	0.000	-	-	-	-	-	-

4.2.3. LISA Aggregation Graphs

In order to show the spatial correlation effect of urban TFCE, ArcGIS software was used in this study to combine the local Moran index and the LISA significance level. Then, LISA aggregation maps were drawn for 2005, 2013 and 2019 as the representative years, as shown in Figure 1[2]. LISA aggregation maps divide local spatial correlation into four types, namely, high–high (H-H) aggregation, high–low (H-L) aggregation, low–high (L-H) aggregation and low–low (L-L) aggregation. Among them, H-H aggregation and L-L aggregation indicate that there is a positive spatial correlation between the TFCE of a given city and the TFCE of other cities in the same agglomeration area. H-L aggregation and L-H aggregation indicate that there is a negative spatial correlation between the TFCE of a given city and the TFCE of other cities in the same agglomeration area.

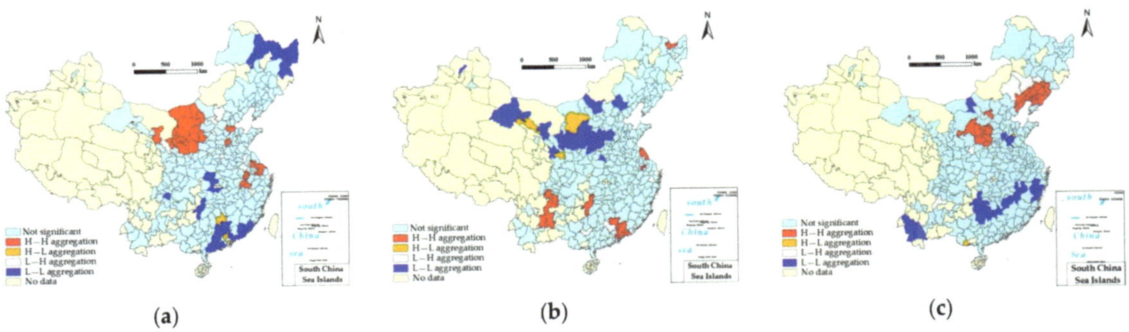

Figure 1. The LISA aggregation maps of urban TFCE in (**a**) 2005, (**b**) 2013 and (**c**) 2019.

As shown by the transfer trends of TFCE local spatial aggregation in Figure 1a–c, the H-H and L-L aggregation areas formed in the spatial correlation of urban TFCE changed greatly over time. Some cities located in H-H aggregation areas in 2005 were transformed into L-L aggregation areas in 2013 and then transformed back into H-H aggregation areas again in 2019, such as Yulin City and Luliang City. Some cities located in L-L aggregation areas in 2005 were transformed into spatially insignificant areas over time, such as Hegang, Jiamusi and Shuangyashan, while other cities located in L-L aggregation areas in 2013 were transformed into H-H aggregation areas in 2019, such as Jiuquan and Jiayuguan. In 2005 and 2013, fewer cities were located in H-L and L-H aggregation areas and the TFCE of these cities was negatively correlated with neighboring areas. For example, Zhangye and Tianshui, which were located in a H-L aggregation area, had high efficiency levels but inhibited the efficiency levels of their neighboring cities. However, other cities, such as

Datong and Zhangjiakou, were located in L-H agglomeration areas, which indicated that the low efficiency levels of these cities did not inhibit TFCE in the surrounding areas but instead promoted it.

In general, compared to 2005 and 2013, the number of cities located in H-H and L-L aggregation areas increased in 2019 and these two areas formed a significant north–south differentiation feature in space. The high aggregation areas were mainly distributed in the northeast, northwest and north of China, such as Shenyang, Beijing, Tangshan and Jiuquan. These cities not only had higher TFCE levels but also had positive spillover effects on the carbon efficiency levels of surrounding cities through the positive diffusion effect. The L-L aggregation areas were mainly located to the south of the Qinling Mountains and the Huaihe River, such as Lishui, Nanchang and Lincang. The TFCE levels of these cities were low, which further inhibited TFCE improvements in surrounding cities and showed the agglomeration effect of inefficient cities in space.

The L-H agglomeration areas were mainly concentrated in the northeastern and northwestern regions, while the number of cities in H-L agglomeration areas was the lowest, indicating that there was a negative correlation between the TFCE of these cities and their neighboring regions and that the low (high) TFCE of one city could promote (inhibit) the TFCE of neighboring cities.

4.2.4. Standard Deviational Ellipse

Figure 2 shows a directional distribution map of urban TFCE based on the standard deviational ellipse, which was drawn using ArcGIS software. The obtained parameters are shown in Table 3. It can be seen from Figure 2 that the TFCE of Chinese cities was centered around Zhumadian City and that the dominant distribution direction extended from northeast to southwest. The barycentric coordinates of the ellipse presented in Table 3 were located between longitude 113°54′–113°93′ E and latitude 32°85′–33°14′ N and were slightly to the southeast of Lanzhou, which is the geographic and geometric center of China (longitude 103°40′ E and latitude 36°03′ N). From the perspective of the shifting trajectory of the center of gravity, the center of gravity of urban TFCE did not change significantly from 2005 to 2019, which meant that the TFCE in eastern and southern China was higher on average than that in other regions. This was consistent with the reality that China's economic development is higher in the east than in the west. During the sample period the azimuth angle of the ellipse ranged between 22° and 24° and increased slightly in 2019 compared to 2005, indicating that the divergence direction of the TFCE of Chinese cities was also relatively stable. The ratio of the minor axis to the major axis gradually decreased over time. The oblateness of the ellipse was the smallest in 2005 and the largest in 2019 i.e., the centripetal force of urban TFCE had an increasing trend from 2005 to 2019. In 2019 the directionality of the urban TFCE distribution was the most significant. As shown in Table 3, the ellipse area showed a fluctuating upward trend and the ellipse area in 2019 decreased by 4.61% compared to the area in 2005, indicating that the spatial aggregation degree of the TFCE of Chinese cities improved slightly. However, in general, this spatial aggregation characteristic was relatively stable during the sample period.

Table 3. Standard deviational ellipse parameters.

Year	Center (Longitude)	Center (Latitude)	Azimuth Angle	Minor Axis/Major Axis	Area (km^2)
2005	113.5349	33.1368	22.7495	0.7239	2,644,508.1170
2010	113.9322	33.0856	22.5042	0.6713	2,747,709.3789
2015	113.7475	32.8490	23.2948	0.6766	2,646,924.5017
2019	113.8674	33.0367	24.0191	0.6579	2,766,434.0552

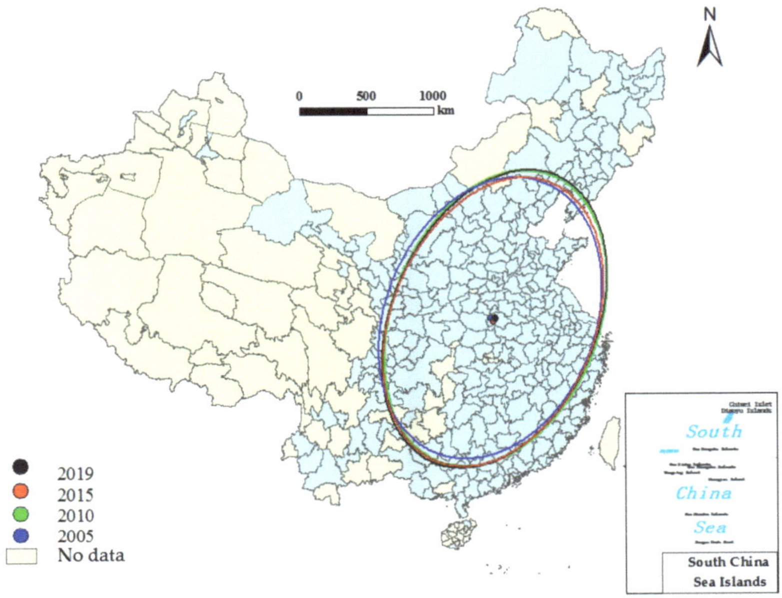

Figure 2. The direction distribution of urban total factor carbon efficiency.

5. Dynamic Evolution Trend Analysis of Urban Total Factor Carbon Efficiency

5.1. Research Method

5.1.1. Kernel Density Estimation

Kernel density estimation is a non-parametric estimation method, which can describe the distribution characteristics of estimation variables well. This method has a low dependence on the model and has a strong robustness, so it is widely used in spatial disequilibrium analysis [39]. Therefore, in order to further explore the distribution differences and evolution rules of China's urban TFCE at the national and regional levels, the kernel density estimation method was adopted to describe the distribution forms of TFCE at the city level using continuous changing curves so as to show the distribution location, form and characteristics of urban TFCE more comprehensively. Let $f(z)$ be the density function of urban TFCE, then

$$f(z) = 1/mv - \sum_{i=1}^{m} K[(\delta_i - \overline{\delta})/v] \quad (21)$$

In Formula (21), m is the number of observations, v is the bandwidth, $K(\cdot)$ is the kernel density function, δ_i is the independently distributed observation value and $\overline{\delta}$ is the mean of the observation values. In this study, the Gaussian kernel density function was selected to estimate the dynamic distribution of TFCE at the national and regional levels. Its formula is as follows:

$$K(z) = 1/\sqrt{2\pi}\exp(-z^2/2) \quad (22)$$

Since there is no exact function expression for non-parametric estimation, it is necessary to observe distribution characteristics by means of graphical comparisons as graphs obtained by kernel density estimation can provide information about different dimensions, such as distribution position and shape and the ductility of the variables.

5.1.2. Spatial Markov Chain Method

Although kernel density curves can better describe the distribution characteristics and convergence degrees of variables, they offer limited internal dynamic information regarding variable distributions and fail to show the relative location change characteristics and future occurrence possibility of TFCE at the city level. The Markov chain method considers random variables as discrete variables, which can make up for these shortcomings. Therefore, we used the Markov chain method to describe the state transition of TFCE in each city, as well as the possibility of transition. A Markov chain is a stochastic process, namely $\{X_t, t \in T\}$, which describes the internal change probability of variables by constructing a Markov transition probability matrix [40]. Markov chains assume that the probability of state j, in which the variable X is in period t, is determined by the state in period t-1 and is independent of the other periods, i.e.:

$$P\{X(t) = j|X_{t-1} = i_{t-1}, X_{t-2} = i_{t-2}, \ldots, X_0 = i_0\} = P\{X(t) = j|X_{t-1} = i_{t-1}\} = p_{ij} \quad (23)$$

In Equation (23), P_{ij} is the transition probability matrix of TFCE in a given city from year t to year t+1. Based on a traditional Markov chain, the concept of "spatial lag" is introduced to obtain a spatial Markov chain. By introducing a spatial weight matrix (W_{ij}), the transition probability matrix of $N \times N$ is decomposed into $N \times N \times N$ so that P_{ij} can reveal the evolution process of the spatial lag effect on urban TFCE. The spatial weight matrix W_{ij} in this study was represented by a spatial adjacency matrix. The formula was as follows:

$$w_{ij} = \begin{cases} 1 & \text{city } i \text{ \& } j \text{ areadjacent} \\ 0 & \text{else} \end{cases} \quad (24)$$

5.2. *Analysis of the Urban TFCE Spatiotemporal Evolution Results*

5.2.1. Analysis of Kernel Density Results

In order to systematically analyze the absolute distribution differences and dynamic evolution laws of urban TFCE, the Gaussian kernel function was selected to depict the distribution laws of TFCE and kernel density curves were drawn using Python 3.8 software. Figure 3 and Table 4 show the distribution dynamics and evolution trends of TFCE in the whole country and the three selected regions.

In terms of the whole country, the main peak position of urban TFCE showed a trend of right–left alternating movement during the investigation period. From 2018, the main peak gradually moved to the right, indicating that the overall TFCE of the whole country effectively improved during this period. This was mainly due to the influence of multiple factors, such as China's green energy transformation and upgrade policies, green scientific and technological innovations and the continuous enhancement of the carbon absorption capacity of terrestrial ecosystems over recent years. In terms of the distribution pattern of the main peak, the height of the main peak first increased and then decreased, while the width of the main peak gradually decreased. As shown in Figure 3a, the height of the main peak in 2019 was higher than that in 2005 and the width narrowed, indicating that the absolute difference in domestic TFCE and the dispersion degree of the distribution both decreased. From the perspective of distribution ductility, the national TFCE distribution curve dragged to the right and showed the characteristics of extended convergence. The higher carbon efficiency levels of some cities led to a right-trailing distribution curve, while the convergence trend indicated that the difference between cities with high carbon efficiency levels and the average level across the country gradually decreased. From the perspective of the polarization trend, the overall TFCE of the country showed a trend of multipolar differentiation, i.e., there as a multipeak phenomenon in addition to the main peak. This was because although the TFCE of some cities showed overall growing trends, they were restricted by various factors, such as differences in energy and resource endowment, economic development and technological level. Therefore, the absolute differences between cities with higher TFCE levels and those with lower TFCE levels were characterized by fluctuating short-term changes.

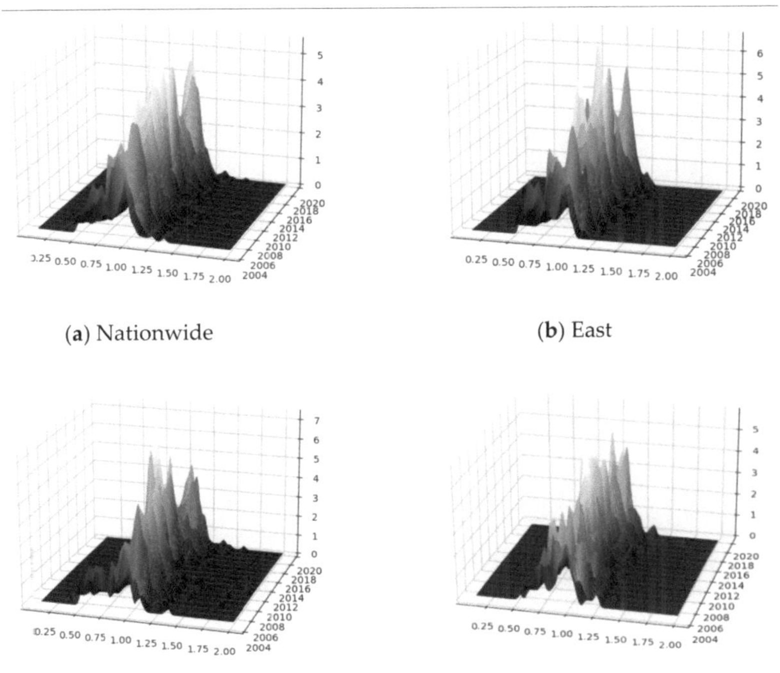

(a) Nationwide (b) East

(c) Central (d) West

Figure 3. Urban TFCE kernel density in the whole country (Nationwide) and the three regions (East; Central; West).

Table 4. The dynamic evolution characteristics of urban total factor carbon efficiency distributions.

Region	Distribution Position	Distribution Main Peak Form	Distribution Ductility	Differentiation Trend
Overall	Moved left and right alternatively	The peak value first increased and then decreased and the width decreased	Right-trailing, convergent extension	Bipolar differentiation
AEast	Moved left and right alternatively	The peak value first increased and then decreased and the width decreased	Right-trailing, convergent extension	Multipolar differentiation
Central	Moved left and right alternatively	The peak value first increased and then decreased and the width first decreased and then increased	Right-trailing, convergent extension	Multipolar differentiation
West	Moved right	The peak increased and the width decreased	Right-trailing, convergent extension	Bipolar differentiation

In terms of the distribution locations and forms of the three selected regions, the positions of the main peaks of TFCE in the eastern and central regions alternated between shifting to the left and right while the position of the main peak in the western region gradually moved to the right, indicating that the TFCE levels in the eastern and central regions fluctuated and changed over time (and even decreased in some years) while the TFCE level in the western region gradually improved. The height of the main peak in the eastern region first increased and then decreased and the width gradually decreased, indicating that the dispersion of its distribution decreased and the absolute difference decreased further. The height of the main peak in the central region also first increased and then decreased and the width increased at the end of the period, which indicated that the dispersion degree and the absolute difference in the TFCE of the central region showed "V" shape patterns. The height of the main peak in the western region increased over time, indicating that the dispersion degree of the TFCE distribution decreased. The peak width also gradually decreased during the sample period, indicating that the absolute TFCE differences within the western region decreased.

From the perspective of distribution extensibility and polarization trends, the TFCE distribution curves for the three regions all trailed to the right and showed extended convergence, indicating that the difference between the cities with high TFCE levels and the average levels in the eastern, central and western regions gradually decreased. There were multiple main peaks in the distribution curves for the eastern and central regions, which showed that the multipolar differentiation in the TFCE levels of cities in the two regions was very significant, thereby confirming that there was also a significant differentiation between cities in the eastern and central regions. The western region showed a polarization trend, with a main peak and a left-hand peak in the early stages and the side peak gradually merging into the main peak in the middle stages, which evolved into a main peak and the polarization phenomenon gradually disappeared. A right-hand peak formed in the later stages.

5.2.2. Analysis of Spatial Markov Chain Results

In order to explore the dynamic evolution trends of urban TFCE levels in China, we used a traditional Markov chain (Table 5) and spatial Markov chain (Table 6) to analyze the transition probability distribution of urban TFCE. In this study, the quartiles were selected to divide TFCE into four types: low level, medium–low level, medium–high level and high level (which are labeled as type I, II, III and IV in the table, respectively). The diagonal elements of the transition probability matrices in the tables represent the probability that TFCE type did not change in the $t + 1$ year, which is also called smooth transition probability. The non-diagonal elements represent the probability of TFCE type changing in the $t + 1$ year, i.e., the probability of a transition up or to the next type. From the traditional Markov chain transfer results presented in Table 3, it can be seen that the probability values of the diagonal elements were significantly greater than those of the non-diagonal elements, which meant that the urban TFCE had strong stationarity. Among these values, the probability of types I, II, III and IV maintaining their original states was 48.90%, 38.10%, 35.60% and 56.30%, respectively, indicating that urban TFCE had a strong path-dependent effect. The probability of the low, medium–low and medium–high levels transitioning to the next level was 21.90%, 27.90% and 25.90%, respectively, indicating that the probability of TFCE level transitioning from the medium–low level to the medium–high level was the highest and the probability of transitioning from the low level to the medium–low level was the lowest. The probability of type I and type II demonstrating a cross-level upward transition was 15.6% and 10.0%, respectively, which was significantly lower than the probability of level-wise upward transition, indicating that the evolution of urban TFCE changed step by step from low to high.

Table 5. The traditional Markov transition probability matrix.

t/t + 1	I	II	III	IV
I	0.4890	0.2190	0.1560	0.1360
II	0.2400	0.3810	0.2790	0.1000
III	0.0970	0.2880	0.3560	0.2590
IV	0.0450	0.1270	0.2650	0.5630

Considering the spatial correlation of TFCE between cities, a spatial lag term was introduced into the traditional Markov chain matrix to establish a spatial Markov transition probability matrix. The results are shown in Table 6. It can be seen that after the spatial lag term was added, the results of the four transition probability matrices presented in Table 6 changed, which meant that there was a certain correlation between the probability of TFCE transitions between neighboring cities within a given geographical space. First of all, when a city had a low or medium–low level of TFCE and its neighboring cities had higher TFCE levels, the probability of the upward transition of that city was greater. As shown in Table 6, when the TFCE of a city was type I and those of the neighboring cities were types II, III or IV, the probability of the upward transition of that city was 24.80%, 20.30% and 27.20%, respectively, which was higher than the transition probability when the neighboring city was type I (19.8%). When the TFCE of a city was type II and the neighboring cities were type III or IV, the probability of the upward transition of that city was 27.60% and 29.20%, respectively, both of which were higher than the probability when the neighboring city was type II. It can also be seen that neighboring cities with higher TFCE levels had better promotion effects on cities with low and medium–low TFCE levels. Secondly, when a city was type II and the neighboring city was type IV, the probability of the downward transition of that city was the smallest. When a city had a medium–high level of TFCE and its neighboring cities were types I, II or III, the probability of the downward transition of that city decreased in turn. To sum up, under the given conditions, the higher the TFCE levels in neighboring cities, the higher the probability of the upward transition of the given city and the lower the probability of downward transition.

Table 6. The spatial Markov transition probability matrix.

Spatial Lag	t/t + 1	I	II	III	IV
I	I	0.4490	0.1980	0.1700	0.1830
	II	0.2230	0.3500	0.3160	0.1120
	III	0.0920	0.2860	0.3280	0.2940
	IV	0.1050	0.1740	0.2330	0.4880
II	I	0.5860	0.2480	0.0950	0.0710
	II	0.2350	0.4050	0.2350	0.1250
	III	0.0950	0.2770	0.3800	0.2480
	IV	0.0620	0.0970	0.2480	0.5930
III	I	0.5270	0.2030	0.1830	0.1080
	II	0.2940	0.3500	0.2760	0.0800
	III	0.1130	0.2580	0.3760	0.2540
	IV	0.0400	0.1340	0.2810	0.5450
IV	I	0.3580	0.2720	0.2220	0.1480
	II	0.2080	0.4310	0.2920	0.0690
	III	0.0840	0.3270	0.3360	0.2520
	IV	0.0260	0.1180	0.2700	0.5860

6. Conclusions and Policy Implications

6.1. Conclusions

The purpose of this paper was to measure and evaluate TFCE at the city level in China as objectively and accurately as possible and comprehensively analyze the regional differences, spatial distribution patterns and dynamic evolution characteristics of urban TFCE. To achieve this, we used a combination of the super-efficiency EBM model and the GML index to measure the TFCE of 283 prefecture-level cities in China from 2004 to 2019. In terms of research methods, the Dagum Gini coefficient, Moran's I index, LISA aggregation and standard deviational ellipse models were used to analyze the spatiotemporal differentiation characteristics of urban TFCE and the kernel density estimation and spatial Markov chain methods were used to analyze the dynamic evolution of urban TFCE. The research conclusions were as follows.

Firstly, the differences in TFCE between Chinese cities decreased, although within the regional differences still constituted the main sources of differences in TFCE across the four regions. Secondly, the TFCE of Chinese cities had significant spatial correlations, the local spatial distributions showed high–high and low–low aggregation and TFCE had obvious directionality and field aggregation. Thirdly, the difference between the cities with the highest TFCE levels and the national average gradually decreased during the sample period and the polarization trends across the whole country and the eastern and central regions showed multipolar polarization. Fourthly, there was a significant path dependence effect of urban TFCE in the transfer process and the probability of TFCE transition between neighboring cities was correlated according to geographical space. The higher the TFCE of a city, the higher the probability of the upward transition of its neighboring cities.

The above research conclusions are not only conducive to comprehensively grasping the development status of the TFCE of Chinese cities but could also provide useful references for local governments to help them to formulate relevant strategies for the coordinated development of regional carbon efficiency. Based on the above research conclusions, the following policy implications are proposed.

6.2. Policy Implications

(i) Technological changes on the energy supply side offer the main technical strategy to achieve carbon neutrality and the development of energy-saving and emission reduction technologies is crucial to improving carbon efficiency. Therefore, firstly, industrial energy consumption and electricity consumption should be reduced and the development of green technologies, such as replacing coal with gas and the carbon neutrality of electric energy, should be vigorously promoted. secondly, the development of clean energy should be accelerated to replace fossil fuels and the total amounts of coal and electricity should be strictly controlled. We need to enhance clean energy power generation and reduce the use of fossil fuels at the source.

(ii) Micro-entities should be encouraged to actively participate in green scientific and technological innovation and support for new energy enterprises should be increased such as preferential tax policies, tax rebates, financial subsidies, etc. At the same time, local governments should increase support for innovation policies and improve the coordination mechanisms of ecological innovation policies. We need to promote the establishment of an eco-innovation system (i.e., a system that integrates science and technology, the environment, energy, industries, construction, transportation and other fields), strengthen the construction of supporting infrastructures, actively guide the regional exchange of knowledge and technology and create sound patterns of innovation and development for science and technology.

(iii) In view of the significant regional differences and the north–south differentiation of China's urban TFCE, the design of carbon efficiency policies should follow the principles of regional policies to avoid further increasing the difference between regions. According to the different resource endowment conditions of the different regions, differentiated policy support should be provided to increase transfer payments for the construction of green innovation resources in underdeveloped regions.

(iv) Considering the positive spatial spillover effect of TFCE to neighboring cities and the spatial correlation of transition probability, each region should pay attention to the exchange and cooperation of carbon emission reduction technologies among different cities while improving its own carbon efficiency level, give full play to regions with high carbon efficiency levels and improve the positive spillover effect of TFCE. It is also necessary to strengthen coordination and interaction between neighboring cities, actively build green, low-carbon and innovative regional collaborative platforms, break through regional boundaries, ensure smooth cross-regional circulation channels for innovative elements and resources and achieve coordinated improvements in regional carbon efficiency levels.

Author Contributions: H.L.: Review and editing and methodology. X.Q.: supervision and software and funding acquisition. All authors have read and agreed to the published version of the manuscript.

Funding: This work was sponsored in part by Chinese Ministry of Education Philosophy and Social Science Research Major Project Fund (18JZD035), Social Science Foundation of Shaanxi Province (2021D054, 2021D053), Social Science Foundation of Shaanxi province Department of Education (2022HZ1118), China Statistical Science Research Project(2022LY061), Fundamental Research Funds of Xi'an Jiaotong University (Special Project of the 20th National Congress of the CPC) (SK2022133).

Institutional Review Board Statement: Not applicable.

Informed Consent Statement: Not applicable.

Data Availability Statement: Not applicable.

Conflicts of Interest: The authors declare no conflict of interest.

Notes

1. According to the IPCC guidelines, the carbon emission coefficients of natural gas, liquefied petroleum gas and urban heating consumption are 1622 $kgCO_2/m^3$, 3.1013 $kgCO_2/kg$ and 2.53 $kgCO_2/kg$, respectively.
2. Due to the limited to space, only the LISA aggregation maps for 2005, 2013 and 2019 are listed here and the remaining maps are retained for retrieval.

References

1. Lin, B.Q. *China Energy Development Report 2019*; Peking University Press: Beijing, China, 2019.
2. Kaya, Y.; Yokobori, K. *Environment, Energy, and Economy: Strategies for Sustainability*; Bookwell Publications: Delhi, India, 1999.
3. Zhou, W.; Nie, M. Research on Regional Differences of China's Industrial Carbon Emission Efficiency—Empirical Analysis Based on Non parametric Frontier. Research on Quantitative Economy. *Tech. Econ.* **2012**, *9*, 58–71.
4. Hu, J.; Yan, S.; Han, J. Research on the implied carbon emission efficiency of China's industrial Sectors—Empirical analysis based on the three-stage DEA model and the non competitive I-O model. *Stat. Res.* **2021**, *38*, 30–43.
5. Gao, Y.; Tian, T. Research on the impact mechanism of carbon trading on carbon efficiency of pilot provinces and cities. *Explor. Econ. Issues* **2022**, *3*, 106–119.
6. Herrala, R.; Coel, R.K. Global CO2 Efficiency:country-wise Estimates using a Stochastic cost frontier. *Energy Policy* **2012**, *45*, 762–770. [CrossRef]
7. Du, K.R.; Zou, C.Y. Regional differences, influencing factors and convergence analysis on China's carbon emission efficiency – an empirical study based on stochastic frontier model and panel unit root. *Zhejiang Soc. Sci.* **2011**, *11*, 32–44.
8. Chen, L.M.; Huang, W. Research on China's Provincial Carbon Emission Efficiency Based on Stochastic Frontier. *Stat. Decis. Mak.* **2013**, *9*, 136–139.
9. Charnes, A.; Cooper, W.; Rhodes, E. Measuring the Efficiency of Decision Making Units. *Eur. J. Oper. Res.* **1978**, *2*, 429–444. [CrossRef]
10. IEA. Tracking Industrial Energy Efficiency and CO2 Emissions. In *Sourceoecd Energy*; OECD Publishing: Paris, France, 2007.
11. Ramanathan, R.A. Multi-factor efficiency perspective to the relationships among world GDP, energy consumption and carbon dioxide emissions. *Technol. Forecast. Soc. Chang.* **2006**, *73*, 483–494. [CrossRef]
12. Tone, K.A. Slacks-Based Measure of Efficiency in Data Envelopment Analysis. *Eur. J. Oper. Res.* **2001**, *130*, 498–509. [CrossRef]
13. Sun, X.; Zhang, H.; Wang, G. Study on regional carbon emission efficiency based on super-efficiency SBM model: A case study of 17 prefecture-level cities in Shandong Province. *Ecol. Econ.* **2016**, *32*, 68–73.
14. Zhang, W.; Zhu, Q.; Li, H. Energy use, carbon emissions and carbon emission reduction total factor efficiency in China. *Econ. Res.* **2013**, *10*, 138–150.
15. Ning, L.; Zheng, W.; Zeng, L. Assessment of China's provincial carbon emission efficiency and analysis on influencing factors from 2007 to 2016—Two-stage analysis based on super efficient SBM Tobit model. *J. Peking Univ. (Nat. Sci. Ed.)* **2021**, *57*, 181–188.
16. Lin, X.Q.; Bian, Y.; Wang, D. Temporal and spatial evolution characteristics and influencing factors of industrial carbon emission efficiency in Beijing Tianjin Hebei region. *Econ. Geogr.* **2021**, *41*, 187–195.
17. Zhang, X.F.; Fan, D.C. Research on the Impact of Carbon Emission Trading Market on Carbon Emission Reduction —Empirical Analysis On the basis of Double Mediation Effect. *Sci. Sci. Manag. Sci. Technol.* **2021**, *42*, 20–38.
18. Jiang, P.; Wang, Y.; Luo, J.; Chu, X.Z.; Wu, B.W. Research on the space-time evolution characteristics and path identification of carbon emission efficiency in the Yangtze River and Yellow River basins. *Environ. Sci. Res.* **2022**, *4*, 1–13.
19. Zhou, X.; Song, L. Low carbon Transformation of China's Industry: Realistic Analysis and Policy Thinking. Research on Quantitative Economy. *Technol. Econ.* **2022**, *39*, 22–41.
20. Mengyao, H.; Weidong, L.; Yitian, X.; Wanbei, J. Regional differences and decoupling trends of carbon emissions in China. *Resour. Sci.* **2021**, *43*, 710–721.
21. Wang, Z.; Su, J.; Peng, K. Calculation and factor decomposition of regional difference of carbon productivity in China. *Stat. Decis.* **2017**, *8*, 116–120.
22. Zhuoqun, Z.; Tao, Z.; Dongfa, F. Regional differences, dynamic evolution and convergence of carbon emission intensity in China. *J. Quant. Technol. Econ.* **2022**, *39*, 67–87.
23. Yang, G.; Te, Z.; Mei, D. Research on agricultural production efficiency and poverty reduction efficiency in China. *J. Quant. Technol. Econ.* **2020**, *4*, 46–64.
24. Tone, K.; Tsutsui, M. An epsilon-based measure of efficiency in DEA-A third pole of technical efficiency. *Eur. J. Oper. Res.* **2010**, *207*, 1554–1563. [CrossRef]
25. Shi, J.G.; Duan, K.F.; Wu, G.D.; Li, J.J.; Xu, K. Efficiency of urban and rural integrated development in the Yangtze River Delta under the constraint of carbon emissions. *Econ. Geogr.* **2021**, *41*, 57–67.

26. Andersen, p.; Petersen, N.C. A Procedure for Ranking Efficient Units in Data Envelopment Analysis. *Manag. Sci.* **1993**, 3[illegible] 1261–1264. [CrossRef]
27. Malmquist, S. Index Numbers and Indifference Surfaces. *Trab. De Estad.* **1953**, *4*, 209–242. [CrossRef]
28. Fukuyama, H.; Weber, W.L. A Directional Slacks-Based Measure of Technical Efficiency. *Soc. Econ. Plan. Sci.* **2009**, *43*, 274–28[illegible] [CrossRef]
29. Oh, D.H. A global Malmquist—Luenberger productivity index. *J. Product. Anal.* **2010**, *34*, 183–197. [CrossRef]
30. Bai, Y.; Deng, X.; Gibson, J.; Zhao, Z.; Xu, H. How does Urbanization affect Residential CO2 Emissions? An Analysis on Urba[illegible] Agglomerations of China. *J. Clean. Prod.* **2019**, *209*, 876–885. [CrossRef]
31. Lin, B.Q.; Jiang, Z. Environmental Kuznets curve prediction and influencing factors analysis of carbon dioxide in China. *Mana[illegible] World* **2009**, *4*, 27–36.
32. Ren, X.; Liu, Y.; Zhao, G. Impact of economic agglomeration on carbon emission intensity and its transmission mechanism. *Chin[illegible] Popul. Resour. Environ.* **2020**, *30*, 95–106.
33. Wu, J.; Guo, Z. Convergence Analysis of China's Carbon Emissions based on Continuous Dynamic Distribution Method. *J. Sta[illegible] Res.* **2016**, *33*, 54–60.
34. Glaeser, L.E.; Kahn, E.M. The Greenness of Cities: Carbon Dioxide Emissions and Urban evelopment. *J. Urban Econ.* **2010**, *6[illegible]* 404–418. [CrossRef]
35. Dagum, C. A New Approach to the Composition of the Gini Income Inequality Ratio. *Empir. Econ.* **1997**, *22*, 515–531. [CrossRef
36. Rey, S.J. Spatial Empirics for Regional Economic Growth and Convergence. *Geogr. Anal.* **2001**, *33*, 195–214. [CrossRef]
37. Yuill, R.S. The standard deviational ellipse: An updated tool for spatial description. *Geogr. Ann.* **1971**, *53*, 28–39. [CrossRef]
38. Gong, J.X. Clarifying the standard deviational ellipse. *Geogr. Anal.* **2002**, *34*, 155–167. [CrossRef]
39. Quah, D. Empirics for Growth and Distribution: Stratification, Polarization and Convergence Clubs. *J. Econ. Growth* **1997**, 2 27–59. [CrossRef]
40. Chen, M.; Zhang, X.; Liu, Y.; Zhong, C. Dynamic Evolution and Trend Prediction of Green TFP Growth: An Empirical Stud[illegible] based on Five Major Urban Agglomerations in China. *Nankai Econ. Res.* **2020**, *1*, 20–44.

land

MDPI

Article

Towards Carbon-Neutral Cities: Urban Classification Based on Physical Environment and Carbon Emission Characteristics

Jiah Lee and Seunghyun Jung *

Department of Building Research, Korea Institute of Civil Engineering and Building Technology, Goyang-si 10223, Republic of Korea
* Correspondence: shjung@kict.re.kr; Tel.: +82-31-910-0377

Abstract: Cities are highly industrialized and populated areas and major sources of greenhouse gas emissions. For carbon neutrality, examining the correlation between urban characteristics and greenhouse gas emissions is necessary. This study aimed to analyze the characteristics of each city from a carbon neutrality perspective. As such, we conducted a carbon-neutral city analysis. First, the physical environmental variables of 250 municipal, county, and district local governments were collected and constructed and then reduced and purified through factor analysis. Second, the type was derived by performing cluster analysis on the reduced factor variables and carbon emissions by analysis unit. Finally, the characteristics of each type were analyzed, and the carbon-neutral city planning and applicable carbon-neutral technology fields were proposed according to the characteristics. After the categorization of carbon-neutral cities throughout Korea, six cluster types were derived; cities in each cluster had similar characteristics. This study suggests that solutions for carbon reduction should be applied by comprehensively considering the social, economic, and environmental characteristics of each city. It concludes that regional physical environmental indicators and energy consumption statistics can be used comprehensively to establish effective policies and apply technologies and techniques at the local government level.

Keywords: greenhouse gas emission; carbon neutrality; urban planning; carbon reduction; city analysis; energy consumption

Citation: Lee, J.; Jung, S. Towards Carbon-Neutral Cities: Urban Classification Based on Physical Environment and Carbon Emission Characteristics. *Land* **2023**, *12*, 968. https://doi.org/10.3390/land12050968

Academic Editor: Yan Li

Received: 2 March 2023
Revised: 10 April 2023
Accepted: 24 April 2023
Published: 26 April 2023

1. Introduction

The global warming caused by the use of fossil fuels has become a major concern for governments worldwide. The challenge of achieving carbon neutrality in cities has attracted significant attention from policymakers, scholars, and practitioners worldwide. The Carbon Neutral Cities Alliance (CNCA) published a framework for action in 2015, which outlines several key strategies for reducing greenhouse gas emissions in cities, including energy efficiency, renewable energy, sustainable transportation, and waste reduction [1]. The Korean government is no exception. The Korean government has taken several steps towards realizing a carbon-neutral society, including enacting the Framework Act on Carbon Neutrality and Green Growth and establishing a roadmap for carbon neutrality. The Korean government's efforts towards carbon neutrality align with the CNCA framework, underscoring the importance of urban policies and plans to reduce greenhouse gas emissions.

However, given that most economic activities generating greenhouse gas emissions occur in cities, it is necessary to prepare measures at the urban level [2]. In particular, as of 2022 in Korea, the urbanization rate—that is, the ratio of the resident population in the urban planning zone to the total population—amounts to 81.4%, which is very high compared to the global urbanization rate of 57.0% [3].

In order to apply policies and technologies to reduce greenhouse gas emissions, it is essential to identify the characteristics of each city first. This is because each city has unique physical and non-physical characteristics, such as size, function, spatial structure, and industrial composition, which affect the type of emissions or absorption [4–11]. Solutions for carbon reduction should be applied by comprehensively considering the social, economic and environmental characteristics of each city. Therefore, this study aimed to analyze the characteristics of each city in Korea from the perspective of carbon neutrality as a pre-stage of urban planning for carbon neutrality.

In particular, the study comprehensively considered various characteristics of the city. The study intended to include low-carbon emission potential by adding renewable energy source values, as well as characteristics of the city's physical environment, such as regional size, economic industry, land uses, and transportation services. By categorizing all cities in Korea based on their characteristics, this study provides direction for the establishment of urban policies to reduce carbon emissions comprehensively. Ultimately, this study contributes to the broader literature on carbon-neutral cities and provides insights for policymakers, scholars, and practitioners working towards achieving carbon neutrality in cities.

2. Review of Theory and Previous Research

2.1. Carbon Neutrality and Carbon-Neutral City

2.1.1. Concept of Carbon Neutrality

The United Nations defines carbon neutrality as a state in which net emissions offset the amount of greenhouse gas absorbed from that emitted or leaked into the atmosphere and become zero (United Nations, UN) [12]. Carbon neutrality does not impose unconditional carbon emission obligations; however, it refers to a comprehensive concept that defines responsible activities by balancing a measured amount of carbon released with an equivalent amount sequestered or offset. In general, the realization of carbon neutrality is divided into two strategies: reducing carbon emissions and increasing carbon absorption.

Policies have been enforced to increase the effect of carbon neutrality by reducing carbon emissions and increasing absorption sources, as well as actively utilizing carbon-neutrality-related plans and technologies.

2.1.2. The Emergence of Carbon-Neutral Cities and Paradigm Changes

The theories on carbon neutrality emerged from the development of "the countryside city" [13], in the early 20th century, to "the ecological city" that emerged in the mid-1970s. Low-carbon, zero-carbon, and carbon-neutral cities appeared to cope with climate change in the 2000s (Table 1). Recently, in response to the demand for carbon neutrality, a project aiming to build a carbon-neutral city with various similar concepts has been actively promoted worldwide.

While a carbon-neutral city can be defined in various ways, it is conceptualized as one with minimal carbon emissions and corresponding measures against carbon inevitably emitted from the economy, society, and leisure activities, resulting in a "zero" carbon net emission [14]. Broadly, carbon-neutral cities comprise eco-friendly and sustainable green cities aiming for carbon neutrality, cities with energy self-sufficiency, cities that absorb enough carbon because of abundant green infrastructure, and cities that use less fossil energy owing to easy walking and the development of public transportation.

In Korea, the legal definition of a carbon-neutral city (Framework Act on Carbon Neutrality and Green Growth) is a city that spatially implements carbon neutrality by actively utilizing plans and technologies related to carbon neutrality. Therefore, considering various urban emission and absorption factors, such as changes in population and industrial structure, urban decline, extinction, and the new development of transportation and IT technology, the urban concept can be expanded to realize carbon neutrality through space and infrastructure. Another characteristic of carbon-neutral cities is that they intend to mitigate climate change by setting a quantitative target of achieving a carbon-neutral state in their spatial unit.

In urban space, various fields such as energy, buildings, transportation, agriculture, and livestock, as well as absorption sources, are mutually influenced; therefore, an integrated approach is needed to view them as a system [15,16]. The advantage of this urban-centered integrated approach is that it can supplement the limitations of one field in another.

Table 1. Urban paradigms and characteristics related to carbon neutrality.

Urban Paradigm	Key Characteristics	Relationship with Carbon Neutrality
Sanitary reform (1840s)	• Removal of dirt and eradication of diseases, especially cholera • Elimination and disposal of sewage and waste, providing clean water	Importance of physical infrastructure
Garden city (1890s)	• New Town Movement in Low-Density, Car-Based Suburbs in the UK with an emphasis on the importance of nature in urban planning • Cities and villages are adjacent to green areas such as gardens, agriculture, etc.	Urban shape change through planning paradigm
Eco city (1990s)	• Integrated city planning and management leveraging the benefits of ecosystem • Urban planning and design in balance with nature	Reflection on the need to meet global and regional ecological needs
Sustainable city (1990s)	• Reducing resource consumption in cities, reducing dependence on cars, and pursuing economic and social sustainability based on ecology	Urban models enabling global and national sustainability models
Green city (2000s)	• Concepts including carbon reduction, sustainability, etc., in existing sustainable cities (eco-friendly cities) • Conservation of ecosystems, eco-friendly cities, cities that minimize carbon emissions, and the pursuit of sustainable development	Starting from qualitative concepts of being eco-friendly and sustainable to quantitative concepts such as carbon
Smart city (2000s)	• Digital-connected cities that measure, manage, and improve the quality of life and the efficiency of urban activities using information and communication technology	Reduction of energy use through smart systems and integration of local solar and wind systems into cities
Low-carbon city (2000s)	• Separating urban economies and activities from fossil fuel use and emphasizing energy efficiency, renewable energy, and green transportation	The main focus on carbon to be addressed
Carbon-neutral city (2020s)	• Reducing all fossil fuel use in larger categories beyond low-carbon cities and creating urban and regional landscapes with carbon sequestration and cyclical economic strategies • Similar concepts include net-zero-carbon city, zero-carbon city, and zero-emission city	Quantitative target setting for carbon neutrality

Reference: [16] Recertifying and modifying.

Cities around the world are increasingly turning to urban planning to achieve carbon neutrality, and a number of projects are underway to support this goal. The CNCA is a global urban cooperative that supports the development of methodologies, planning, standardization, and governance tools for measuring and improving carbon neutrality in cities [17]. Major eco-friendly cities in Europe, such as Copenhagen and Helsinki, as well as cities with high urbanization rates, such as New York in the United States and London in the United Kingdom, are participating. The Zero Cities Project, launched by the Urban Sustainability Directors Network (USDN) in 2017, is another example of an effort to develop carbon-free building sectors [18].

Meanwhile, the European Union has selected 100 cities to achieve climate neutrality by 2030, and launched the "100 Climate-Neutral Cities by 2030" campaign to support less capable cities. The EU is providing financial support and activity guidelines for cities t prepare and operate visions and roadmaps according to their size and capabilities [19]. I addition, the EU has several other initiatives to promote carbon-neutral urban plannin the initiative named the Covenant of Mayors for Climate and Energy, Smart Cities an Communities, the European Green Capital Award, and the EU's largest program, name Horizon 2020 [20–23].

One additional project worth noting is the C40 Cities Climate Leadership Grou a network of the world's megacities committed to addressing climate change [24]. C4 connects cities with the expertise and resources that they need to take bold climate actio and provides a platform for peer-to-peer sharing and collaborative problem solving. Th group has more than 90 member cities worldwide, including Tokyo, Paris, and Los Angele and works to implement policies and projects to reduce greenhouse gas emissions an improve air quality.

Overall, these and other urban planning projects are critical in the fight against climat change, as cities are responsible for a significant share of the global carbon emissions. B implementing strategies for achieving carbon neutrality, cities can help to reduce their envi ronmental impact and create more sustainable, livable communities for future generation

2.2. The Relationship between Carbon Neutrality and the Urban Spatial Structure

Cities are highly industrialized and populated areas that are significant sources o greenhouse gas (GHG) emissions. Reducing GHG emissions is essential for mitigatin climate change and achieving carbon neutrality. As such, understanding the correlatio between urban characteristics and GHG emissions is crucial for developing effective policie and strategies to reduce emissions in urban areas.

To establish a carbon-neutral city, cities should be categorized based on the discus sions regarding the components, functions, and structures of carbon-neutral cities. T analyze and understand cities correctly, attempts have been previously made to classif them according to such criteria as size, function, structure, shape, and location [25–28 Analyzing and understanding cities in this manner is crucial as it enables the identificatio of appropriate carbon-neutrality-related projects and technologies based on the uniqu components and functions of each city. Thus, sustainable standards for zero-energy citie have been raised, and the development of simplified methodological evaluation tools ha been required [29,30].

Studies have been conducted in earnest to understand the characteristics of energ consumption by country and city since the 1980s, when urbanization intensified. Earl studies were conducted to analyze the energy demands of cities. Owens (1984a, 1984b compared and analyzed urban change and energy demands (emissions) at the time of rapi urbanization [31,32]. As cities become more complex and diverse, he proved that not onl simple land use patterns but also population movement between regions and the resultin increase in employment and economic power had an impact on the energy demand.

Urban form and density are important factors that play a role in GHG emissions. Com pact, mixed-use urban areas with good public transportation systems can reduce emissions by encouraging active transportation and reducing car dependency [4,33,34]. Studies have also found that increasing the population density in urban areas can reduce emissions by promoting the efficient use of resources and encouraging sustainable transportation modes [35,36]. Newman and Kenworthy (1989; 2006) used regression analysis to examine the impact of the urban form on emissions in cities in Australia, Canada, and the United States [37,38]. The study found that cities with higher levels of density and mixed land use had lower levels of emissions. Studies by Heinonen and Junnila (2011) also compared and analyzed carbon consumption in the suburbs to the dense metropolitan center [7], deriving urban density as a major factor affecting greenhouse gas emissions. Liu et al. (2017) also explored the impact of the population density, land use mix, and public transportation accessibility on carbon emissions in Chinese cities [39]. The study found that cities with higher levels of population density and land use mix had lower levels of emissions.

One of the most active scholars, Frank (2000), analyzed how land use and vehicle traffic characteristics relate to automobile exhaust, a major factor in carbon emissions in Seattle [4]. As a result, he proved that population density and employment density, which are the main measurements of land use variables, have positive (+) relationships with exhaust gas. More specifically, his 2006 study demonstrated the influence of factors such as the residential density, number of intersections, and commercial areas on land characteristics [5]. Xia et al. (2018) proved that changing land use characteristics from agriculture to industry negatively affect carbon neutrality [9]. This is consistent with the findings of Gao et al. (2019) that urban green spaces such as parks and gardens can improve the air quality and reduce emissions by isolating carbon dioxide [40].

Transportation is also a key factor that influences GHG emissions in urban areas. Studies have consistently shown that cities with high levels of car dependency have higher levels of GHG emissions compared to cities with well-developed public transportation systems [41,42]. Mavroeidi et al. (2013) examined the impact of transportation mode choice on emissions in European cities [43]. The study found that cities with higher levels of public transportation use and cycling had lower levels of emissions. In a study by Dou et al. (2016), access to transportation facilities was noted as a key strategy for low-carbon urban planning [9]. Another study by Zhou et al. (2017) used regression analysis to explore the relationship between the transportation mode share, population density, and carbon emissions in Chinese cities [44]. The study found that cities with higher levels of public transportation use and population density had lower levels of carbon emissions. Liu et al. (2022) analyzed the relationship between carbon emissions and comprehensive urban environmental variables such as urban density, transportation network, industrial structure, and location [11].

Studies discussing greenhouse gas reduction based on the shape and use of buildings in cities have also emerged [32,33]. Buildings account for a significant portion of GHG emissions in urban areas, and energy-efficient buildings have been identified as an effective means to reduce emissions [45–48]. Studies have found that improvements in building design and materials, such as insulation, lighting, and ventilation systems, can significantly reduce energy consumption and emissions [49,50]. Building codes and standards that mandate energy efficiency measures have also been shown to be effective in reducing GHG emissions in urban areas [51]. The studies by Du et al. (2018) and Wang et al. (2020) used regression analysis to explore the impact of building energy efficiency on carbon emissions in Chinese cities and the United States [52,53]. The studies found that cities with more energy-efficient buildings had lower levels of emissions.

These previous research works suggest that transportation, building energy consumption, and urban form and density are important factors to consider when attempting to reduce GHG emissions in urban areas. Policies and strategies that promote sustainable transportation modes, energy-efficient buildings, and compact, mixed-use urban areas can effectively reduce emissions. In summary, the physical environment of cities plays a crucial role in carbon emissions, and urban design and planning are essential to achieving carbon neutrality. However, more research is needed to identify the most effective strategies and tools for carbon-neutral urban development. To support urban planning for carbon neutrality, this study categorizes cities based on their unique characteristics and greenhouse gas emissions. The results contribute to the development of targeted policy alternatives for achieving carbon-neutral cities.

3. Methodology

Previous studies on energy consumption analysis have been conducted mainly in a single city or a few specific cities, and there is a limit when viewing the various environments of the city comprehensively by analyzing some variables such as transportation and land use. Therefore, to reflect the characteristics of each city in the carbon-neutral city plan, cities should be categorized using variables that are more specific. As such, this study conducts a carbon-neutral city analysis. Consequently, regional physical environmental indicators and energy consumption statistics can be used comprehensively to establish effective policies and apply technologies and techniques at the local government level.

3.1. Cities to Be Analyzed

The spatial scope of the analysis included all 250 local governments in Korea.

The temporal range was set to 2019, when the statistics on carbon emissions and regional characteristic indicators obtained simultaneously in all regions analyzed were established.

For 2019's carbon emissions, the 2019 energy supply and demand statistics issued by the KESIS National Energy Statistics Comprehensive Information System were used as shown in Figure 1, and for regional characteristic indicators, the 2019 statistics of local governments were used [34]. KESIS's energy supply and demand statistics for cities, counties, and districts are used for preparing annual energy statistics and providing energy statistics necessary for regional energy planning and energy supply prospects considering regional characteristics. The main contents include measurements of final energy consumption by city, county, and district, and energy consumption by sector, such as industry, transportation, home, commerce, and public.

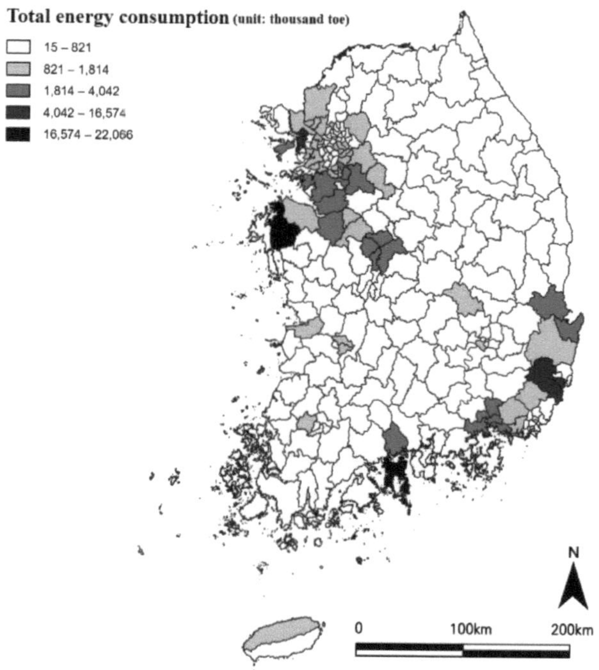

Figure 1. Total energy consumption of local governments nationwide in 2019.

The analysis procedure for the categorization of carbon-neutral cities was as follows. First, after collecting and constructing the physical environmental variables of 250 municipal, county, and district local governments, the analysis variables were reduced and purified through factor analysis.

Second, the type was derived by performing cluster analysis on the reduced factor variables and carbon emissions by analysis unit. Finally, the characteristics of each type were analyzed, and the carbon-neutral city planning and applicable carbon-neutral technology fields were proposed according to the characteristics.

Table 3. *Cont.*

Statistics Data			Average	Max	Min	Stv
Economy Industry Characteristics	Economy Industry	Number of businesses (unit)	21,369.64	86,643.00	1308.00	19,409.01
		Number of businesses per 1000 people	88.89	476.60	49.90	38.75
		Number of manufacturers (unit)	334.26	4260.00	2.00	492.88
		Financial independence (%)	21.73	68.90	4.00	13.61
	Special-Purpose Area	Land use_residential ratio	21.63	91.50	0.12	15.74
		Land use_commercial ratio	3.06	44.24	0.00	4.88
		Land use_industrial ratio	6.77	51.38	0.00	8.41
		Land use_green area ratio	65.03	92.84	0.00	20.53
Traffic Service		Road ratio	0.05	0.74	0.00	0.08
		Paved road	502,674.04	2,045,533.00	55,276.00	335,497.26
		Number of cars registered per person (unit)	0.51	2.45	0.11	0.22
Production Volume by Renewable Energy Sources		Solar power	2183.77	4023.00	122.00	1241.18
		Photovoltaics	1,114,623.37	2,768,303.00	69,194.00	816,313.91
		Wind power	184,342.43	774,049.00	2.00	286,381.55
		Hydropower	251,147.73	708,305.00	101.00	250,306.44
		Marine power	452,760.97	474,321.00	0.25	98,800.49
		Geothermal power	19,306.90	48,650.00	1771.00	14,078.94
		Hydrothermal energy	2212.53	11,954.00	8.00	3741.24
		Fuel cell	230,104.90	793,295.00	50.00	28,7512.71
Energy Consumption Characteristics		Total energy consumption	965.79	22,066.00	15.00	2367.23
	Industry	Industrial energy consumption	497.68	21,230.00	1.00	2224.91
	Transportation	Transportation sector energy consumption	222.82	2313.00	4.00	277.44
	Household	Household energy consumption	121.29	509.00	3.00	124.29
	Commercial Sector	Commercial sector energy consumption	94.66	534.00	4.00	99.32
	Public Sector	Public sector energy consumption	29.41	419.00	1.00	41.59

4.2. *Refining and Reducing Analysis Variables*

Based on the interrelationships between the variables related to regional characteristics, variables were reduced through exploratory factor analysis. First, the measured values of all variables were standardized and used. In the first attempt at factor analysis,

22 variables, excluding energy consumption by region, were used to perform Verimax principal component analysis. In the process, the commonality of the land use–industrial ratio variable and the number of automobile registrations per person was 0.3 or less, and the two variables were deleted.

In the next attempt, the remaining 20 variables were input and re-analyzed. Consequently, five factors with an intrinsic value of 1 or more were extracted; they have a 100% cumulative explanatory power for the whole, and the load on each factor is shown in Table 4.

Table 4. Factors extracted by factor analysis and the load on factors.

Names of Factors	Names of Variables	Factor Rotation 1	Factor Rotation 2	Factor Rotation 3	Factor Rotation 4	Factor Rotation 5
Space Size and Land Use Characteristics	Population density	**0.88**	0.01	0.33	0.02	0.01
	Residential area per person	**0.81**	0.19	0.12	0.04	0.35
	Land use_green area ratio	**0.81**	0.02	0.24	0.08	0.09
	Road ratio	**0.65**	0.28	0.08	0.28	0.04
	Land use_residential ratio	**0.58**	0.23	0.38	0.22	0.19
	Urban area ratio	**0.55**	0.14	0.63	0.16	0.07
Economy Industry Characteristics 1	Number of businesses (unit)	0.28	0.27	0.27	**0.79**	0.02
	Paved road (m)	0.25	0.18	0.16	**0.77**	0.09
	Financial independence (%)	0.22	0.52	0.30	**0.60**	0.07
	Number of manufacturers (unit)	0.06	0.33	0.08	**0.70**	0.05
Economy Industry Characteristics 2	Land use_commercial area ratio	0.32	0.01	0.13	0.04	**0.84**
	Number of businesses per 1000 people (unit)	0.04	0.11	0.03	0.03	**0.91**
Region Renewable Energy Production 1	Hydropower	0.44	**0.63**	0.08	0.12	0.09
	Fuel cell	0.26	**0.87**	0.18	0.07	0.11
	Geo-power	0.07	**0.95**	0.11	0.13	0.09
	Marine power	0.03	**0.95**	0.12	0.15	0.09
Region Renewable Energy Production 2	Solar power	0.36	0.26	**0.72**	0.09	0.02
	Photovoltaics	0.27	0.07	**0.84**	0.06	0.13
	Wind power	0.25	0.16	**0.66**	0.11	0.11
	Hydrothermal power	0.04	0.17	**0.76**	0.07	0.18
SS loadings		4.65	3.80	3.27	2.33	1.94
Proportion Explained		0.29	0.24	0.20	0.15	0.12
Cumulative Proportion		0.29	0.53	0.73	0.88	1.00

The first factor comprises variables related to space size and land use characteristics. The characteristics of space size correspond to the population density and residential area per person. The characteristics related to land use correspond to the ratio of urban areas, the ratio of housing and green areas for land use, and the ratio of roads. The second factors are related to the economic industry characteristics and comprise the variables related to absolute figures related to the industry, such as the number of businesses, the number of manufacturers, and the length of paved roads. The third factor comprises variables related to the relative figures related to the ratio of land use commercial areas and the distribution ratio of industries by the number of businesses per 1000 people. The fourth and fifth factors are largely classified into two factors for each type of renewable energy. Solar power, photovoltaics, wind power, and hydrothermal power are divided into relatively generalized renewable energy groups and other groups of hydropower, fuel cell, geo-power, and marine energy.

The above five factors were standardized, converted, and used as regional characteristic variables for urbanization.

4.3. Analysis of Typification of Carbon-Neutral Cities

After standardizing the five regional characteristic variables and the energy consumption characteristic variables of cities and counties derived through factor analysis, cities were categorized by cluster analysis. After adjusting and analyzing the number of clusters, six clusters were found to be the most appropriate within the significance level of 0.01. However, very few cases included in two clusters, Clusters 5 and 6, were derived from three regions and one region. These can be interpreted as representing clusters with minimal cases.

Table 5 shows the regional classification values for each of the six cluster types according to the cluster analysis results. Cluster 1 has 70 cases, including most areas in metropolitan cities. Cluster 2 includes most of the local small and medium-sized cities; therefore, it is the largest case. Cluster 3 includes most of Gyeonggi-do, the metropolitan area. Cluster 4 includes fewer cases among local small and medium-sized cities, similar to Cluster 2. Clusters 5 and 6 have fewer cases and cover three areas, Jung-gu in Seoul, Jung-gu in Busan, and Jung-gu in Daegu, and the latter includes one area in Hwaseong.

Table 5. Types of carbon-neutral cities by region.

Regional Classification *	Cluster Classification					
	Cluster 1 (N = 70)	Cluster 2 (N = 109)	Cluster 3 (N = 41)	Cluster 4 (N = 26)	Cluster 5 (N = 3)	Cluster 6 (N = 1)
Seoul	24	-	-	-	1	-
Busan	15	-	-	-	1	-
Daegu	7	-	-	-	1	-
Incheon	8	2	-	-	-	-
Gwangju	5	-	-	-	-	-
Daejeon	5	-	-	-	-	-
Ulsan	5	-	-	-	-	-
Sejong	1	-	-	-	-	-
Gyeonggi-do	-	-	41	-	-	1 (Hwaseong-si)
Gangwon-do	-	16	-	2	-	-
Chungbuk	-	9	-	5	-	-
Chungnam	-	13	-	3	-	-
Jeonbuk	-	13	-	2	-	-
Jeonnam	-	22	-	-	-	-
Gyeongbuk	-	19	-	5	-	-
Gyeongnam	-	14	-	8	-	-
Jeju	-	1	-	1	-	-

* Representative target site (partial). Cluster 1: (Seoul) Jongno-gu, Yongsan-gu, Seongdong-gu, (Busan) Yeongdo-gu, Suyeong-gu, Gijang-gun, (Daegu) Dong-gu, Seo-gu, (Incheon) Yeonsu-gu, Bupyeong-gu, (Gwangju) Seo-gu, Buk-gu, (Daejeon) Yuseong-gu, Daedeok-gu, Jung-gu, Nam-gu, Ulju-gun, (Sejong) Sejong-si, etc. Cluster 2: (Incheon) Ganghwa-gun, Ongjin-gun, (Gangwon) Gangneung-si, Donghae-si, Hoengseong-gun, (Chungbuk) Jecheon-si, Jeungpyeong-gun, Jincheon-gun, (Chungnam) Boryeong-si, Seosan-si, Dangjin-si, (Jeonbuk) Gunsan-si, Namwon-si, Buan-gun, (Jeonnam) Mokpo-si, Yeosu-si, Gwangyang-si, (Gyeongbuk) Gimcheon-si, Goryeong-gun, Uljin-gun, (Gyeongnam) Jinju-si, Tongyeong-si, and Hamyang-gun, (Jeju), Seogwipo-si, etc. Cluster 3: (Gyeonggi-do) Suwon Gwonseon-gu, Seongnam Bundang-gu, Uijeongbu-si, Anyang Dongan-gu, Bucheon-si, etc. Cluster 4: (Gangwon-do) Chuncheon-si, Wonju-si, (Chungbuk) Cheongju Sangdang-gu, Chungju-si, (Chungnam) Cheonan Dongnam-gu, Asan-si, (Jeonbuk) Jeonju Wansan-gu, (Gyeongbuk) Pohangnam-gu, Gyeongju-si, Gumi-si, (Gyeongnam) Changwon Uijang-gu, Changwon Jinhae-gu, Sacheon-si, Geoje-si, (Jeju) Jeju-si, etc. Cluster 5: (Seoul) Jung-gu, (Busan) Jung-gu, (Daegu) Jung-gu. Cluster 6: (Gyeonggi-do) Hwaseong City.

4.4. Characteristics of Carbon-Neutral Cities by Type

The average values of all statistical data used in the analysis were compared to confirm the characteristics of the Si, Gun, and Gu regions by six clusters derived from the cluster analysis (Table 6). The characteristics of each cluster type were compared with the average value of each variable, and the name of the type was given accordingly (Figure 2).

- Cluster 1 has the following regional characteristics and energy consumption characteristics. It has a high population density in terms of space size and land use, with the highest proportion of residential areas and the lowest proportion of green areas. Conversely, the residential area per capita is relatively low because the population density is very high and the proportion of roads is 0.08, which is high. The economic industry is characterized by areas where manufacturing industries are less distributed and the proportion of paved roads is relatively low. The proportion of commercial areas is high; however, among the six clusters, the number of businesses per 1000 people and financial independence are among the upper–middle ranks. Renewable energy is not produced adjacent to the sea; therefore, there are no marine energy production facilities. The area is considered to have insufficient capacity to produce new and renewable energy as the energy produced by hydropower, geo-power, solar power, photovoltaics, wind power, and hydrothermal power is the lowest. Total energy consumption was found to be in the mid-range—that is, the early 1000 toe range. Consumption was high in the order of industry, transportation, household, commercial and public sectors, and consumption was found to be moderate compared to other clusters. Based on these characteristics, Type 1 of the carbon-neutral city can be named as a dense residential center type (energy-small and medium-sized cost).
- Cluster 2 has the lowest population density, with a very small urban area ratio of 0.09, and green areas account for the majority. Despite the low ratio of the residential area, the per capita residential area is high because of the small population density. The proportion of roads is 0.01, which is low. The manufacturing industry is the least distributed, and the number of businesses, the ratio of commercial areas to land use, and financial independence are very low; therefore, it appears to be an area with no economic activity. Conversely, the production of new and renewable energy generated by solar power, photovoltaics, wind power, and hydrothermal power is significantly higher than that of other clusters. This may be because solar and wind power generation equipment is furnished while considering small-sized local cities' characteristics of spatial structure and climate. Energy consumption characteristics are the lowest at 600,000 toe, most of which is occupied by the industrial sector. This reflects the characteristics of local cities with a small population as, compared to other regions, the energy consumption in daily activities such as home, transportation, commerce, and the public is insignificant. Based on these characteristics, Type 2 carbon-neutral cities can be named low-density local small cities (low energy consumption).
- Cluster 3 has the following regional characteristics and energy consumption characteristics. The population density is in the middle-class area, comprising a high ratio of urban areas and residential areas, and the green area ratio is the highest. The proportion of roads is 0.08, which is high. In this area, a large number of manufacturers are distributed, and the number of businesses, paved road areas, and financial independence are all high. These regions have high population density and an active economy. Most areas near the metropolitan area have very high hydro-energy production related to new and renewable energy. The production of fuel cells, geo-power, and marine energy is also the highest. Conversely, the energy production of photovoltaics and wind power is low. The total energy consumption is moderate at around 1,000,000 toe. By sector, unlike other clusters, energy consumption in the transportation sector is the highest, followed by home, industry, commerce, and public. Based on these characteristics, Type 3 of the carbon-neutral city can be named the metropolitan area type (energy-small and medium-sized cost).

- Cluster 4 has a low population density, urban area ratio, and residential area ratio, and a low road ratio of 0.03. Conversely, the per capita housing area and green area ratio are high. Manufacturers are most distributed here, and the number of businesses and paved roads is the highest. While the proportion of commercial areas is relatively small, the degree of financial independence and the number of businesses per 1000 are high; therefore, it seems that the manufacturing-oriented industry is outstandingly active. The production of new and renewable hydropower, solar power, photovoltaics, wind power, and hydrothermal power energy is very high, indicating that many energy production facilities are located in the region. The total energy consumption is the highest at 1,800,000 toe. Considering that industries and transportation sectors account for the highest consumption, the area can be considered a transportation hub with substantial traffic and a manufacturing center with active industrial activities. In addition, energy consumption is high in the order of home, commerce, and public. Based on these characteristics, Type 4 of the carbon-neutral city can be viewed as a low-density manufacturing industry terrain (energy high cost).
- Cluster 5 is an overcrowded urban area with a very high proportion of population density, urban area ratio, and residential area for land use; the smallest proportion of green areas; and the highest proportion of roads. It has the lowest distribution of manufacturers, and the smallest proportion of paved roads. Conversely, as the proportion of commercial areas and the number of businesses per 1000 people is the highest, this area has active commercial activities. The production of new and renewable energy is the lowest among all types; accordingly, the production capacity of new and renewable energy is insufficient. Total energy consumption is the lowest at approximately 500,000 toe. By sector, consumption is high in the order of transportation, home, commerce, industry, and public, and consumption is lower than most types except for the commercial sector. Based on these characteristics, Type 5 of carbon-neutral cities is the urban type (low energy consumption).
- Regarding the last type, Cluster 6 has the highest population density. The green area ratio is high, while the residential area ratio is low. Conversely, the ratio of roads is 0.74, which is very high compared to other types. In particular, the manufacturing industry is at least 6 to 42 times higher than in other types. The number of businesses, paved roads, and financial independence is also the highest, and it can be considered a manufacturing-oriented industrial dense area. Compared to other types, the production of hydropower, fuel cells, geo-power, marine power, solar power, and photovoltaics is very high. On the other hand, wind and hydrothermal energy production is low, which seems to be due to the absence of related energy production facilities. Total energy consumption is the highest at 3,314,000 toe. Most energy is consumed in the industrial and transportation sectors, which seems to result from the regional characteristics of manufacturing-specialized areas. In addition, energy consumption is relatively high in the home, commercial, and public sectors. Based on these characteristics, Type 6 of the carbon-neutral city can be called the manufacturing industry specialization type (maximum energy consumption).

Table 6. Cluster analysis results summary and the average values of variables by type of carbon neutral city.

Variables		Cluster Classification						F-Value (p-Value)
		1 (N = 70)	2 (N = 109)	3 (N = 41)	4 (N = 26)	5 (N = 3)	6 (N = 1)	
Space Size and Land Use Characteristics	Population density (per/ha)	100.07	1.90	37.35	9.68	139.69	151.03	6.175 ($p < 0.001$)
	Urban area ratio	0.92	0.09	0.69	0.37	0.93	0.40	
	Green area ratio	49.59	71.98	74.26	70.11	37.18	75.70	
	Residential area per person (per/m^2)	28.38	33.40	28.74	31.09	27.55	29.30	
	Land use_residential ratio	35.19	14.90	19.79	14.01	44.53	16.66	
	Road ratio	0.08	0.01	0.08	0.03	0.12	0.74	
Economy Industry Characteristics 1	Number of businesses (unit)	25,361.90	6732.57	38,148.12	42,547.35	26,724.71	66,767.00	85.117 ($p < 0.001$)
	Paved road (m)	339,321.64	462,457.59	489,513.98	1,172,214.31	281,173.71	662,864.00	
	Financial independence (%)	23.75	11.79	38.53	28.14	25.28	68.90	
	Number of manufacturers (unit)	287.11	108.60	687.63	726.15	199.32	4260.00	
Economy Industry Characteristics 2	Land use_commercial Ratio	3.87	1.99	2.42	1.89	6.75	1.90	92.677 ($p < 0.001$)
	Number of businesses per 1000 people	84.26	91.51	72.04	85.12	103.25	81.90	
Region Renewable Energy Production 1	Hydropower	6272.94	278,309.36	566,631.00	306,274.04	1157.05	566,631.00	42.346 ($p < 0.001$)
	Fuel cell	218,488.76	68,104.99	793,295.00	21,412.40	262,097.71	793,295.00	
	Geo-power	8441.76	16,331.39	48,650.00	14,908.00	10,086.80	48,650.00	
	Marine power	-	0.25	474,321.00	0.25	262,097.71	793,295.00	
Region Renewable Energy Production 2	Solar power	710.74	2808.93	2696.00	2858.54	757.05	2696.00	7.541 ($p < 0.001$)
	Photovoltaics	173,962.51	1,715,357.38	1,071,664.00	1,305,998.15	196,942.02	1,071,664.00	
	Wind power	5418.84	356,132.54	4834.00	250,205.08	245.49	4834.00	
	Hydrothermal power	60.22	3693.90	8.00	2029.05	23.00	8.00	
Energy Consumption Characteristics *	Total energy consumption	1017.90	686.60	1077.61	1807.00	500.07	3314.00	-
	Industrial energy Consumption	501.20	519.39	217.15	841.00	63.54	1817.00	-
	Transportation sector energy consumption	245.00	90.92	374.61	473.42	157.51	781.00	-
	Household energy consumption	137.27	35.39	244.07	246.12	136.80	335.00	-
	Commercial sector energy consumption	106.19	29.39	189.17	174.50	113.17	342.00	-
	Public sector energy consumption	28.21	11.68	52.41	72.12	28.90	39.00	-

* These variables are not used for cluster analysis.

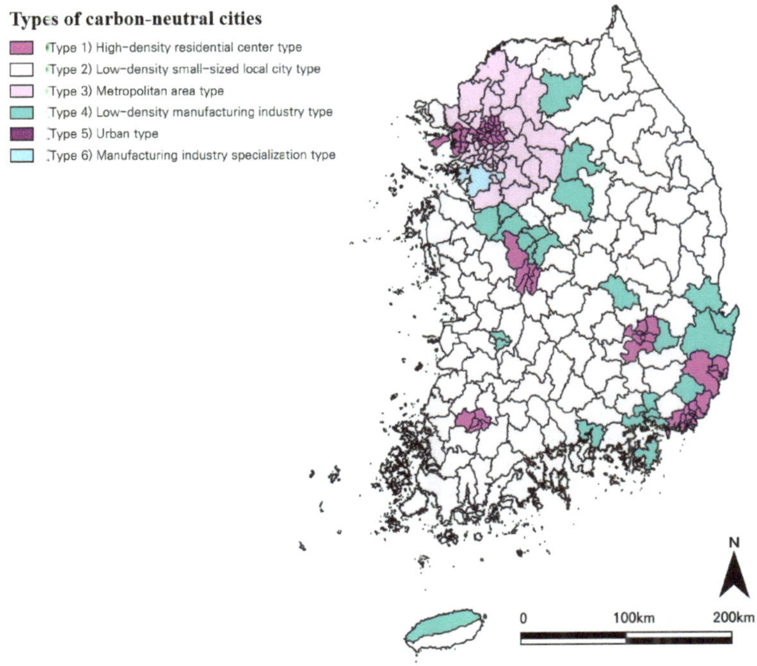

Figure 2. Types of carbon-neutral cities.

5. Discussion and Conclusions

After the categorization of carbon-neutral cities throughout Korea, six cluster types were derived. Cities in each cluster have similar characteristics, and there are many homogeneous factors in greenhouse gas emissions. These results are consistent with the arguments of previous studies that various characteristics of cities and energy consumption characteristics are closely related [4–11,37,38].

In other words, the results suggest that the classification of cluster types by city should be preceded by planning and managing carbon-neutral cities and that policies and technologies related to carbon neutrality need to be applied differently according to the characteristics of the cluster type.

The analysis identified five key planning elements that are central to carbon-neutral urban planning in Korea: transportation (mobility), buildings, energy, absorption sources, and industries. Each planning element has specific characteristics that are related to the physical and non-physical features of a city. The technology in the mobility/transportation sector can be applied mainly to urban areas with high traffic demands and high efficiency due to mobility conversion [9,43,44]. Buildings are easy to apply when constructing and renovating buildings for an entire area [45–48]. Energy can be utilized for information on areas suitable for new and renewable energy production. For example, an energy absorption source can be applied to improve the absorption performance of park areas in established cities. Industry can be applied to the type of industrial manufacturing-centered city with technology dissemination. Accordingly, we propose the following recommendations for each cluster type.

For Type 1 (high-density residential center type), the direction of expanding low-carbon means of transportation in cities through mobility conversion should be considered first, considering the characteristics of regions with high urban activities. Type 2 (low-density local small-town type) is an area where public transportation is not developed compared to large cities, and policies are needed to improve transportation to meet carbon neutrality through its expansion. Conversely, as Type 3 (metropolitan area type) has the characteristics of a bed town outside of Seoul, it is important to establish a policy to convert the means of commuting to the city center from passenger cars to public transportation. As Type 4 (low-density manufacturing industry terrain) and Type 6 (manufacturing industry specialization type) have high energy demands in the industry, the application of renewable energy production technology should be actively considered. Additionally, direct carbon absorption facilities such as Carbon Capture, Utilization, and Storage (CCUS) should be introduced. Type 5 (urban type) is a traditional urban area in Korea; therefore, it will not be easy to secure absorption sources in the city because of the characteristics already developed. Instead, an integrated system such as a Building Energy Management System (BEMS) can be introduced for the energy management of existing commercial buildings. The carbon-neutral performance of the cluster can be improved by increasing the ratio of new and renewable energy when renovating buildings.

In sum, this study demonstrates that key regional characteristic indicators can be derived according to the classification of carbon-neutral cities, and related policies can be promoted around them. Our findings have important implications for policymakers and practitioners involved in carbon-neutral urban planning, highlighting the need to apply policies and technologies differently based on the characteristics of each cluster type. By adopting an efficient approach to carbon neutrality planning, cities in Korea can achieve sustainable and effective reductions in greenhouse gas emissions. This study identified the major variables that affect greenhouse gas emissions by city type, but did not analyze the factors in detail or consider the weight of each variable. In future studies, we plan to identify the key factors that have a significant impact on greenhouse gas emissions by region type. We will also develop mitigation policies and guidelines for applying relevant technologies in consideration of these factors. Additionally, we aim to classify other cities and establish international standards for carbon-neutral cities, based on a comparison with the Korean case.

Author Contributions: Conceptualization, J.L. and S.J.; methodology, J.L. and S.J.; software, J.L.; formal analysis, J.L.; data curation, J.L.; original draft preparation, J.L.; writing—review and editing, J.L. and S.J.; supervision, S.J. All authors have read and agreed to the published version of the manuscript.

Funding: This research was supported by the National Research Foundation of Korea (NRF), funded by the Ministry of Science and ICT (MSIT) (NRF-2022R1C1C2013464).

Informed Consent Statement: Informed consent was obtained from all subjects in the study.

Data Availability Statement: The data presented in this study are available upon request from the corresponding author.

Acknowledgments: This work has utilized a certain part of the results of the research project 'Planning Research on a Pilot Model of the 2050 Carbon Neutral City (21RDPP-C165367-01)', funded by the Korea Agency for Infrastructure Technology Advancement (KAIA) via a grant funded by the Ministry of Land, Infrastructure and Transport.

Conflicts of Interest: The authors declare no conflict of interest. The funders had no role in the design of the study; in the collection, analyses, or interpretation of data; in the writing of the manuscript; or in the decision to publish the results.

References

1. Carbon Neutral Cities Alliance. Carbon Neutral Cities Alliance: A Framework for Action. 2015. Available online: https://carbonneutralcities.org/wp-content/uploads/2015/04/CNCA-Framework-for-Action.pdf (accessed on 5 December 2022).
2. IPCC. *Climate Change 2007—The Physical Science Basis: Working Group I Contribution to the Fourth Assessment Report of the IPCC*; Solomon, S.D., Qin, M., Manning, Z., Chen, M., Marquis, K.B., Averyt, M., Tignor, H.L.M., Eds.; Cambridge University Press: Cambridge, UK; New York, NY, USA, 2007.
3. KOSIS (Korea Statistical Information Service). Available online: https://kosis.kr/statHtml/statHtml.do?orgId=101&tblId=DT_2KAA204&conn_path=I2,2022urbanizationrate (accessed on 29 September 2022).
4. Ewing, R. Is Los Angeles-style sprawl desirable? *J. Am. Plann. Assoc.* **1997**, *63*, 107–126. [CrossRef]
5. Frank, L.D.; Stone, B., Jr.; Bachman, W. Linking land use with household vehicle emissions in the central Puget Sound: Methodological framework and findings. *Transp. Res. D Transp. Environ.* **2000**, *5*, 173–196. [CrossRef]

6. Frank, L.D.; Sallis, J.F.; Conway, T.L.; Chapman, J.E.; Saelens, B.E.; Bachman, W. Many pathways from land use to health: Associations between neighborhood walkability and active transportation, body mass index, and air quality. *J. Am. Plann. Assoc.* **2006**, *72*, 75–87. [CrossRef]
7. Stone, B., Jr.; Mednick, A.C.; Holloway, T.; Spak, S.N. Is compact growth good for air quality? *J. Am. Plann. Assoc.* **2007**, *73*, 404–418. [CrossRef]
8. Heinonen, J.; Junnila, S. Implications of urban structure on carbon consumption in metropolitan areas. *Environ. Res. Lett.* **2011**, *6*, 014018. [CrossRef]
9. Dou, Y.; Luo, X.; Dong, L.; Wu, C.; Liang, H.; Ren, J. An empirical study on transit-oriented low-carbon urban land use planning: Exploratory Spatial Data Analysis (ESDA) on Shanghai, China. *Habitat. Int.* **2016**, *53*, 379–389. [CrossRef]
10. Xia, C.; Li, Y.; Xu, T.; Ye, Y.; Shi, Z.; Peng, Y.; Liu, J. Quantifying the spatial patterns of urban carbon metabolism: A case study of Hangzhou, China. *Ecol. Indic.* **2018**, *95*, 474–484. [CrossRef]
11. Liu, K.; Ni, Z.; Ren, M.; Zhang, X. Spatial differences and influential factors of urban carbon emissions in China under the target of carbon neutrality. *Int. J. Environ. Res. Public Health* **2022**, *19*, 6427. [CrossRef]
12. United Nations. Paris Agreement. 2015. Available online: https://unfccc.int/process-and-meetings/the-paris-agreement (accessed on 12 January 2023).
13. Howard, E. *Garden Cities of Tomorrow*; MIT Press: Cambridge, MA, USA, 1965; Volume 23.
14. KAIA (Korea Agency for Infrastructure Technology Advancement). Carbon Neutral City. *Land Transp. Technol. Issue Rep.* **2021**, *1*, 3. Available online: http://www.kaia.re.kr/portal/bbs/view/B0000070/10703.do?menuNo=201225&cate1=2 (accessed on 10 December 2022).
15. Seto, K.C.; Davis, S.J.; Mitchell, R.B.; Stokes, E.C.; Unruh, G.; Ürge-Vorsatz, D. Carbon lock-in: Types, causes, and policy implications. *Annu. Rev. Environ. Resour.* **2016**, *41*, 425–452. [CrossRef]
16. Seto, K.C.; Churkina, G.; Hsu, A.; Keller, M.; Newman, P.W.; Qin, B.; Ramaswami, A. From low-to net-zero carbon cities: The next global agenda. *Annu. Rev. Environ. Resour.* **2021**, *46*, 377–415. [CrossRef]
17. Carbon Neutral Cities Alliance. Carbon Neutral Cities Alliance 2019 Annual Report. 2019. Available online: https://carbonneutralcities.org/tile/cnca-2019-annual-report (accessed on 5 December 2022).
18. USDN (Urban Sustainability Directors Network). Zero Cities Project. 2021. Available online: https://www.usdn.org/projects/zero-cities-project.html (accessed on 3 March 2023).
19. Directorate-General for Research and Innovation. Proposed Mission: 100 Climate-neutral Cities by 2030—By and for the Citizens. European Commission. 2020. Available online: https://research-and-innovation.ec.europa.eu/knowledge-publications-tools-and-data/publications/all-publications/100-climate-neutral-cities-2030-and-citizens_en (accessed on 3 March 2023).
20. Covenant of Mayors. (n.d.). Available online: https://www.covenantofmayors.eu/about.html (accessed on 10 April 2023).
21. European Commission. (n.d.). Smart Cities and Communities. Available online: https://ec.europa.eu/digital-single-market/en/smart-cities-communities (accessed on 10 April 2023).
22. European Commission. (n.d.). European Green Capital Award. Available online: https://ec.europa.eu/environment/europeangreencapital/about-award/index.html (accessed on 10 April 2023).
23. European Commission. (n.d.). Horizon 2020. Available online: https://ec.europa.eu/programmes/horizon2020 (accessed on 10 April 2023).
24. C40 Cities Climate Leadership Group. (n.d.). About C40. Available online: https://www.c40.org/about (accessed on 10 April 2023).
25. Suarez-Villa, L. Metropolitan evolution, sectoral economic change, and the city size distribution. *Urban Stud.* **1988**, *25*, 1–20. [CrossRef]
26. Batty, M.; Kim, S.K. Form follows function: Reformulating urban population density functions. *Urban Stud.* **1992**, *29*, 1043–1069. [CrossRef]
27. Conzen, M.P. The study of urban form in the United States. *Urban Morphol.* **2001**, *5*, 3–14. [CrossRef]
28. Waddell, P.; Borning, A.; Noth, M.; Freier, N.; Becke, M.; Ulfarsson, G. Microsimulation of urban development and location choices: Design and implementation of UrbanSim. *Netw. Spat. Econ.* **2003**, *3*, 43–67. [CrossRef]
29. Koutra, S.; Becue, V.; Gallas, M.A.; Ioakimidis, C.S. Towards the development of a net-zero energy district evaluation approach: A review of sustainable approaches and assessment tools. *Sustain. Cities Soc.* **2018**, *39*, 784–800. [CrossRef]
30. Urrutia-Azcona, K.; Sorensen, S.; Molina-Costa, P.; Flores-Abascal, I. Smart zero carbon city: Key factors towards smart urban decarbonisation. *Dyna* **2019**, *94*, 676–683. [CrossRef]
31. Owens, S. Spatial structure and energy demand. Energy policy and land use planning: An international perspective. In *Energy Policy and Land-Use Planning: An International Perspective*, 1st ed.; Cope, D.R., Hills, P., James, P., Eds.; Pergamon Press: Oxford, UK, 1984; Volume 32, pp. 215–240.
32. Owens, S.E. Energy and spatial structure: A rural example. *Environ. Plan A* **1984**, *16*, 1319–1337. [CrossRef]
33. Crane, R.; Crepeau, R. Does neighborhood design influence travel? A behavioral analysis of travel diary and GIS data. *Transp. Res. Part D Transp. Environ.* **1998**, *3*, 225–238. [CrossRef]
34. Ewing, R.; Cervero, R. Travel and the built environment: A meta-analysis. *J. Am. Plan. Assoc.* **2010**, *76*, 265–294. [CrossRef]
35. Jiang, Y.; Zhang, L.; Wang, Y.; Gao, J.; Zhang, Y. Research on the relationship between urban form and CO_2 emissions based on China's cities. *J. Clean. Prod.* **2018**, *179*, 251–261. [CrossRef]

36. Zhao, Y.; Long, R.; Li, H.; Wang, Q.; Cheng, X. Urbanization, population density and CO_2 emissions in Chinese cities. *J. Clean Prod.* **2020**, *244*, 118901. [CrossRef]
37. Newman, P.W.; Kenworthy, J.R. Gasoline consumption and cities: A comparison of US cities with a global survey. *J. Am. Plann Assoc.* **1989**, *55*, 24–37. [CrossRef]
38. Newman, P.; Kenworthy, J. Urban design to reduce automobile dependence. *Opolis Int. J. Suburb. Metrop. Stud.* **2006**, *2*, 9–22.
39. Liu, L.; Li, X.; Liu, C.; Wu, H. Urban form, population density and public transportation connectivity: What matters most for urban CO_2 emissions in China? *Habitat Int.* **2017**, *60*, 70–78.
40. Gao, J.; Sun, Y.; Hu, T.; Liu, Y.; Liu, L.; Wang, H.; Wu, J.; Zhou, J. Can urban parks and gardens reduce carbon emissions? Evidence from a Chinese city. *J. Clean. Prod.* **2019**, *219*, 649–657. [CrossRef]
41. Newman, P.; Kenworthy, J. *Cities and Automobile Dependence: An International Sourcebook*; Gower Technical: Brookfield, VT USA, 1989.
42. Banister, D.; Hickman, R. *Transport Futures: Thinking the Unthinkable*; Routledge: London, UK, 2006.
43. Mavroeidi, E.; Kostopoulos, K.; Yannis, G. Exploring the impact of transport mode choice on CO_2 emissions in urban areas *Procedia Soc. Behav. Sci.* **2013**, *48*, 1507–1516. [CrossRef]
44. Zhou, K.; Yang, S.; Zhang, J. The relationship between transportation mode share, population density, and carbon emissions Evidence from Chinese cities. *Habitat Int.* **2017**, *66*, 54–68. [CrossRef]
45. Huang, W.; Li, F.; Cui, S.H.; Huang, L.; Lin, J.Y. Carbon footprint and carbon emission reduction of urban buildings: A case in Xiamen City, China. *Procedia Eng.* **2017**, *198*, 1007–1017. [CrossRef]
46. Huang, B.; Xing, K.; Ness, D.; Liao, L.; Huang, K.; Xie, P.; Huang, J. Rethinking carbon-neutral built environment: Urban dynamics and scenario analysis. *Energy Build.* **2022**, *255*, 111672. [CrossRef]
47. Eom, J.; Shin, Y.; Lee, M. Estimation of GHG emission reduction potential of building energy efficiency retrofit measures: A case study of Seoul, Korea. *Sustainability* **2018**, *10*, 3868. [CrossRef]
48. Sayan, S.; Fouseki, K.; Nijkamp, P. The effect of energy efficiency measures on urban carbon emissions: A review of the literature *Renew. Sustain. Energy Rev.* **2019**, *107*, 207–218. [CrossRef]
49. Balaras, C.A.; Dascalaki, E.G.; Gaglia, A.G. Energy efficiency of buildings: A review. *J. Build. Apprais.* **2007**, *3*, 307–332. [CrossRef]
50. Hong, T.; Taylor-Lange, S.C.; D'Oca, S. Building energy benchmarking: Methods and comparison with a case study. *Build. Environ* **2019**, *150*, 82–98. [CrossRef]
51. Zhou, W.; Fu, X.; Chen, H. Impacts of building energy efficiency codes on greenhouse gas emissions in Chinese cities. *Build. Environ.* **2020**, *178*, 107100. [CrossRef]
52. Du, L.; Wu, J.; Wang, H. The impact of building energy efficiency on carbon emissions in Chinese cities: Evidence from panel data analysis. *Renew. Sustain. Energy Rev.* **2018**, *94*, 1006–1014. [CrossRef]
53. Wang, J.; Zou, Y.; Huang, Y.; Gu, J. The impact of energy efficiency on carbon emissions: Evidence from US cities. *Energy Policy* **2020**, *139*, 111338. [CrossRef]
54. KESIS (Korea Energy Statistical Information System). 2019 Energy Supply and Demand Statistics for Cities, Counties, and Districts. 2019. Available online: http://www.kesis.net/sub/sub_0003.jsp?CATEGORY_ID=C_007&SEQ=17&LOW_SEQ=1&M_MENU_ID=M_M_002&S_MENU_ID=S_M_013 (accessed on 10 December 2022).

land

MDPI

Article

Evaluating the Impact of Smart City Policy on Carbon Emission Efficiency

Xingneng Xia, Ruoxi Yu and Sheng Zhang *

School of Public Policy and Administration, Xi'an Jiaotong University, Xi'an 710049, China; xiaxingneng1991@stu.xjtu.edu.cn (X.X.); yuruoxi@stu.xjtu.edu.cn (R.Y.)
* Correspondence: zhangsheng@xjtu.edu.cn

Abstract: Smart city policy (SCP) is crucial to addressing climate change and achieving sustainable urban development with low carbon emissions. The purpose of this paper is to investigate the mechanisms through which smart city policies have an impact on carbon emission efficiency (CEE). In terms of research methodology, we construct a quasi-natural experiment on smart city policies in China and use the time-varying DID approach to study this issue. The DEA method was used to measure the CEE. For the data sample, panel data from 281 cities in China between 2007 and 2020 was used in this study. The findings are as follows: ① SCP has a significant impact on CEE. This conclusion remains valid after introducing parallel trend tests, placebo tests, and other robustness tests. ② The mechanism test result reveals that SCP has a positive impact on urban CEE through three main channels: promoting industrial upgrading, increasing public environmental attention, and enhancing marketization. ③ The analysis of heterogeneity reveals that the impact of SCP on CEE is noticeable in cities that belong to well-developed economic regions with a lower intensity of environmental regulations, higher levels of green finance, and fewer official changes. This research contributes to the existing literature on the environmental assessment of SCP and offers valuable policy insights for cities to tackle climate change and sustainable urban planning.

Keywords: smart city; carbon emission efficiency; environmental assessment; sustainable urban development; urban planning

Citation: Xia, X.; Yu, R.; Zhang, S. Evaluating the Impact of Smart City Policy on Carbon Emission Efficiency. *Land* **2023**, *12*, 1292. https://doi.org/10.3390/land12071292

Academic Editor: Yan Li

Received: 22 May 2023
Revised: 20 June 2023
Accepted: 24 June 2023
Published: 26 June 2023

1. Introduction

As human economic activities continue to expand and globalization accelerates, the severity of global climate change and environmental issues is increasing. It is crucial to address the challenges of reducing energy consumption, improving carbon emission efficiency, and promoting sustainable urban development while pursuing economic growth [1,2]. Achieving a green and low-carbon development model is a pressing global concern [3,4]. In order to combat the pressing issues of climate change and environmental pollution, the United Nations has implemented the 2030 Agenda for Sustainable Development. Furthermore, during COP27 in 2022, the UN established a set of practical agreements aimed at reducing carbon emissions. Meanwhile, China has taken strategic action to reduce carbon emissions, pledging to peak carbon emissions by 2030 and achieve carbon neutrality by 2060 [5,6].

Smart cities are considered an important policy practice to reduce urban carbon emissions and achieve sustainable urban development [7]. It refers to the intelligent management and transformation of a city's infrastructure, public services, government governance, business operations, and residents' lives through the usage of advanced information and communications technology (ICT), big data, the Internet of Things (IoT), and cloud computing [8–11]. The European Commission considers energy, transport, and ICT to be the core elements of smart city construction. For example, Francesco Russo et al. found that smart cities are the main path to pursuing sustainability goals in mobile planning tools [12]. Additionally, Paola Panuccio found that smart planning is an important tool in the pursuit of urban sustainability in complex territorial systems, based on the Italian cas study [13]. The ultimate purpose of a smart city is to create more sustainable and efficier urban environments [14].

Since the 1990s, numerous countries and regions, such as the European Union, th United States, and Singapore, have implemented smart city projects. In 2012, Chin launched its own nationwide smart city program, with over 100 cities approved as nationa smart city pilot cities in 2012, 2013, and 2014. These initiatives have yielded valuable polic experience for the study of smart cities.

In recent years, scholars have conducted many studies on sustainable developmen The research covers several aspects, including ecosystem services, the circular econom climate change, and the interaction between social development and natural ecology. Fo instance, Yan et al. examined the main EU nations' energy efficiency using panel dat collected between 2010 and 2018 and found that countries with higher ecological securit outperformed others in terms of energy efficiency [15]. Moreover, Lagodiienko et al. foun that the external economic activity of enterprises has an important role in sustainabl development based on Ukrainian data [16].

Academics have also conducted extensive research on smart cities and carbon emissio issues. Research on smart cities is conducted in three main aspects: theoretical research practical research, and methodological research. Theoretical research focuses on the cor ceptual connotation, explanatory framework, and functional role of smart cities [17–19 Practical research includes the digital economy, green innovation, and communicatior technology [20–22]. Methodological research encompasses policy evaluation, practica case studies, and development trend research [23–25]. Research on carbon emissions ha focused on identifying the factors that influence carbon emissions, developing effectiv measurement methods, and evaluating the performance of carbon emissions [26–29].

The current body of literature serves as a strong foundation for this paper, howeve there is a noticeable gap in research regarding the effect of smart city policy (SCP) or carbon emissions, particularly in terms of how these policies can improve carbon emissior efficiency (CEE). So, can SCP, a sustainable urban development policy that is widely usec around the world, have an impact on CEE? What is the mechanism of its impact? What ar the factors associated with this impact on cities? Therefore, this paper builds on previou research to propose our research topic, which is to evaluate the impact of smart city policie on carbon emission efficiency. The research goal is to explore how smart city policies affec carbon emission efficiency and give constructive policy recommendations for improving urban energy use efficiency and promoting sustainable urban development. The research tasks of this paper are three: first, to construct a quasi-natural experiment on the impact o smart city policies on carbon emission efficiency based on China's smart city constructior experience using the time-varying DID method; second, to measure urban carbon emissior efficiency based on the DEA model; and third, to identify the action channels of smart city policies on carbon emission efficiency.

This paper offers a unique contribution to the existing studies by highlighting three distinct aspects that have not been previously explored. Firstly, from a research perspec tive, this paper presents a quasi-natural experiment on SCP using practical experience from China. It investigates the impact of SCP on CEE, providing empirical evidence to enrich the literature on SCP evaluation and improve CEE. Secondly, this paper employs two different measurement methods to assess CEE. The baseline model uses SBM-DEA while the robustness test utilizes EBM-DEA. This approach helps to prevent any bias in the estimation results that may have arisen from relying solely on a single measurement method, as was done in prior studies. Lastly, from an analytical perspective, the mechanism of action is examined through three aspects: promoting industrial upgrading, increasing public environmental attention, and enhancing marketization. To analyze the heterogeneity of policy impact, four perspectives are selected: regional differences among cities, environmental regulation intensity, green finance level, and official change cycles. This approach offers a fresh analytical perspective for evaluating the impact of SCP on CEE and provides theoretical guidance for the scientific construction of SCP.

The remainder of the article is structured as follows: Section 2 is the policy background and mechanism analysis. Section 3 displays the research design. Section 4 presents the benchmark analysis of SCP on CEE. Section 5 is the robustness test. Section 6 outlines the heterogeneous analysis of SCP. The following Section 7 draws on the mechanism analysis. The final section gives the conclusions and policy implications.

2. Policy Background and Mechanism Analysis

2.1. Policy Background

Numerous countries have already embarked on practical explorations of smart city construction. One such example is the Intelligent Transportation System (ITS) project, launched by the city of Seattle in 1996, aimed at resolving the issue of urban traffic congestion. The ITS project uses intelligent transportation technology to dynamically monitor and analyze traffic flow, optimizing carbon emissions by reducing traffic congestion and improving traffic efficiency. Another instance is the Songdo International Business District (SIBD) project, implemented in Incheon, South Korea, in 2003, which investigated the potential of urban science and technology parks and sought to optimize their operational efficiency. One of the features of the SIBD project is the adoption of green building design principles. The project involves the extensive use of energy-efficient equipment and technologies in the main buildings. High-efficiency lighting systems, solar panels, and geothermal energy are used in buildings to reduce energy consumption and carbon emissions. Additionally, in 2009, the Amsterdam Smart City (ACS) project was launched by the city of Amsterdam in the Netherlands. This project developed an open data sharing platform that facilitates data sharing and applications to improve the efficiency of city management and the quality of life of residents. Similarly, the 'Virtual Singapore' (VS) project aims to build a large-scale 3D city model and a multidimensional data collaboration platform based on the concept of data-driven digitization. The goal of this project is to improve the efficiency of city operations. A common feature of the ACS and VS projects is the use of big data and smart technologies to collect and analyze a variety of urban data to support data-based decision-making. This helps identify areas with high energy consumption and carbon emissions and develop corresponding emission reduction measures.

The development of smart cities in China was officially initiated in 2012 with the announcement of the *Notice on National Smart City Pilot Work* (NNSCPW) and the *National Smart City Pilot Index System* (NSCPIS) by the Chinese Ministry of Housing and Urban-Rural Development. This policy practice marked the beginning of smart city construction in China. NNSCPW and NSCPIS mainly provide guidelines for the direction of urban carbon reduction in four aspects: urban infrastructure protection, green building, urban intelligent management, and sustainable industrial development. Furthermore, the construction direction and target for smart city construction have been indicated in two documents: *Several Opinions on Promoting Information Consumption to Expand Domestic Demand* (SOPICEDD) from 2013 and the *National New Type Urbanization Plan (2014–2020)* (NNTUP) from 2014. SOPICEDD and NNTUP point out that information infrastructure construction and information consumption are the focus of smart city construction. Information-driven smart city construction based on information can promote sustainable urban development and efficient allocation of energy use. Meanwhile, the following release of two documents, namely *Guidance on Promoting the Healthy Development of Smart Cities* and *Guidance on the Construction and Application Implementation of the Smart City Standard System and Evaluation Index System*, has established evaluation standards and operational guidelines for the development of smart cities. This provides a unified analytical framework for scientifically evaluating the efficiency of urban operations, measuring energy consumption, and identifying carbon emissions footprints. Notably, in 2017, the inclusion of smart city construction in the report of the 19th National Congress of the Communist Party of China marked the formal recognition of smart cities as a national strategy and top-level design.

Following this, the Chinese government issued a series of guidance documents to promote the integration and development of smart cities with various industries. This indicates that the construction of smart cities is moving towards specialization and sophistication.

During the development process of SCP, it became evident that these policies have a direct impact on carbon emissions. As environmental issues continue to escalate, improving CEE has become a key focus of SCP. As a result, it is crucial to investigate the impact of SCP on CEE and explore the mechanisms behind this impact.

2.2. Mechanism Analysis

2.2.1. SCP Can Improve CEE by Promoting Industrial Upgrading

The effect of industrial upgrading can be observed in three main areas: the aggregation of industrial factors, the transformation of traditional industries, and the development of emerging industries. In terms of the aggregation of industrial factors, SCP aims to optimize industrial layout and enhance industrial agglomeration, resulting in efficient allocation and utilization of resources [21]. By sharing infrastructure, technical resources and talents, enterprises in industrial agglomeration areas can reduce production costs and improve economic efficiency [30]. Industrial agglomeration not only facilitates the formation of the scale effect but also reduces energy consumption and carbon emissions during transportation by optimizing the supply chain and logistics [31]. Furthermore, the synergic effect and technological innovation brought by industrial agglomeration can promote the development and application of green production technologies, enhancing the efficiency of energy utilization and carbon emissions in the production process. In terms of the transformation of traditional industries, the SCP aims to promote technological transformation in traditional industries by introducing advanced energy-saving and low-carbon technologies, thereby reducing energy consumption and carbon emissions [32]. To achieve this, the policy offers R&D support, financial subsidies, and tax incentives to reduce the cost of renovation and stimulate the reform motivation of enterprises. The policy not only encourages enterprises to reduce carbon emissions but also aims to improve their efficiency by implementing clean production standards, carbon emission quotas, and market mechanisms. This can be achieved through the adoption of intelligent, networked and environmentally friendly production methods, which can help traditional industries achieve their goals of high efficiency, low carbon emissions, and environmental protection [33]. In terms of the development of emerging industries, the SCP aims to optimize and upgrade the economic structure by supporting the development of emerging industries such as new energy, clean technology, and green building [34]. These low-carbon industries have high added value and environmental performance, which can effectively reduce carbon emission intensity [35]. The smart city policies offer financial support, talent training, market development, and other measures to attract companies to invest in emerging industries. This leads to an increase in the proportion of emerging industries in the economy, resulting in a significant improvement in CEE. Additionally, the development of emerging industries drives the green transformation of related industrial chains, creating a low-carbon industrial ecology.

2.2.2. SCP Can Improve CEE by Raising Public Environmental Attention

The public environmental concern effect of smart city policies is mainly reflected in three aspects: raising environmental awareness, promoting green consumption, and influencing policy implementation. From the perspective of raising environmental awareness, SCP aims to increase public awareness of environmental protection through enhanced environmental education and public participation [36]. A highly environmentally conscious public is more likely to adopt a low-carbon lifestyle, which includes green travel, energy savings, and waste reduction. Moreover, improving environmental awareness helps to create a positive social atmosphere and encourages enterprises to fulfill their social responsibility by paying attention to carbon emission issues, thus improving CEE [37]. From the perspective of promoting green consumption, SCP aims to encourage environmentally conscious consumption by promoting green products and services as well as establishing green labeling and certification systems. This approach has proven effective in reducing carbon emissions as consumers opt for energy-efficient home appliances, clean energy, and public transportation [38]. From the perspective of influencing policy implementation, SCP enables the government to obtain real-time information and policy recommendations on carbon emissions by creating a data analysis platform to analyze, evaluate, and monitor data related to public environmental concerns [39]. This information helps the government adjust policy direction and measures in a timely manner, optimize resource allocation, and achieve CEE.

2.2.3. SCP Can Improve CEE by Increasing Marketization

The marketization effects of SCP are mainly reflected in three aspects: improving resource allocation efficiency, stimulating innovation-driven development, and promoting carbon emissions trading. Regarding improving resource allocation efficiency, the SCP aims to efficiently allocate and utilize resources through market-oriented operations, optimize industrial layouts, guide capital investment, and adjust energy structures [40]. As part of this goal, the policy encourages the development of clean and renewable energy sources to reduce reliance on high-carbon energy. Additionally, the policy not only encourages the adoption of circular economy and green development strategies but also introduces market competition mechanisms to stimulate the growth of the renewable energy market. Thus, it helps reduce resource waste and improve CEE. Regarding stimulating innovation-driven development, the SCP aims to enhance talent competition and innovation development mechanisms through market-based approaches. It supports investments in research and development, talent training, and technological innovation to stimulate innovation. The policy promotes healthy competition, reduces costs and energy consumption, and fosters green technology innovation and research [41]. Additionally, it encourages the research and application of clean, low-carbon, and energy-saving technologies to reduce carbon emissions. Regarding promoting carbon emissions trading, the SCP encourages businesses to engage in the trading of carbon emission rights by creating a carbon emission trading market and quota system [42]. This approach facilitates a market-based allocation of carbon emissions, which incentivizes companies to reduce their carbon emissions and improve their CEE. Consequently, enterprises can use market-based trading to buy or sell carbon emission rights, which can help them achieve their carbon emission reduction targets. Additionally, the carbon emission trading market can aid in the rational allocation of resources and ultimately reduce the overall cost of carbon emission reduction.

3. Research Design

3.1. *Multiple Time-Varing DID Model*

This paper utilizes the three batches of Chinese smart city construction lists released by the Ministry of Housing and Urban-Rural Development (MHUD) in 2012, 2013, and 2014 to construct a quasi-natural experiment for Chinese smart city policy. The selected cities for smart city construction are considered the treatment group, while the cities not selected are the control group. Following the classical methodology [43,44], a multiple-time-varying DID model is used to identify the policy impact of smart city construction on urban carbon emission efficiency.

$$Eff_{Carbon\,i,t} = \alpha + \beta Smart_{Policy\,i,t} + \gamma Controls + CityFE + YearFE + \varepsilon_{i,t} \tag{1}$$

In the above Equation (1), Eff_{Carbon} denotes the efficiency of urban carbon emissions; $Smart_{Policy}$ denotes the smart city policy; $Controls$ denotes the control variables; $CityFE$ denotes control of city fixed effects; $YearFE$ denotes control of year fixed effects; ε denotes the random disturbance term. The estimated coefficient β measures the average difference in carbon efficiency before and after being subjected to smart city policy implementation.

3.2. DEA Model and CEE

Data envelopment analysis (DEA) is a method that uses linear programming to evaluate the relative efficiency of decision units in a multi-input, multi-output system. This method is widely used in efficiency evaluation and is one of the common models used to calculate carbon emission efficiency. In addition, the DEA model, as a non-parametric method, has also been widely used in urban system planning [45]. However, the traditional DEA model ignores the measurement error caused by the slack variables. To overcome this problem, Tone introduces a non-radial and non-angular SBM model based on the measurement of slack variables, which solves the slackness problem of the traditional DEA model. It is worth noting that the model cannot further differentiate and rank the efficiency of multiple decision-making units (DMUs) with efficiency values equal to 1 (effective frontier surface). Consequentially, Tone proposes a super-efficient SBM model that combines the super-efficiency model with the SBM model. The model incorporates undesirable output variables, corrects slack variables, and could be able to decompose multiple decision units with an efficiency value of 1. This enables comparison and ranking among efficient decision units and improves the accuracy of the model. Therefore, the super-efficiency SBM-DEA model, considering non-desired outputs, is chosen to measure urban carbon emission efficiency in this paper. The model is set as follows:

$$Min\rho = \frac{\frac{1}{m}\sum_{i=1}^{m}\frac{\overline{x_i}}{x_{ij}}}{\left[\frac{1}{S_1+S_2}\left(\sum_{q=1}^{S_1}\frac{\overline{y_q^w}}{y_{qj}^w}+\sum_{q=1}^{S_2}\frac{y_q^b}{y_{qj}^b}\right)\right]}$$

$$s.t.\begin{cases}\overline{x} \geq \sum_{j=1,\neq k}^{n}\varnothing_j x_j \\ \overline{y^w} \leq \sum_{j=1,\neq k}^{n}\varnothing_j y_j^w \\ \overline{y^b} \geq \sum_{j=1,\neq k}^{n}\varnothing_j y_j^b \\ \overline{x} \geq x_k, 0 \leq \overline{y^w} \leq y_k^w \\ \overline{y^b} \geq y_k^b, \varnothing \geq 0\end{cases} \quad (2)$$

In the above model: ρ denotes the Eff_{Carbon}; n denotes the numbers of city decision units; m denotes the numbers of city inputs; S_1 denotes the numbers of city desirable outputs; S_2 denotes the numbers of city undesirable outputs; x_{ij} denotes city inputs; y_{qj}^w denotes city desirable outputs; y_{qj}^b denotes city undesirable outputs; $\overline{x_i}$ denotes the slack variable of city inputs; $\overline{y^w}$ denotes the slack variable of city desirable outputs; $\overline{y^b}$ denotes the slack variable of city undesirable outputs; x_k denotes the maximum city inputs; y_k^w denotes the maximum slack variable of city desirable outputs; $\varnothing$ denotes the model vector weighting.

Consequently, in the selection of indicators, the number of urban labor forces employed, the stock of urban fixed capital investment, and coal consumption are used as input indicators. Regional GDP is the desired output. Urban CO_2 emissions are used as undesirable outputs. Finally, the indicators are brought into the SBM-DEA model to measure CEE.

3.3. Variable Selection

Explained variables: CEE (Eff_{Carbon}). In this paper, the SBM-DEA model [46] based on non-desired outputs is used to measure the CEE. Meanwhile, in order to ensure the robustness of the results, the EBM-DEA model [47] is also used in this paper to measure the CEE in the robustness test.

Explanatory variables: SCP (*Smart_Policy*). In this paper, we consider SCP as a quasi-natural experiment and construct interaction terms for the city dummy variable and the policy implementation time dummy variable to characterize the policy treatment effect of smart city policy. Specifically, if a city is classified as a smart city, let $Smart_{Policy} = 1$, otherwise take the value 0.

Control variables. In this paper, the following control variables are selected based on existing studies to consider other possible factors that may affect CEE: ① Urban economic development level (*economy*) is expressed by the logarithm of GDP per capita. ② Urban information infrastructure level (*infrastructure*) is expressed by the logarithm of the number of international internet users. ③ Urban financial development level (*finance*) is expressed by the logarithm of the balance of deposits and loans of financial institutions at the end of the year. ④ Urban environmental pollution intensity (*pollution*) is expressed by the logarithm of industrial sulfur dioxide emissions.

3.4. Data Description

The base data for this paper is the panel data of 281 cities in China from 2007 to 2020. The data sources are the China Statistical Yearbook, the China Urban Statistical Yearbook, and the China Environmental Statistical Yearbook. For data processing, linear interpolation was used to fill in individual missing values according to local government websites and statistical bulletins. The descriptive statistics of the main variables in this paper are shown in Table 1.

Table 1. Descriptive statistics.

Variables	N	Mean	SD	Min	Max
eff_carbon	3633	0.363	0.177	0.0713	1.484
economy	3628	10.52	0.685	4.595	13.06
finance	3633	17.15	1.198	13.82	21.69
infrastructure	3351	13.04	1.111	5.468	17.76
polution	3591	10.08	1.284	0.693	13.43
baidu index	3633	24.61	27.05	0	174
IU index	3633	6.479	0.359	5.517	7.657
marketization index	3633	10.95	2.697	3.351	19.69

4. Benchmark Analysis

Table 2 presents the regression results for the impact of SCP on CEE. Column (1) displays the estimation results without considering control variables or fixed effects. In Column (2), we observe the estimation results without considering control variables but controlling for city and year fixed effects. Finally, Column (3) shows the estimation results that consider the control variables as well as city and year fixed effects. According to the regression results, the coefficient of the *smart_policy* is positively significant at the 1% level in both scenarios. This indicates that SCP has an elevated effect on CEE of about 3.72% without considering control variables and fixed effects. In the case where only city-fixed effects and time-fixed effects are considered, the SCP boost to CEE is about 2.04%. Accordingly, the net policy effect of SCP on CEE improvement is 1.61% after excluding the confounding factors of city, time, and other variables. This suggests that SCP has a strong impact on CEE, indicating its robustness. Therefore, implementing SCP can greatly enhance CEE.

Table 2. Benchmark regression estimates.

Variables	(1)	(2)	(3)
smart_policy	0.0372 ***	0.0204 ***	0.0161 ***
	(5.802)	(3.157)	(2.616)
economy			0.0868 ***
			(3.456)
finance			0.0200
			(0.852)
infrastructure			−0.0191 ***
			(−2.590)

Table 2. Cont.

Variables	(1)	(2)	(3)
pollution			−0.0162 ***
			(−3.902)
City FE	NO	YES	YES
Year FE	NO	YES	YES
Observations	3633	3633	3329
R-squared	0.010	0.692	0.754

Notes: *** denotes significant at the 1% level.

5. Robustness Test

5.1. Parallel Trend Test

The multiple time-varying DID model assumes that there are no significant differences between the experimental and control groups prior to the implementation of the policy. In other words, the policy sample must satisfy parallel trends. Table 3 presents the hypothesis test results for the parallel trend analysis of SCP. The *policy_year* variable denotes the relative year of policy implementation, with *pre** indicating the period before policy implementation, *post** indicating the period after policy implementation, *policy_test* indicating the policy effect test coefficient, and *current* indicating the baseline period of policy implementation. Table 3 shows that the coefficients for the policy effect test were not significant or negative prior to the implementation of SCP. This indicates that there was no significant difference between the control and experimental groups before the policy was put in place. Therefore, the research sample passed the parallel trend hypothesis test.

Table 3. Parallel trend test.

Policy_Year	Policy_Test	Policy_Year	Policy_Test
pre5	−0.0079	*post3*	0.0030 **
	(−0.652)		(2.351)
pre4	−0.0134	*post4*	0.0110 *
	(−1.084)		(1.867)
pre3	−0.0105	*post5*	0.0149 **
	(−0.843)		(2.197)
pre2	−0.0106	*post6*	0.0273 **
	(−0.798)		(2.169)
Current	0.0143	*post7*	0.0306 *
	(1.213)		(1.801)
post1	0.0020 *	*post8*	0.0487 *
	(1.766)		(1.883)
post2	0.0055 *		
	(1.858)		
Observations	3633	Observations	3633
R-squared	0.693	R-squared	0.693

Notes: * denotes significant at the 10% level; ** denotes significant at the 5% level; The values in parentheses are the *t*-statistics of the parameters.

5.2. Placebo Test

To exclude bias in the estimation results of this paper due to the chance factors of policy sample setting, a placebo test was conducted. This study employs a placebo test by randomly selecting a sample of cities from the test area of the smart city policy. Additionally, the bootstrap method is utilized to perform 500 iterations of the random process to simulate the model and derive the model estimation results. Moreover, the paper presents a kernel density distribution plot of the estimated coefficients measuring the impact of SCP on CEE. The placebo test plots for the randomized policy sample are presented in Figure 1, respectively.

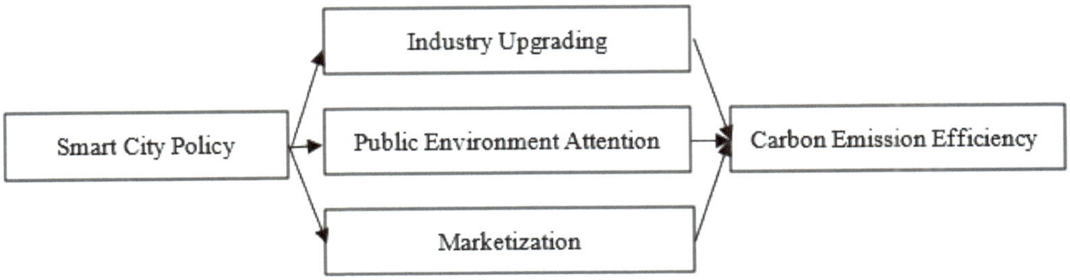

Figure 1. The mechanism effect of SCP on CEE.

Figure 2 displays a normal distribution of the estimated coefficients, where the mean value is insignificant and close to 0. Additionally, most of the *p*-values are higher than 0.1. Notably, the true estimated coefficient of SCP (0.0161) falls within the small probability range of the kernel density plot of the coefficients for the placebo test shown above. In other words, the impact of SCP on CEE is not a random occurrence. Therefore, the results presented in this paper remain reliable and valid.

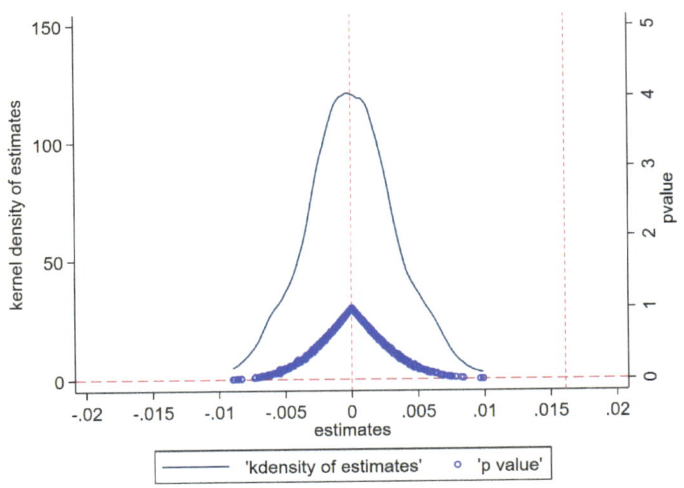

Figure 2. SCP placebo test.

5.3. Other Robustness Tests

To enhance the credibility and dependability of the findings in this research paper, we conducted four additional robustness tests: ① In order to exclude any interference caused by the different measurement methods of explanatory variables, the EBM-DEA model is used to estimate CEE, replacing the estimation obtained through the SBM-DEA model; ② Winsorizing at the 1% level for continuous variables involves excluding the effect of sample extremes to ensure that the analysis is not skewed by outliers; ③ To account for possible omitted variables that may affect CEE, two additional variables are included in the model: government fiscal intervention (*gfi*) and science and technology expenditure (*ste*). ④ In order to exclude the interference of model estimation errors due to policy lags, the explanatory variables are lagged separately in one and two periods using a lagged period model.

The results of the four robustness tests are presented in Table 4. The estimated coefficients remain largely consistent in direction and significance with the original benchmark model. Hence, the model findings in this paper are robust.

Table 4. Other Robustness Tests Estimation.

Variables	(1) EBM-DEA	(2) Winsorization	(3) *gfi*	(4) *ste*	(5) Lag 1 Period	(6) Lag 2 Periods
smart_policy	0.0215 ***	0.0134 **	0.0163 ***	0.0150 **	0.0177 ***	0.0161 **
	(3.812)	(2.274)	(2.667)	(2.435)	(2.783)	(2.562)
economy	0.106 ***	0.123 ***	0.0818 ***	0.0827 ***	0.0673 ***	0.0470 **
	(3.739)	(8.438)	(3.320)	(3.198)	(2.874)	(2.404)
finance	0.0251	−0.00353	0.00911	0.0168	−0.0239	−0.0448
	(1.243)	(−0.175)	(0.375)	(0.730)	(−0.965)	(−1.638)
infrastructure	−0.0147 **	−0.0232 ***	−0.0195 ***	−0.0194 ***	−0.00629	0.00626
	(−2.372)	(−3.297)	(−2.645)	(−2.634)	(−1.058)	(0.913)
pollution	−0.0136 ***	−0.0172 ***	−0.0161 ***	−0.0158 ***	−0.0118 ***	−0.00910 **
	(−3.687)	(−3.702)	(−3.853)	(−3.779)	(−3.070)	(−2.385)
City FE	YES	YES	YES	YES	YES	YES
Year FE	YES	YES	YES	YES	YES	YES
Observations	3329	3329	3329	3329	3028	2731
R-squared	0.755	0.745	0.755	0.755	0.782	0.800

Notes: ** denotes significant at the 5% level; *** denotes significant at the 1% level.

6. Heterogeneity Analysis

6.1. Regional Heterogeneity

Resource dependency theory suggests that the economic growth and development paths of cities often depend on the material resources (natural resources and land resources) and immaterial resources (human resources and technological resources) they possess. The resource endowments of cities in different regions of China vary widely. Cities located in various regions have distinct economic conditions and resources, which consequently impact CEE in diverse ways. Hence, to investigate potential differences in the policy effects of SCP in different regions, we divided our sample of cities into three regions based on the geographical and economic regions in China: eastern, central, and western. We then conducted a group regression test to analyze the policy effects within each region. Table 5 displays the estimated impact of SCP on CEE in the eastern, central, and western regions. The results are shown in columns (1), (2), and (3) for each respective region. The results show that the coefficient estimate of *smart_policy* in the sample of cities in the eastern region is 0.0322 and is significant at the 1% level. However, in the sample of cities in the central and western regions, the coefficient estimates of *smart_policy* are 0.00751 and 0.0106, respectively, and both fail to pass the significance test. The findings show that the impact of SCP on CEE varies across different regions. Specifically, SCP has a significant positive effect on CEE in the eastern region, while its impact on the central and western regions is insignificant. This shows a boosting effect of SCP on the CEE of the eastern regional cities of about 3.2%. One possible explanation for this phenomenon is that cities located in eastern China tend to be more economically developed and have a higher concentration of technology and skilled professionals, which allows them to effectively reduce carbon emissions [48]. Conversely, cities in central and western China may be less economically advanced and have limited access to technology and talent, which could result in less motivation to prioritize carbon reduction efforts.

Table 5. Heterogeneity analysis in different regions.

Variables	(1) East	(2) Central	(3) West
smart_policy	0.0322 ***	0.00751	0.0106
	(3.420)	(0.791)	(0.830)
economy	0.105 ***	0.108 ***	0.0814 *
	(4.569)	(4.109)	(1.683)
finance	0.0159	0.0363	0.0101
	(0.667)	(0.882)	(0.233)

Table 5. *Cont.*

Variables	(1) East	(2) Central	(3) West
infrastructure	−0.00535	−0.0332 ***	−0.0245
	(−0.579)	(−2.599)	(−1.610)
pollution	−0.0181 ***	−0.0163 **	−0.0159 **
	(−3.047)	(−2.346)	(−2.086)
City FE	YES	YES	YES
Year FE	YES	YES	YES
Observations	1188	1156	985
R-squared	0.800	0.735	0.722

Notes: * denotes significant at the 10% level; ** denotes significant at the 5% level; *** denotes significant at the 1% level.

6.2. Heterogeneity of Government Environmental Regulation

Different cities differ in their economic incentives, environmental enforcement efforts, carbon reduction policy goals, and technological innovation capabilities, thus leading to different government environmental regulation conditions. Disparities in government environmental regulations may result in varied approaches by firms to addressing environmental protection, ultimately influencing the effectiveness of carbon reduction efforts [49]. Therefore, to investigate the potential heterogeneous differences in the policy impact of SCP on cities with varying levels of environmental regulatory intensities, the city sample is divided into two groups based on their respective environmental regulatory intensities: a high intensity group and a low-intensity group. Table 6 displays the estimated impact of SCP on CEE for the two groups of cities. The results indicate that the estimated coefficient of *smart_policy* for cities with low environmental regulation intensity is 0.0297 and statistically significant at the 1% level. In other words, SCP improves the CEE of low-environmental-regulation cities by about 2.97%. However, the estimated coefficient for cities with high environmental regulatory intensity is not significant. This suggests that the high intensity of government environmental regulation has not contributed to the improvement of CEE, while the appropriate relaxation of environmental regulation has contributed to the improvement of the carbon emission situation. Hence, the implementation of strict environmental regulations can lead to a 'crowding-out effect' on regional environmental improvements, as local entities may respond strategically to environmental inspections. On the other hand, relaxed environmental regulations can have a 'compensating effect' on regional environmental improvements, as local entities may focus more on substantive clean technology upgrades to improve CEE.

Table 6. Heterogeneity analysis in different intensity of environmental regulation.

Variables	(1) Low Environmental Regulation	(2) High Environmental Regulation
smart_policy	0.0297 ***	0.00399
	(2.797)	(0.466)
economy	0.0845 ***	0.124 ***
	(3.709)	(5.225)
finance	0.0119	0.0128
	(0.345)	(0.320)
infrastructure	−0.0187	−0.0139 **
	(−1.607)	(−2.452)
pollution	−0.0138 **	−0.0164 ***
	(−2.286)	(−2.998)
City FE	YES	YES
Year FE	YES	YES
Observations	1617	1677
R-squared	0.744	0.830

Notes: ** denotes significant at the 5% level; *** denotes significant at the 1% level.

6.3. Heterogeneity of Green Finance

Green finance plays a crucial role in driving green technology innovation and has a profound impact on the proliferation and acceptance of clean technologies [50]. Cities with higher levels of green finance typically have access to more investment and financing support for green projects. These funds can be used to promote the development and application of renewable energy, energy efficiency improvements, and carbon reduction technologies. In contrast, cities with lower levels of green finance may face a shortage of funds and difficulty accessing the investment and financing support needed for green projects, thus limiting the implementation of carbon emission reductions. In order to analyze the impact of SCP on CEE in cities with varying levels of green finance, this study categorizes the sample into two groups: cities with low levels of green finance and cities with high levels of green finance. This paper utilizes the entropy value method to compute the green finance index of prefecture-level cities and employs it as a substitute variable to determine the level of green finance in cities based on existing studies. Table 7 displays the regression coefficients for the impact of SCP on CEE for the two city groups. The regression coefficient value for *smart_policy* in the low green finance level city group was −0.00356 which was not significant. However, the regression coefficient value for *smart_policy* in the high green finance level city group was 0.03000 and passed the significance test at the 1% level. Hence, the level of green finance in cities has a significant impact on the effectiveness of SCP in reducing carbon emissions. In terms of policy effects, the SCP has a CEE enhancement effect of about 3% for cities with high green finance levels. It is likely that a favorable green financial environment enables urban entities to have greater financial resources to invest in green innovation, which in turn increases their motivation and capacity for green innovation. As a result, CEE is improved.

Table 7. Heterogeneity analysis in different level of green finance.

Variables	(1) Low Green Finance	(2) High Green Finance
smart_policy	−0.00356	0.0300 ***
	(−0.429)	(2.693)
economy	0.0544 *	0.0885 ***
	(1.858)	(4.377)
finance	0.0725 *	−0.0397
	(1.935)	(−1.504)
infrastructure	−0.00821	−0.0272 *
	(−1.334)	(−1.882)
pollution	−0.00565	−0.0227 ***
	(−0.872)	(−3.639)
City FE	YES	YES
Year FE	YES	YES
Observations	1687	1632
R−squared	0.841	0.740

Notes: * denotes significant at the 10% level; *** denotes significant at the 1% level.

6.4. Heterogeneity of Officer Change

The turnover of officials is a critical component of a city's administrative system. It directly impacts the efficiency of the administration and, consequently, the effectiveness of policies implemented by the city [51]. Changes in officials may lead to policy adjustments and changes, especially in policy measures for environmental protection and carbon emission reduction. If each change of officials is accompanied by policy reformulation or changes, it will affect the planning and investment decisions of enterprises and residents on carbon emission reduction, thus reducing carbon emission efficiency. Hence, to investigate the impact of changes in city officials on the carbon reduction effect of SCP, this study categorizes cities into two groups: those with changes in officials and those without. Referring to existing studies, whether a change in mayor occurs is used as a proxy variable to examine the heterogeneity of change in city officials. Table 8 displays the regression coefficients for both the officer-changed city group and the officer-not-changed city group in columns (1) and (2), respectively. According to Table 8, the coefficient of *smart_policy* for cities without official changes is 0.0183, which is statistically significant at the 5% level. However, for cities with official changes, the value of the *smart_policy* coefficient is 0.00999 and does not pass the significance test. It can be observed that SCP exerts a CEE-enhancing policy effect on office-not-changed cities, with an enhancement effect of 1.83%. The findings indicate that SCP has a positive impact on CEE in cities without official changes but does not result in significant improvements in CEE in cities with official changes. The reason for the importance of a city's principal officials is that they have a significant impact on the policy stability and continuity of the entire city. A stable policy environment enhances governing efficiency and policy effectiveness, which in turn strengthens the policy effect of SCP. This ultimately leads to an improvement in the city's CEE.

Table 8. Heterogeneity analysis in different turnovers of officer change.

Variables	(1) Officer-Not-Changed	(2) Officerchanged
smart_policy	0.0183 ** (2.538)	0.00999 (0.706)
economy	0.0723 *** (2.610)	0.131 *** (4.611)
finance	0.0420 (1.295)	−0.0397 (−1.146)
infrastructure	−0.0191 ** (−2.393)	−0.0194 (−1.010)
pollution	−0.0148 *** (−2.857)	−0.0208 ** (−2.368)
City FE	YES	YES
Year FE	YES	YES
Observations	2290	1003
R-squared	0.769	0.764

Notes: ** denotes significant at the 5% level; *** denotes significant at the 1% level.

7. Mechanism Analysis

7.1. Industrial Upgrading

According to the mechanism analysis in the previous section, SCP affects CEE through industrial upgrading. This paper utilizes the industrial upgrading index (IU index) as a proxy for the industrial upgrading mechanism and examines the relationship between *smart_policy* and the *IU index* by constructing the interaction term. Table 9 presents the regression results of the industrial upgrading mechanism. In column (1), the regression coefficient of the *smart_policy*IU index* is 0.217 and is statistically significant at the 1% level. The model estimation results show that SCP has led to a 21.7% increase in CEE because of the effect of industrial upgrading. This indicates that the smart city policy promotes carbon efficiency through the industrial upgrading mechanism. Therefore, the industrial upgrading mechanism is validated.

Table 9. Mechanism Analysis of SCP on CEE.

Variables	(1) Industry Upgrading	(2) Public Environmental Attention	(3) Marketization
smart_policy	−0.438 *** (−4.647)	−0.0123 (−1.600)	−0.0405 (−1.555)
*smart_policy*IU index*	0.217 *** (4.822)		

Table 9. *Cont.*

Variables	(1) Industry Upgrading	(2) Public Environmental Attention	(3) Marketization
*smart_policy*baidu index*		0.0233 ***	
		(6.220)	
*smart_policy*marketization index*			0.0288 **
			(2.269)
economy	0.0900 ***	0.0875 ***	0.0872 ***
	(3.465)	(3.475)	(3.464)
finance	0.0260	0.0229	0.0183
	(1.106)	(0.980)	(0.781)
infrastructure	−0.0140 **	−0.0114 *	−0.0186 **
	(−1.969)	(−1.656)	(−2.512)
polution	−0.0153 ***	−0.0138 ***	−0.0162 ***
	(−3.711)	(−3.383)	(−3.888)
City FE	YES	YES	YES
Year FE	YES	YES	YES
Observations	3329	3329	3329
R-squared	0.757	0.758	0.755

Notes: * denotes significant at the 10% level; ** denotes significant at the 5% level; *** denotes significant at the 1% level.

7.2. Public Environmental Attention

In order to verify the role of public environmental attention in the mechanism, this paper uses the *Baidu index* as a proxy variable for public environmental attention. To measure the public environmental concern of each city, we utilized China's largest Chinese search engine and searched for the term 'environmental protection'. We set the search time and location and used Python web crawler technology to capture the frequency of urban environmental protection words.

Column (2) of Table 9 reports the regression results of the mechanism of action of public environmental attention. It can be shown that the regression coefficient value for the *smart_policy*baidu index* is 0.0233 and is significant at the 1% level. From the results of the estimated parameters of the model, it is easy to conclude that SCP has increased the CEE by 2.33% through the mechanism of public environmental attention. In other words, SCP has a catalytic effect on CEE by increasing public environmental concern. Hence, the public environmental attention mechanism passed validation.

7.3. Marketization

To investigate the impact of SCP on CEE using marketization mechanisms, this study constructs interaction terms between marketization mechanism variables and *smart_policy* and includes them in the benchmark regression. According to previous studies, the *marketization index* is used as a proxy variable for the marketization mechanism to measure the marketization level of each city.

Column (3) of Table 9 reports the regression results for the market-based mechanism. The regression coefficient of the *smart_policy*marketization index* is 0.0288 and is significant at the 5% level. From the results of the Marketization mechanism test, the increase in the level of urban marketization contributes about 2.88% to the CEE. This shows that marketization mechanisms do play a catalytic role in the carbon reduction process of SCP. In other words, the marketization mechanism passed validation.

8. Conclusions and Policy Implications

8.1. Conclusions

This paper evaluates the impact of China's SCP on CEE and its role in promoting green development and sustainable cities. The study uses Chinese city-level data from 2007 to 2020 and a quasi-natural experiment approach, employing a multiple-time-varying DID method. The findings of the study are presented below:

First, SCP significantly improves CEE. From the model estimation results, the implementation of SCP resulted in a 1.61% improvement in CEE after excluding relevant confounding factors. This finding holds after parallel trend tests, placebo tests, and other robustness tests.

Second, there is heterogeneity in the impact of SCP on CEE. In terms of city regions, SCP significantly contributes to CEE in eastern cities while having no impact on western and central cities. In terms of the environmental regulatory intensity of cities, SCP promotes CEE in cities with low environmental regulatory intensity. Additionally, among cities with different levels of green finance, the pilot policy exerts a significant carbon reduction effect in cities with high levels of green finance. Moreover, the implementation of SCP led to a greater improvement in CEE in cities without official changes compared to those with official changes.

Third, SCP acts on CEE through three mechanisms: industrial upgrading, public environmental attention, and marketization. Specifically, the impact of SCP on CEE is positively correlated with industrial upgrading, public environmental attention, and the level of marketization.

To our best knowledge, this study is the first to examine how SCP affects the CEE. Furthermore, we identify the paths (industry upgrading, public environmental attention, and marketization) of action for the policy. The results of this study may provide policymakers with clear operational directions and policy suggestions on how to increase the efficiency of urban carbon emissions.

8.2. Policy Implications

Firstly, Policymakers should prioritize summarizing and promoting the positive effects of carbon reduction in SCP. SCP plays a crucial role in improving energy efficiency, promoting eco-friendly practices, and achieving sustainable and innovative development. Therefore, it is important to leverage the potential of smart city policies to address environmental challenges and promote green growth.

Moreover, to enhance CEE, policymakers must prioritize industrial upgrading, public environmental attention, and marketization in smart city policies. It is crucial to acknowledge the significant role played by these factors and actively create channels for smart city policies to contribute to this goal. Specifically, to drive economic green growth, we recommend promoting the consolidation of industrial elements, fostering new industries while revamping traditional ones, and prioritizing industrial upgrading. Additionally, we suggest increasing public awareness of environmental protection, encouraging green consumption, facilitating access to information, and promoting environmental concerns. To achieve efficient resource allocation, we propose stimulating innovation-driven development, implementing a carbon emission trading system, and enhancing marketization.

Lastly, to ensure effective implementation of SCP, policymakers must consider the varying effects of such policies on different cities. It is crucial to optimize the implementation details and operational guidelines of SCP based on the unique characteristics of each city. City managers should focus on classification management and policy relevance when promoting smart city construction. The first point is to allocate more resources to central and western cities, focusing on improving the CEE of these cities with poor economies and green technologies. The second point is to promote greater autonomy among enterprises and other market players by reducing government administrative intervention in environmental regulation. The third objective is to strengthen the development of urban green finance, increase the funding available for environmental innovation markets, and encourage green innovation. The fourth point is to minimize the turnover of officials in order to maintain policy stability and continuity, ultimately improving the efficiency of policy operations.

8.3. Future Research Perspectives

This paper is the first to investigate the impact of smart city policies on urban carbon emission efficiency. However, there are still some shortcomings. Future researchers can start their research in the following two directions: ① Expand the measurement of carbon emission efficiency indicators. In this paper, when measuring the input indicators of carbon emission efficiency, coal consumption was mainly selected among the fuel types. Other fuel sources, such as natural gas, are not included in the analysis. Scholars can consider more fuel sources in the measurement of input indicators in order to obtain more accurate measurement results. ② Extending the scope of the city sample. The sample of smart city policies in this paper is from Chinese cities, and whether the findings are equally applicable to cities in other developing and developed countries is subject to further validation.

Author Contributions: Conceptualization, X.X. and S.Z.; Methodology, X.X.; Software, X.X.; Formal analysis, X.X., R.Y. and S.Z.; Investigation, R.Y.; Resources, S.Z.; Data curation, R.Y.; Writing—original draft, X.X.; Visualization, R.Y.; Supervision, S.Z.; Project administration, S.Z. All authors have read and agreed to the published version of the manuscript.

Funding: This research received no external funding.

Data Availability Statement: Data is unavailable due to privacy.

Conflicts of Interest: The authors declare no conflict of interest.

References

1. Tonne, C.; Adair, L.; Adlakha, D.; Anguelovski, I.; Belesova, K.; Berger, M.; Brelsford, C.; Dadvand, P.; Dimitrova, A.; Giles-Corti, B. Defining pathways to healthy sustainable urban development. *Environ. Int.* **2021**, *146*, 106236. [CrossRef] [PubMed]
2. Wen, H.; Liang, W.; Lee, C.-C. China's progress toward sustainable development in pursuit of carbon neutrality: Regional differences and dynamic evolution. *Environ. Impact Assess. Rev.* **2023**, *98*, 106959. [CrossRef]
3. Lin, B.; Li, Z. Towards world's low carbon development: The role of clean energy. *Appl. Energy* **2022**, *307*, 118160. [CrossRef]
4. Nie, Y.; Zhang, G.; Duan, H.; Su, B.; Feng, Y.; Zhang, K.; Gao, X. Trends in energy policy coordination research on supporting low-carbon energy development. *Environ. Impact Assess. Rev.* **2022**, *97*, 106903. [CrossRef]
5. Liu, Z.; Deng, Z.; He, G.; Wang, H.; Zhang, X.; Lin, J.; Qi, Y.; Liang, X. Challenges and opportunities for carbon neutrality in China. *Nat. Rev. Earth Environ.* **2022**, *3*, 141–155. [CrossRef]
6. He, J.; Li, Z.; Zhang, X.; Wang, H.; Dong, W.; Du, E.; Chang, S.; Ou, X.; Guo, S.; Tian, Z. Towards carbon neutrality: A study on China's long-term low-carbon transition pathways and strategies. *Environ. Sci. Ecotechnol.* **2022**, *9*, 100134. [CrossRef]
7. Blasi, S.; Ganzaroli, A.; De Noni, I. Smartening sustainable development in cities: Strengthening the theoretical linkage between smart cities and SDGs. *Sustain. Cities Soc.* **2022**, *80*, 103793. [CrossRef]
8. Zubizarreta, I.; Seravalli, A.; Arrizabalaga, S. Smart city concept: What it is and what it should be. *J. Urban Plan. Dev.* **2016**, *142*, 04015005. [CrossRef]
9. Caragliu, A.; Del Bo, C.F. Smart innovative cities: The impact of Smart City policies on urban innovation. *Technol. Forecast. Soc. Chang.* **2019**, *142*, 373–383. [CrossRef]
10. Chu, Z.; Cheng, M.; Yu, N.N. A smart city is a less polluted city. *Technol. Forecast. Soc. Chang.* **2021**, *172*, 121037. [CrossRef]
11. Al Sharif, R.; Pokharel, S. Smart city dimensions and associated risks: Review of literature. *Sustain. Cities Soc.* **2022**, *77*, 103542. [CrossRef]
12. Russo, F.; Rindone, C. Smart city for sustainable development: Applied processes from SUMP to MaaS at European level. *Appl. Sci.* **2023**, *13*, 1773. [CrossRef]
13. Panuccio, P. Smart planning: From city to territorial system. *Sustainability* **2019**, *11*, 7184. [CrossRef]
14. Duygan, M.; Fischer, M.; Pärli, R.; Ingold, K. Where do Smart Cities grow? The spatial and socio-economic configurations of smart city development. *Sustain. Cities Soc.* **2022**, *77*, 103578. [CrossRef]
15. Yan, D.; Liu, H.; Yao, P. Assessing Energy Efficiency for Economic and Sustainable Development in the Region of European Union Countries. *Front. Environ. Sci.* **2021**, *9*, 779163. [CrossRef]
16. Lagodiienko, V.; Popelo, O.; Zybareva, O.; Samiilenko, H.; Mykytyuk, Y.; Ahmed Saif Alsawwafi, F.M. Peculiarities of the Management of the Foreign Economic Activity of Enterprises in Current Conditions of Sustainability. *Int. J. Sustain. Dev. Plan.* **2022**, *17*, 1215–1223. [CrossRef]
17. Baraniewicz-Kotasińska, S. Smart city. Four approaches to the concept of understanding. *Urban Res. Pract.* **2022**, *15*, 397–420. [CrossRef]
18. Westraadt, L.; Calitz, A. A modelling framework for integrated smart city planning and management. *Sustain. Cities Soc.* **2020**, *63*, 102444. [CrossRef]

19. Mondschein, J.; Clark-Ginsberg, A.; Kuehn, A. Smart cities as large technological systems: Overcoming organizational challenges in smart cities through collective action. *Sustain. Cities Soc.* **2021**, *67*, 102730. [CrossRef]
20. Ma, D.; Zhu, Q. Innovation in emerging economies: Research on the digital economy driving high-quality green development. *J. Bus. Res.* **2022**, *145*, 801–813. [CrossRef]
21. Gao, K.; Yuan, Y. Is the sky of smart city bluer? Evidence from satellite monitoring data. *J. Environ. Manag.* **2022**, *317*, 115483. [CrossRef] [PubMed]
22. Batabyal, A.A.; Beladi, H. The optimal provision of information and communication technologies in smart cities. *Technol. Forecast. Soc. Chang.* **2019**, *147*, 216–220. [CrossRef]
23. Liu, K.; Meng, C.; Tan, J.; Zhang, G. Do smart cities promote a green economy? Evidence from a quasi-experiment of 253 cities in China. *Environ. Impact Assess. Rev.* **2023**, *99*, 107009. [CrossRef]
24. Zhang, X.; Wang, D.; Zhang, Y.; Manokaran, K.B.; Antony, A.B. IoT driven framework based efficient green energy management in smart cities using multi-objective distributed dispatching algorithm. *Environ. Impact Assess. Rev.* **2021**, *88*, 106567.
25. Kim, J. Smart city trends: A focus on 5 countries and 15 companies. *Cities* **2022**, *123*, 103551. [CrossRef]
26. Quan, C.; Cheng, X.; Yu, S.; Ye, X. Analysis on the influencing factors of carbon emission in China's logistics industry based on LMDI method. *Sci. Total Environ.* **2020**, *734*, 138473. [CrossRef]
27. Gao, M.; Yang, H.; Xiao, Q.; Goh, M. A novel method for carbon emission forecasting based on Gompertz's law and fractional grey model: Evidence from American industrial sector. *Renew. Energy* **2022**, *181*, 803–819. [CrossRef]
28. Wang, Q.; Su, M. Drivers of decoupling economic growth from carbon emission–an empirical analysis of 192 countries using decoupling model and decomposition method. *Environ. Impact Assess. Rev.* **2020**, *81*, 106356. [CrossRef]
29. Ropo, M.; Mustonen, H.; Knuutila, M.; Luoranen, M.; Kosonen, A. Considering embodied CO_2 emissions and carbon compensation cost in life cycle cost optimization of carbon-neutral building energy systems. *Environ. Impact Assess. Rev.* **2023**, *101*, 107100. [CrossRef]
30. Xiao, X.; Xie, C. Rational planning and urban governance based on smart cities and big data. *Environ. Technol. Innov.* **2021**, *21*, 101381. [CrossRef]
31. Xu, N.; Ding, Y.; Guo, J. Do Smart City policies make cities more innovative: Evidence from China. *J. Asian Public Policy* **2022**, *15*, 1–17. [CrossRef]
32. Kamyab, H.; Klemeš, J.J.; Van Fan, Y.; Lee, C.T. Transition to sustainable energy system for smart cities and industries. *Energy* **2020**, *207*, 118104. [CrossRef]
33. Qi, W.; Shen, Z.J.M. A smart-city scope of operations management. *Prod. Oper. Manag.* **2019**, *28*, 393–406. [CrossRef]
34. Razmjoo, A.; Gandomi, A.H.; Pazhoohesh, M.; Mirjalili, S.; Rezaei, M. The key role of clean energy and technology in smart cities development. *Energy Strategy Rev.* **2022**, *44*, 100943. [CrossRef]
35. Liu, X.-M.; Zhang, Y.-Q. Study on the impact of intelligent city pilot on green and low-carbon development. *Environ. Sci. Pollut. Res.* **2023**, *30*, 57882–57897. [CrossRef]
36. Heitlinger, S.; Bryan-Kinns, N.; Comber, R. The right to the sustainable smart city. In Proceedings of the 2019 CHI Conference on Human Factors in Computing Systems, New York, NY, USA, 4–9 May 2019; pp. 1–13.
37. Razmjoo, A.; Østergaard, P.A.; Denai, M.; Nezhad, M.M.; Mirjalili, S. Effective policies to overcome barriers in the development of smart cities. *Energy Res. Soc. Sci.* **2021**, *79*, 102175. [CrossRef]
38. Tang, C.; Han, Y.; Ng, P. Green consumption intention and behavior of tourists in urban and rural destinations. *J. Environ. Plan. Manag.* **2022**, *66*, 2126–2150. [CrossRef]
39. Gu, Y.; Ho, K.-C.; Xia, S.; Yan, C. Do public environmental concerns promote new energy enterprises' development? Evidence from a quasi-natural experiment. *Energy Econ.* **2022**, *109*, 105967. [CrossRef]
40. Huang, K.; Luo, W.; Zhang, W.; Li, J. Characteristics and problems of smart city development in China. *Smart Cities* **2021**, *4*, 1403–1419. [CrossRef]
41. Gao, K.; Yuan, Y. The effect of innovation-driven development on pollution reduction: Empirical evidence from a quasi-natural experiment in China. *Technol. Forecast. Soc. Chang.* **2021**, *172*, 121047. [CrossRef]
42. Zhang, X.; Fan, D. Collaborative emission reduction research on dual-pilot policies of the low-carbon city and smart city from the perspective of multiple innovations. *Urban Clim.* **2023**, *47*, 101364.
43. Beck, T.; Levine, R.; Levkov, A. Big bad banks? The winners and losers from bank deregulation in the United States. *J. Financ.* **2010**, *65*, 1637–1667. [CrossRef]
44. Liu, X.; Li, Y.; Chen, X.; Liu, J. Evaluation of low carbon city pilot policy effect on carbon abatement in China: An empirical evidence based on time-varying DID model. *Cities* **2022**, *123*, 103582. [CrossRef]
45. Russo, F.; Rindone, C. *Data Envelopment Analysis (DEA) for Evacuation Planning*; WIT Press: Southampton, UK, 2010.
46. Tone, K. A slacks-based measure of efficiency in data envelopment analysis. *Eur. J. Oper. Res.* **2001**, *130*, 498–509. [CrossRef]
47. Tone, K.; Tsutsui, M. An epsilon-based measure of efficiency in DEA–A third pole of technical efficiency. *Eur. J. Oper. Res.* **2010**, *207*, 1554–1563. [CrossRef]
48. Tao, Y.; Li, F.; Crittenden, J.; Lu, Z.; Ou, W.; Song, Y. Measuring urban environmental sustainability performance in China: A multi-scale comparison among different cities, urban clusters, and geographic regions. *Cities* **2019**, *94*, 200–210. [CrossRef]
49. Han, Y. Impact of environmental regulation policy on environmental regulation level: A quasi-natural experiment based on carbon emission trading pilot. *Environ. Sci. Pollut. Res.* **2020**, *27*, 23602–23615. [CrossRef]

50. Irfan, M.; Razzaq, A.; Sharif, A.; Yang, X. Influence mechanism between green finance and green innovation: Exploring regiona policy intervention effects in China. *Technol. Forecast. Soc. Chang.* **2022**, *182*, 121882. [CrossRef]
51. Ji, X.; Liu, S.; Lang, J. Assessing the impact of officials' turnover on urban economic efficiency: From the perspective of politica promotion incentive and power rent-seeking incentive. *Socio-Econ. Plan. Sci.* **2022**, *82*, 101264. [CrossRef]

land

Article

The Effect of Public Traffic Accessibility on the Low-Carbon Awareness of Residents in Guangzhou: The Perspective of Travel Behavior

Qingyin Li †, Meilin Dai †, Yongli Zhang and Rong Wu *

School of Architecture and Urban Planning, Guangdong University of Technology, Guangzhou 510062, China; 2112110025@mail2.gdut.edu.cn (Q.L.); 3219008409@mail2.gdut.edu.cn (M.D.); 2112110048@mail2.gdut.edu.cn (Y.Z.)
* Correspondence: wurong@gdut.edu.cn
† These authors contributed equally to this work.

Abstract: The demand for transportation among urban residents in China is increasing in tandem with the nation's population growth, rising consumption levels, and increasing car ownership rates. Breaking the existing high-carbon travel practices and reshaping positive low-carbon awareness represents an inevitable way to change existing transportation structures and reduce urban traffic congestion and carbon emissions. A mediating effect model was employed and we found that community satisfaction is an essential variable in the effect of traffic accessibility and travel behavior on low-carbon awareness. First, the impact of residents' zero and low-carbon actions on their low-carbon awareness is mediated by community satisfaction. Furthermore, compared to high-income groups, community satisfaction exerts a robust mediating influence on low-income groups. The mediating effect of community satisfaction on the relationship between residential proximity to commercial centers and low-carbon awareness among individuals with low incomes is evident. Based on these findings, this paper explores the heterogeneity and associated measures of low-carbon awareness among residents. The conclusion of this study provides suggestions to promote residents' low-carbon awareness by improving their travel experience from the perspective of community construction, providing scientific reference and a basis for the formulation of transportation policies for low-carbon city construction.

Keywords: low-carbon awareness; travel behavior; public traffic accessibility; community satisfaction; mediating effect

Citation: Li, Q.; Dai, M.; Zhang, Y.; Wu, R. The Effect of Public Traffic Accessibility on the Low-Carbon Awareness of Residents in Guangzhou: The Perspective of Travel Behavior. *Land* **2023**, *12*, 1910. https://doi.org/10.3390/land12101910

Academic Editor: Yan Li

Received: 18 August 2023
Revised: 8 September 2023
Accepted: 25 September 2023
Published: 11 October 2023

1. Introduction

The population, consumption level, and car ownership level of Chinese cities are rapidly increasing, and the transportation demand and car travel proportion of urban residents are also increasing [1]. However, the rapid growth of urban transportation demand and the widespread use of cars have introduced many problems and challenges to Chinese cities, such as environmental pollution, traffic congestion, and carbon emissions [2–4]. The level of carbon emissions attributed to the transportation sector is closely related to the level of economic development. The proportion of carbon emissions in the transportation sector of developed countries to total emissions is generally high, e.g., 33% in the United States and 25% in the United Kingdom [5]. However, developing countries such as China are experiencing rapid development, and their transportation demand and carbon emissions from the transportation sector are rapidly increasing [1]. In September 2020, at the 75th United Nations General Assembly, the Chinese government announced that China would strive to achieve a national peak in carbon emissions by 2030 and achieve "carbon neutrality" by 2060 [6]. In order to avoid sprawl and car-related pollution, residents need to be guided to shift from a high-carbon car-oriented mode of travel to a low-carbon mode of transport oriented towards public transport, walking, and cycling [7]. The United Nations decided to designate 3 June as World Bicycle Day [8], and the original intention of this decision is that promoting cycling can help residents understand the benefits of low-carbon action and increase their low-carbon awareness. Therefore, methods to enhance residents' low-carbon awareness and guide them to use low-carbon transportation have become issues of concern for policymakers. To formulate effective policies to enhance residents' low-carbon awareness and promote low-carbon travel methods, it is necessary to deeply understand the factors that affect low-carbon awareness, which will help further reduce carbon emissions in a reasonable and effective manner.

Previous research on the relationship between low-carbon awareness and behavior mainly focused on the micro scales, involving aspects such as travel demand [9], travel modes [10], urban form [11], and individual characteristics [12]. Meanwhile, theories from various disciplines [13–16] have been used to explore the relationship between awareness and behavior. Many studies have shown that residents' low-carbon awareness, to some extent, affects their travel modes [17–19]. For example, residents with strong low-carbon awareness tend to choose more electric vehicles and promote sustainable consumption [20,21]. Meanwhile, in recent years, scholars have noted that behaviors may influence residents' subjective awareness to a certain extent and that the more that residents perceive low-carbon travel to be convenient, the greater the likelihood that they will choose low-carbon travel, with a consequent increase in low-carbon awareness [9]. Although scholars have conducted a series of explorations on the impact mechanism between low-carbon awareness and travel behavior [18,22,23], there are still some limitations. First, there is insufficient discussion on the mediating effect between travel behavior and low-carbon awareness. Research has found that improving the traffic environment and community services can increase residents' community satisfaction and further affect residents' travel methods and low-carbon awareness [9,24]. Second, considering the influencing factors of low-carbon awareness is not comprehensive enough; there is a lack of extensive exploration of low-carbon awareness from both internal and external factors. In addition, the specific definition of low-carbon awareness still needs to be improved. In an endeavor to fill these gaps, on the one hand, this study provides a concrete definition of low-carbon awareness from the perspective of values, attitudes and knowledge. On the other hand, in order to comprehensively explore the impact mechanism between low-carbon awareness and travel behavior, this paper considers the impact of public traffic accessibility and explores the potential mediating effect of community satisfaction. In addition, this paper provides a scientific and reasonable reference for adjusting low-carbon policies, low-carbon-oriented community construction, and optimizing traffic management measures in Guangzhou. The findings in this paper have significant theoretical and practical significance for promoting low-carbon awareness among urban residents and constructing a low-carbon environment.

The rest of this paper is organized as follows: Section 2 provides a review of the existing literature pertaining to low-carbon behavior, low-carbon awareness, and influencing factors. Section 3 highlights the data resources and modeling methods used in this paper. Section 4 presents a mediating effect model of the effect of public traffic accessibility and travel behavior on low-carbon awareness. Section 5 presents the discussion and conclusion derived from the findings and policy proposals for the future.

2. Literature Review

2.1. Low-Carbon Awareness and Low-Carbon Behavior

Low-carbon awareness is an abstract concept lacking a specific definition, but some scholars try to describe environmental awareness as a multidimensional structure [25,26]. It is generally believed that there are three prerequisites for low-carbon awareness, namely values, attitudes, and knowledge. Values mainly represent environmental values, that is, a person's ecological worldview [27], which can be divided into biospheric, anthropocentric, and self-worth [28,29]. Low-carbon attitudes are mainly aimed at low-carbon issues and issue-related things. According to previous studies, people with good low-carbon attitudes are more inclined to participate in environmental protection activities [30,31]. Low-carbon knowledge is considered to be an important basis for low-carbon awareness, and a large number of studies have explored the important role of low-carbon knowledge in the formation of low-carbon awareness [32–34]. Based on previous research, this study defines low-carbon awareness as a state of the combined action of values, attitudes, and knowledge that tends to reduce energy consumption and carbon emissions, ultimately promoting environmental sustainability.

Low-carbon behaviors include private low-carbon behaviors and public low-carbon behaviors. Private low-carbon behaviors mainly include low-carbon consumption behaviors and the use and processing of products and services that have a positive impact on the environment [14]. Public low-carbon behaviors mainly affect the environment indirectly through public policies or the low-carbon behaviors of others, such as supporting environmental policies and encouraging others to participate in environmental protection activities [21]. Based on the definition and classification of low-carbon behaviors in the literature, low-carbon behaviors in this study specifically refer to low-carbon travel behaviors, that is, actively adopting transportation modes that can reduce CO_2 emissions during travel, such as taking buses, subways, cycling, and walking. A study has shown that urban residents can usually take three measures to achieve the purpose of low-carbon travel: changing travel modes, shortening travel distances, and reducing the number of long-distance travel events [35]. The low-carbon travel behavior of urban residents is an important prerequisite to realize low-carbon transportation and sustainable development in the urban sector.

Low-carbon awareness and low-carbon behaviors play an important role in promoting environmentally friendly development, but there have been debates on the correlation between the two factors [36–38]. Scholars in many fields have used the theory of planned behaviour [13], value–belief theory [14], the attitude–behavior–external conditions model [15], and the distributed cognition theory [16] to explore the relationship between traveling behaviors and low-carbon awareness. Residents' low-carbon awareness is vital in travel mode choice [20,39]. Increased environmental awareness, such as recognizing the dangers of environmental pollution, can reduce residents' choice of private car travel [37]. Most individuals think lowering carbon emissions is a far-off issue since they do not think it will immediately influence their lives [22,23]. We aim to investigate what role a community's built environment plays in influencing awareness of low-carbon behavior and to promote a positive public understanding of carbon reduction from the perspective of behavior guidance from low-carbon-oriented community construction. In addition, most studies focus on awareness determinism, with relatively little discussion on the impact of behavior on awareness. More research needs to be conducted on the effect of travel behavior on low-carbon awareness. Some scholars have noted that behavior may influence residents' subjective awareness to a certain extent, and when residents realize that low-carbon travel is more convenient, they are more likely to choose low-carbon travel, thus increasing low-carbon awareness [9]. This paper considers the influence of travel behavior on low-carbon awareness to explore further the deeper relationship between travel behavior and low-carbon awareness.

2.2. Influencing Factors of Low-Carbon Awareness

The factors affecting residents' low-carbon awareness can be categorized into internal and external factors. Internal factors include the individual's psychological state, values, low-carbon knowledge, and personal norms. Some studies believe that an individual's low-carbon awareness is shaped by their values and beliefs about the environment [38]. Varela-Candamio et al. show that improving citizens' knowledge of global warming can effectively improve citizens' low-carbon awareness [40]. External factors include the natural environment (e.g., geographical location, natural resources, and climatic conditions) [41] and the social environment (e.g., infrastructure, policy guidelines, cultural history, and interpersonal relationships) [42,43]. It is generally believed that external factors have a stimulating effect on the low-carbon awareness of the residents [44,45]. Taking interpersonal relationships as an example, some studies have shown that people are generally beneficially influenced by the low-carbon awareness of family members, friends, and colleagues [46,47]. In addition, traffic accessibility significantly affects residents' low-carbon awareness. Urban infrastructure and spatial patterns guide residents' transportation and housing choices and influence their travel behavior patterns [48]. Zhang et al. found that the closer a residential area is to a bus stop, the more residents in the area tend to adopt public transportation and low-carbon lifestyles [49].

Theodori believes that community satisfaction refers to residents' satisfaction with community functions and the quality of commercial, childcare, medical, and other public services around their homes [50]. It has been shown that community satisfaction affects residents' travel behavior and low-carbon awareness to a certain extent. Improving the transport environment and public transport support can encourage urban residents to choose low-carbon traveling modes [9]. Community services and a high-quality environment can improve residents' low-carbon awareness and knowledge [24].

Significantly, in previous studies, researchers mainly conducted an in-depth analysis of the influencing factors of low-carbon awareness in the fields of psychology and social economics. However, there are few studies on the impact of the external environment such as an environment built upon low-carbon awareness. Enough attention should be paid to comprehensively measure the influencing factors of low-carbon awareness in the future to improve residents' low-carbon awareness and promote the sustainable development of cities. In addition, existing studies have yet to explore whether there is a mediating role between low-carbon awareness and travel behavior. Therefore, this study investigated whether the community satisfaction variable mediates public traffic accessibility and travel behavior in low-carbon awareness.

3. Materials and Methodology

3.1. Case Study

Guangzhou, in the south-central region of Guangdong Province, enjoys a prime urban location. It is one of the major cities in the Greater Bay Area of Guangdong, Hong Kong and Macao. In this study, 30 sample communities located in the urban core of Guangzhou were selected. These communities are Huangpu, Haizhu, Baiyun, Liwan, Yuexiu, Tianhe and Panyu (Figure 1). Guangzhou's urbanization rate increased from 25.71% at the end of 1984 to 86.48% in 2022 due to the city's rapid economic development [51]. In recent years, Guangzhou has controlled the growth rate of automobile ownership to gain the time and space needed to implement a transit-priority strategy. According to recent data, as of the year 2022, the population of permanent residents in Guangzhou reached a total of 10.12 million individuals. Additionally, it has been projected that the number of cars in Guangzhou will amount to 3,309,000 units [52]. Since 2012, the operational distance of the Guangzhou Metro has witnessed a notable surge, escalating from 236 to 611 km, thereby surpassing the objective of doubling the mileage [52]. The central city of Guangzhou possesses a more concentrated transport road network. It assumes a more significant number of transport functions, rendering the study of the central city more representative compared to the peripheral areas. This study explores the specific influence mechanisms of public traffic accessibility, travel behavior, low-carbon awareness, and community satisfaction on carbon emissions from transportation. As such, it can be a valuable resource for government departments in formulating carbon reduction strategies.

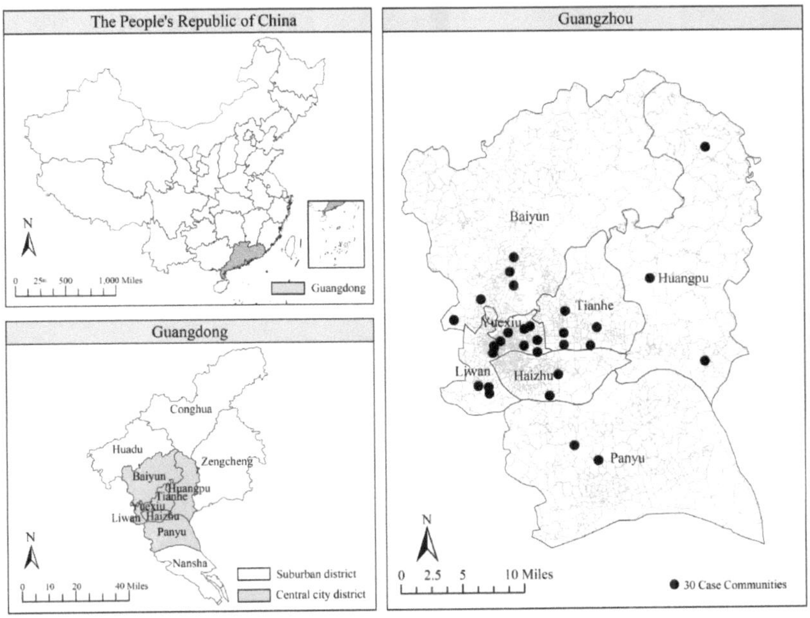

Figure 1. The locations of Guangzhou and their studied communities in the central city district.

3.2. Data Sources

The primary data source in this study was derived from the "Guangzhou Residents' Green and Low-Carbon Travel Survey", a questionnaire-based research initiative conducted between July and September 2022 across 30 communities in Guangzhou. A stratified probability-proportional scale sampling (PPS) technique was employed to ensure the representativeness and typicality of the research. This method involved randomly selecting 30 sample communities that possess the characteristics of the central urban area of Guangzhou. These districts encompass the six major types of residential areas found in China. A random selection of 30–70 households was made from each chosen neighborhood using equidistant sampling based on the door number information gathered during the pre-survey—the survey instrument comprised two distinct sections. The initial section of the questionnaire gathered fundamental data regarding the sociodemographic attributes of the participants, encompassing details such as age, education, gender, and income. Furthermore, the questionnaire encompassed an examination of the travel behavior of the residents, their level of satisfaction with the community, and their awareness of low-carbon practices. This included an assessment of the frequency of residents' travel modes, their attitudes toward low-carbon policies, and their satisfaction with the community. Ultimately, 1496 valid questionnaires were collected as a part of this survey.

Table 1 presents the descriptive statistical data derived from the collected samples. The distribution of the respondents' gender is relatively equitable, with males and females accounting for 47.6% and 52.4% of the sample, respectively. The mean age of the participants is 33.69 years. The data show that 82.6% of respondents are non-Communist Party members. There are notable disparities in marital status, employment, and lower educational attainment between individuals belonging to high-income and low-income categories. The high-income demographic consists primarily of married individuals, most of whom are employed in office-based occupations and possess advanced educational credentials.

Table 1. Sociodemographic characteristics of the sample.

Variables	Total	Low Income	High Income
Age (years)			
Mean value	33.69	34.14	33.41
Gender (%)			
Female	52.4	62.6	41.1
Male	47.6	37.4	58.9
Marital status (%)			
Single	43.4	30.94	37.90
Married	55.7	27.30	60.49
Divorced	0.9	0.75	0.64
Political affiliation (%)			
Communist Party member	17.4	10.1	21.8
Non-Communist Party member	82.6	89.9	78.2
Employment (%)			
Unemployed	76.2	56.0	4.4
Employed	23.8	44.0	95.6
Education (%)			
Lower education	19.05	36.30	9.10
Middle and lower education	17.31	23.96	13.60
Middle and higher education	26.60	22.87	29.12
Higher education	29.68	16.88	37.58
Hukou status (%)			
Nonlocal	31.0	41.1	24.9
Local	69.0	58.9	75.1

Note: (1) Low income means monthly income <5000 RMB and high income means monthly income ≥5000 RMB (2) Hukou status means the Chinese household registration system.

3.3. Research Methods

3.3.1. The Research Framework

This paper investigates the effect of public traffic accessibility and travel behavior on citizens' low-carbon awareness. In addition, we propose a framework to address the relationship between public traffic accessibility, travel behavior, community satisfaction, and low-carbon awareness, as shown in Figure 2. The underlying logic is that travel behavior influences residents' low-carbon awareness and may be mediated by community satisfaction, which influences low-carbon awareness as well, i.e., community satisfaction plays a mediating effect in the relationship between travel behavior and low-carbon awareness.

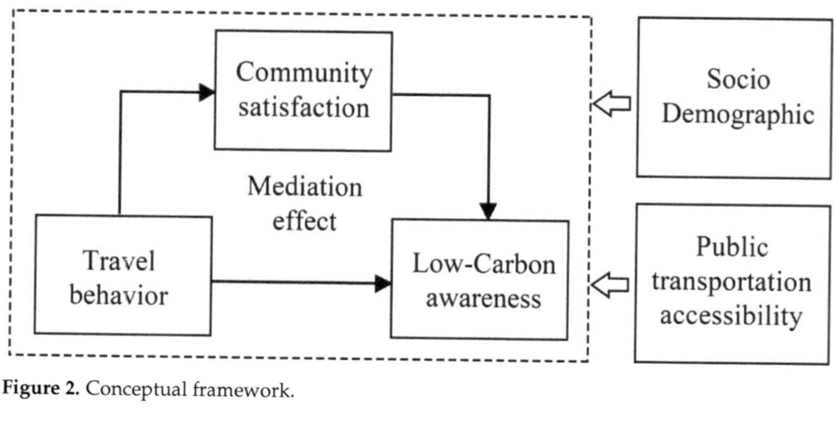

Figure 2. Conceptual framework.

3.3.2. Model Construction

Based on the proposed conceptual framework, a regression model was used to quantify the effects of individual factors, travel behavior, and public traffic accessibility on citizens' low-carbon awareness. The functional form of the linear regression model is as follows:

$$Y_{ij} = \alpha_1 + \eta X_j + \beta_1 Z_{ij} + \gamma_1 W_{ij} + \mu_{ij} + \varepsilon_{1ij}\,, \tag{1}$$

where Y_{ij} represents the low-carbon awareness of individual i in community j; X_j represents the variables for the public traffic accessibility of community j; Z_{ij} represents the variables for the sociodemographic characteristics of resident i of city j; W_{ij} represents the variable for the travel behavior of resident i of community j; η represents the total effect of the independent variables; β_1 represents the coefficient of the sociodemographic characteristics; γ_1 represents the coefficient of the travel behavior; α_1 represents the intercept; μ_{ij} represents the residual of the sociodemographic characteristics; and ε_{1ij} represents the residual of the travel behavior.

Based on the theoretical framework, we applied a stepwise approach [53] and bootstrap [54] in stata16.0 to test the mediating role of community satisfaction between travel behavior and low-carbon awareness. We first used linear probability modeling (LPM) to regress the dependent variable (low-carbon awareness) on the independent variable (travel behavior and public traffic accessibility). Next, we used LPM to regress the mediating variable (community satisfaction) on the independent variable (travel behavior). We then used LPM to regress the dependent variable onto both the independent and mediating variables. Mediation occurs when the dependent variable is influenced by both the independent variable and the mediator and when the mediator is influenced by the independent variables [55]. In addition, we used the bootstrap method to test whether other indicators have a mediating effect.

3.4 *Variables and Measures*

3.4.1. Travel Behavior

We defined the behavior indicators by referring to Bai's article on individual low-carbon behavior [56]. This study employs the concepts of zero-carbon action, low-carbon action, and high-carbon action to examine travel behavior (Table 2). Zero-carbon action is measured by "I walk when I go out" and "I travel by bicycle"; low-carbon action is measured by "I travel by subway" and "I travel by bus"; high-carbon action is measured by "I travel by cab" and "I travel by car." For each item, a 5-point Likert scale was used to measure the degree of frequency, where five means "frequently" and one means "never", to measure whether residents' travel behavior is characterized as low-carbon.

Table 2. Measured variables of travel behavior.

Constructed Dimensions	Variables	Specific Variables
Travel behavior	Zero-carbon action	I walk when I go out I travel by bicycle
	Low-carbon action	I travel by subway I travel by bus
	High-carbon action	I travel by cab I travel by car

3.4.2. Community Satisfaction

Community satisfaction was measured by five parameters: community services, school childcare, shopping and commercial facilities, transportation conditions, and property management. This measurement aims to assess the satisfaction level among residents regarding the community facilities that provide support. The Kaiser–Meyer–Olkin (KMO) value of the scale in this work is 0.818, and the Bartlett sphere test is significant. The Cronbach's α of community satisfaction is 0.802 (Table 3), which indicates good reliability of the questionnaire. In this study, the responses were measured on a scale of 1 to 5, representing the following levels of agreement: strongly disagree, disagree, neutral, agree, and strongly agree, respectively. There is a positive correlation between the scale score and the level of safety reported by the respondent.

Table 3. Measured variables of community satisfaction and low-carbon awareness.

Constructed Dimensions	Cronbach's α	Cronbach's α Based on Standardized Terms	Variables
Community satisfaction	0.802	0.803	Community services
			Schools and childcare
			Shopping and commercial facilities
			Transportation conditions
			Property management
Low-carbon awareness	0.859	0.865	Low-carbon consumption is a healthy way
			Low carbon is important to our living environment
			Outraged at the waste of energy
			Low-carbon consumption is something to be proud of
			Wasting energy is shameful behavior
			Know how to reduce carbon emissions
			To protect the environment, you will use a more environmentally friendly way of traveling
			I'll focus more on low-carbon consumption if there's a carbon tax

3.4.3. Low-Carbon Awareness

Jia et al. developed a specific scale to measure low-carbon awareness [22]; this work draws on their research scale on low-carbon awareness, and the scale was strictly revised according to the program for use. The residents' level of low-carbon awareness was assessed using a set of questions that were designed to measure their understanding and beliefs regarding low-carbon consumption. The KMO value of the scale in this work is 0.874, and the Bartlett sphere test is significant. The Cronbach's α of low-carbon awareness is 0.859 (Table 3), which indicates good reliability of the questionnaire. These questions included statements such as "Low-carbon consumption is a healthy way", and "Low carbon is essential to our living environment". Additionally, the questionnaire included items that aimed to gauge respondents' attitudes toward energy waste, such as "Being indignant about wasting energy behavior", "Low-carbon consumption is something to be proud of", and "Wasting energy is a shameful behavior". Furthermore, participants were asked about their knowledge of carbon emissions reduction strategies and their willingness to adopt more environmentally friendly modes of transportation to protect the environment. For each question, the available responses ranged from 1 = "Strongly Disagree" to 5 = "Strongly Unified". Subsequently, the scores of these projects were summarized and low-carbon awareness scores were calculated.

3.4.4. Public Traffic Accessibility

We referred to the research conclusion of Zhang et al. [49] and selected the distance to the commercial center, the distance to the nearest subway station, and the nearest bus stop as indicators of public traffic accessibility. These indicators were used to measure the traffic accessibility of the community, as shown in Table 4. Distance to the commercial center refers to the distance of the community from the commercial center of the administrative district, which reflects the community transportation location to a certain extent. The distance to the nearest subway station and the distance to the nearest bus stop can help determine whether the community has convenient transportation.

Table 4. Measured variables of public traffic accessibility.

Constructed Dimensions	Variables
Public traffic Accessibility	Distance to the business left Distance to the nearest subway station Distance to the nearest bus stop

Note: the data are from AMAP (https://www.amap.com (accessed on 7 October 2022)).

4. Results and Analysis

4.1. Descriptive Analysis of Travel Behavior and Low-Carbon Awareness

Regarding travel behavior, the mean values for zero-carbon action, low-carbon action and high-carbon action among residents of Guangzhou are 6.810, 6.978, and 5.756, respectively. In Table 5, the total score for zero-carbon, low-carbon, and high-carbon action reached a maximum value of 10. The average total score of community satisfaction of Guangzhou residents is 18.455 (out of a maximum of 25 points); there are five items, and the average value of each item is 3.691 (in the questionnaire, 4 = "satisfied"). This result indicates that actual residents are satisfied with the community's amenities.

Table 5. Score of each variables.

	Low-Carbon Awareness	Zero-Carbon Action	Low-Carbon Action	High-Carbon Action	Community Satisfaction
Mean	33.382	6.810	6.979	5.756	18.455
Min	8.000	2.000	2.000	2.000	6.000
Max	40.000	10.000	10.000	10.000	25.000
S.E.	4.456	1.821	1.877	1.323	3.078

In contrast to others, Guangzhou residents' low-carbon awareness had a higher mean score of 33.382 for each item (out of a maximum of 40 points), there were eight items, and the mean for each item was 4.172 (in the questionnaire, 4 = "unified"). The mean of the total scores for the high-income and low-income groups are 32.520 and 33.446, respectively. The *t*-test results for the differences between high-income and low-income groups are statistically significant, as the significance (Sig.) of income is higher than 0.5 for Levene's test and less than 0.05 in the first row of the *t*-test (Table 6). The low-carbon awareness of low-income groups seems stronger. Research has found that residents prefer shorter travel times and lower transportation costs, while low-income groups prefer low-carbon travel methods [57]. The following section further delves into the impact mechanism of low-carbon awareness and explores whether travel behavior affects low-carbon awareness through community satisfaction by considering other control variables in the regression model, revealing the impact mechanism of behavior on awareness.

Table 6. Independent sample *t*-test results for low-carbon awareness.

Dependent Variable: Low-Carbon Awareness					Levene's Test		*t*-Test			
Demographic		Mean	S.E.	Mean of S.E.	F	Sig.	t	Sig. (Two Tail)	Mean	S.E.
Age	<60 years	33.402	4.463	0.118	0.266	0.606	0.811	0.417	0.465	0.574
	≥60 years	32.937	4.299	0.542			0.840	0.404	0.465	0.554
Gender ***	Female	33.736	4.354	0.156	0.000	0.995	3.231	0.001	0.743	0.230
	Male	32.993	4.537	0.170			3.225	0.001	0.743	0.230

Table 6. *Cont.*

Dependent Variable: Low-Carbon Awareness					Levene's Test		*t*-Test			
Demographic		Mean	S.E.	Mean of S.E.	F	Sig.	t	Sig. (Two Tail)	Mean	S.E.
Marital status **	Unmarried	33.658	4.369	0.168	0.031	0.861	2.177	0.030	0.503	0.231
	Married	33.155	4.516	0.158			2.184	0.029	0.503	0.230
Political affiliation **	Non-Communist Party member	33.273	4.465	0.127	1.816	0.178	−2.069	0.039	−0.628	0.303
	Members of Chinese Communist Party	33.900	4.384	0.271			−2.094	0.037	−0.628	0.300
Employment	Unemployed	33.183	4.577	0.243	0.155	0.694	−0.969	0.333	−0.262	0.271
	Employed	33.445	4.418	0.131			−0.951	0.342	−0.262	0.276
Education	Lower education	33.171	4.337	0.170	2.813	0.094	−1.616	0.106	−0.375	0.232
	Higher education	33.546	4.542	0.157			−1.625	0.104	−0.375	0.231
Hukou status	Nonlocal	33.226	4.562	0.212	1.738	0.188	−0.908	0.364	−0.226	0.249
	Local	33.453	4.408	0.137			−0.896	0.370	−0.226	0.252
Income **	Low-income	33.446	4.407	0.118	0.115	0.734	2.170	0.030	0.516	0.238
	High-income	32.520	5.028	0.498			2.166	0.030	0.516	0.238

Note: **, and *** represent significance levels of 1%, and 0.1%, respectively.

4.2. Effects of Travel Behavior and Community Satisfaction on Low-Carbon Awareness

A three-step methodology was employed to investigate the initial research inquiry [53,58] In Model 1, without a mediation variable, the relationship between travel behavior and residents' low-carbon awareness is strong and statistically significant (Table 7). A positive correlation exists between zero-carbon action and low-carbon action and residents' low-carbon awareness. Conversely, a negative correlation exists between high-carbon action and residents' low-carbon awareness. For every point increase in low-carbon action, residents' awareness of low-carbon practices demonstrates a corresponding increase of 0.489 points.

Table 7. Regression models for low-carbon awareness for the entire sample.

	Low-Carbon Awareness	Community Satisfaction	Low-Carbon Awareness
Public traffic accessibility			
Distance to the business center	−0.200	−0.957 **	0.217
Distance to the nearest subway station	−1.123 **	−0.418	−0.937 **
Distance to the nearest bus stop	0.378	0.434	0.188
Travel behavior			
Zero-carbon action	0.489 ***	0.223 ***	0.390 ***
Low-carbon action	0.472 ***	0.158 ***	0.403 ***
High-carbon action	−0.463 ***	−0.087	−0.424 ***
Sociodemographic			
Age	−0.007	−0.004	−0.005
Gender (reference group: male)			
Female	0.865 ***	0.228	0.766 ***
Marital status (reference group: single)			
Married	−0.568 *	−0.072	−0.537 *
Divorced	−1.991 ***	0.192	−2.076 ***
Political affiliation (reference group: Communist Party member)			
Non-Communist Party member	−0.504	−0.056	−0.481

Table 7. *Cont.*

	Low-Carbon Awareness	Community Satisfaction	Low-Carbon Awareness
Employment (reference group: Unemployed)			
Employed	−0.180	0.333	−0.326
Education (reference group: lower education)			
Middle and lower education	−0.047	−0.094	−0.007
Middle and higher education	−0.320	0.078	−0.355
Higher education	0.387	−0.232	0.489 *
Hukou status (reference group: local)			
Nonlocal	0.090	0.252	−0.022
Income (reference group: low-income)			
High income	0.790 **	−0.039	0.805 **
Mediator variable			
Community satisfaction			0.442 ***
Constant	29.620 ***	16.350 ***	22.390 ***

Note: *, **, and *** represent significance levels of 5%, 1%, and 0.1%, respectively.

Similarly, an increase in low-carbon actions causes a rise of 0.472 points in residents' low-carbon awareness. Conversely, an escalation in high-carbon actions yields a decrease of 0 463 points in residents' low-carbon awareness. Hence, promoting zero-carbon behavior can effectively enhance residents' awareness of low-carbon practices, surpassing the impact of both low-carbon and high-carbon behaviors.

In Model 2, the regression analysis reveals that the independent variables of zero-carbon action and low-carbon action, which represent travel behavior, significantly reflect community satisfaction, as measured by the mediation factor. A quantitative analysis shows that an increment of 1 point in zero-carbon action is associated with a corresponding increase of 0.223 points in community satisfaction. Similarly, a 1-point increase in low-carbon action is linked to a 0.158-point increase in community satisfaction. The findings of this study indicate that individuals who engage in lower-carbon travel behaviors tend to exhibit higher levels of satisfaction with their community, evidenced by the data presented in Table 7.

Compared with Model 1, when travel behavior and community satisfaction are included in the same model (Model 3, the dependent variable is low-carbon awareness), the zero-carbon action and low-carbon action in travel behavior are still significant. However, their effects on low-carbon awareness decrease: the value of the coefficient of zero-carbon action decreases from 0.489 to 0.390, and the coefficient of low-carbon action decreases from 0.472 to 0.403. This result confirms that the intermediary indicator of community satisfaction influences the relationship between travel behavior and low-carbon awareness [59].

Using the bootstrap method to test whether other indicators have a mediating effect [54], it was found that the distance from the commercial center has a mediating effect on low-carbon awareness. Although the regression results of this indicator are not significant in Model 1, mediation validation found that the distance from the commercial center has a mediating effect on low-carbon awareness (Table 8), which is due to the suppression of community satisfaction [60]. Specifically, community satisfaction has an inhibiting effect, with each unit of increasing commercial center distance resulting in a corresponding decrease of 0.957 in the residents' community satisfaction, positively correlated with low-carbon attitudes. Although the distance from the commercial center may not directly impact low-carbon awareness on the surface, it has been found through testing that the longer this distance, the lower the community satisfaction of residents, thereby resulting in a negative impact on low-carbon awareness. This conclusion is consistent with previous research findings; residents living in areas with sufficient facilities and a high land-use mix often have low-carbon daily travel, and their low-carbon awareness level is higher [49]. Therefore, residents can often complete short commutes by walking or cycling to meet their daily needs. At the same time, zero-carbon and low-carbon transportation is more economical than high-carbon transportation. Therefore, this convenient and economica approach can better promote their positive emotions toward low-carbon awareness. I the community, if the facilities are less than perfect, residents may find it challenging t complete their daily life more conveniently through zero-carbon and low-carbon action an even have a poor experience with zero-carbon and low-carbon travel. These factors ma increase their resistance to low-carbon awareness, thus promoting their choice to travel b car to reduce their travel time cost.

Table 8. Bootstrap results for the entire sample.

Variables	Coefficient	Estimate	S.E.	Z	$p > z$
Distance to business left	Indirect effect	−0.587	0.183	−3.210	0.001
	Direct effect	−0.134	0.491	−0.272	0.785
	Total effect	−0.720	0.524	−1.375	0.169
Distance to the nearest subway station	Indirect effect	0.248	0.207	−1.530	0.126
	Direct effect	1.312	0.507	−2.589	0.010
	Total effect	1.560	0.542	−2.878	0.004
Zero-carbon action	Indirect effect	0.132	0.026	4.980	0.000
	Direct effect	0.455	0.059	7.733	0.000
	Total effect	0.586	0.061	9.537	0.000
Low-carbon action	Indirect effect	0.099	0.025	3.920	0.000
	Direct effect	0.463	0.057	8.178	0.000
	Total effect	0.561	0.060	9.407	0.000
High-carbon action	Indirect effect	−0.003	0.033	−0.100	0.924
	Direct effect	−0.220	0.081	−2.711	0.007
	Total effect	−0.223	0.087	−2.568	0.010
Gender	Indirect effect	0.101	0.078	1.290	0.198
	Direct effect	0.642	0.215	2.986	0.003
	Total effect	0.743	0.230	3.231	0.001
Married	Indirect effect	−0.114	0.084	−1.360	0.173
	Direct effect	−0.389	0.216	−1.799	0.072
	Total effect	−0.503	0.231	−2.177	0.029
Divorced	Indirect effect	0.204	0.425	0.480	0.631
	Direct effect	−1.211	1.160	−1.044	0.296
	Total effect	−1.006	1.241	−0.811	0.417
High income	Indirect effect	0.028	0.082	0.340	0.734
	Direct effect	0.488	0.222	2.198	0.028
	Total effect	0.516	0.238	2.170	0.030

This conclusion suggests that travel behavior directly affects residents' low-carbon awareness and indirectly affects residents' low-carbon awareness through the mediating effect of community satisfaction. The effect of residents' community satisfaction on low-carbon awareness displays a positive feedback effect: the higher the residents' satisfaction with the community, the stronger their low-carbon awareness. In addition, research has found through testing that the relationship between the distance between communities and business centers and low-carbon awareness is influenced by community satisfaction. The farther the distance, the lower the residents' community satisfaction, which negatively impacts low-carbon awareness.

We also found that the distance to the nearest subway station negatively impacts residents' low-carbon awareness. For every additional point of subway station distance, their low-carbon awareness point decreases by 1.123. This result may be attributed to the reduced public traffic accessibility in the community with increased distance. Residents' willingness to choose low-carbon travel modes thus decreases, which leads to a decrease in low-carbon awareness [9]. In addition, the study also found that females have higher low-carbon awareness, which is consistent with previous studies [10].

Regarding education, groups with higher education levels have higher low-carbon awareness, which may be because low-carbon publicity in schools gives residents a deeper understanding of environmental pollution, thus a higher low-carbon awareness [56,61]. Previous studies have found that residents with high levels of education are more inclined to use more energy-efficient and low-carbon electric vehicles [12]. In addition, the study also found that the higher income of the residents, the higher their low-carbon awareness. This result is understandable; environmental attitudes and situational factors can impact individual environmental behavior [15] and an increase in income may increase their environmental demand, so low-carbon awareness also increases.

4.3. Travel Behavior and Low-Carbon Awareness: The Income Division

The empirical results comparing the two subsample models using the same methodology as the entire sample are presented in Table 9. This study presents a comparative analysis examining the correlation between travel behavior and low-carbon awareness among individuals of high-income and low-income backgrounds.

Table 9. Regression subsample for income division.

	Low-Income			High-Income		
	Low-Carbon Awareness	Community Satisfaction	Low-Carbon Awareness	Low-Carbon Awareness	Community Satisfaction	Low-Carbon Awareness
Public traffic accessibility						
Distance to business center	−1.288	−1.166 *	−0.831	0.462	−0.869	0.818
Distance to the nearest subway station	−1.962 ***	−0.328	−1.847 ***	−0.397	−0.545	−0.131
Distance to the nearest bus stop	1.201 **	0.469	1.022 *	−0.348	0.331	−0.498
Travel behavior						
Zero-carbon action	0.581 ***	0.285 ***	0.463 ***	0.440 ***	0.180 ***	0.360 ***
Low-carbon action	0.575 ***	0.199 ***	0.494 ***	0.436 ***	0.146 **	0.368 ***
High-carbon action	−0.443 ***	−0.123	−0.393 ***	−0.460 ***	−0.062	−0.431 ***
Sociodemographic						
Age	−0.010	−0.001	−0.010	0.010	−0.012	0.0153
Gender (reference group: male)						
Female	0.532 *	0.213	0.454 *	1.074 ***	0.222	0.980 ***
Marital status (reference group: single)						
Married	−0.807	−0.139	−0.758	−0.590 *	0.005	-0.594 *
Divorced	−1.962 **	−0.565	−1.708 *	−2.293 ***	1.334 *	−2.909 ***
Political affiliation (reference group: Communist Party member)						
Non-Communist Party member	−0.953	−0.309	−0.841	−0.353	0.032	−0.371

Table 9. *Cont.*

	Low-Income			High-Income		
	Low-Carbon Awareness	**Community Satisfaction**	**Low-Carbon Awareness**	**Low-Carbon Awareness**	**Community Satisfaction**	**Low-Carbon Awareness**
Employment (reference group: unemployed)						
Employed	−0.521	0.196	−0.603 *	0.697	0.541	0.458
Education (reference group: lower education)						
Middle and lower education	0.165	−0.032	0.177	−0.296	−0.090	−0.272
Middle and higher education	−0.243	0.284	−0.359	−0.456	0.001	−0.467
Higher education	−0.355	−0.167	−0.280	0.502	−0.277	0.615 *
Hukou status (reference group: local)						
Nonlocal	0.355	0.406	0.180	−0.090	0.065	−0.117
Mediator variable						
Community satisfaction			0.411 ***			0.454 ***
Constant	29.270 ***	15.950 ***	22.750 ***	29.390 ***	16.630 ***	21.850 ***

Note: *, **, and *** represent significance levels of 5%, 1%, and 0.1%, respectively.

For the model without a mediation factor (community satisfaction), travel behavior strongly and significantly predicted residents' low-carbon awareness (Table 9): zero-carbon action, low-carbon action, and low-carbon awareness were positively correlated, while high-carbon action was negatively correlated. In the model with the mediation factor as the dependent variable, zero-carbon action and low-carbon action were also significantly positively correlated with community satisfaction: residents with lower carbon travel behavior had higher satisfaction with the community. When the independent variable (travel behavior) and mediation factor (community satisfaction) were included in the same model as the dependent variable, the impact of zero-carbon action and low-carbon action was significantly reduced, although it remained significant. These results indicate that the mediating effect of community satisfaction seems reasonable for both groups, consistent with the findings of the entire sample: travel behavior not only directly affects respondents' low-carbon awareness but also indirectly affects their low-carbon awareness through the mediating effect of community satisfaction.

The zero-carbon action coefficient for low-income groups decreased from 0.581 to 0.463, and the zero-carbon action coefficient for high-income groups decreased from 0.440 to 0.360 when the mediation factor was included (Table 9). This result indicates that community satisfaction has a more significant impact on low-income groups. In addition, the bootstrap method was used to test whether other indicators in the high-income and low-income group models have a mediating effect. It was found that the distance from the commercial center has a mediating effect on the low-carbon awareness of low-income groups (Table 10), which is consistent with the overall situation and is even more apparent considering that the mediating proportion reached 97%. However, this phenomenon only exists in the model of low-income groups. The zero-carbon action, low-carbon action, and distance from commercial centers of low-income groups are mediated by community satisfaction. In contrast, only the zero-carbon behavior and low-carbon behavior of high-income groups are mediated by community satisfaction (Table 11). Therefore, community satisfaction is essential to low-carbon awareness, especially among low-income groups.

Table 10. Bootstrap results for the low-income group.

Variables	Coefficient	Estimate	S.E.	z	$p > z$
Distance to business left	Indirect effect	−0.589	0.282	−2.090	0.036
	Direct effect	−0.015	0.786	−0.019	0.985
	Total effect	−0.605	0.839	−0.721	0.471
Distance to the nearest subway station	Indirect effect	−0.145	0.341	−0.430	0.671
	Direct effect	−1.991	0.763	−2.609	0.009
	Total effect	−2.136	0.818	−2.613	0.009
Distance to the nearest bus stop	Indirect effect	0.217	0.283	0.770	0.444
	Direct effect	0.595	0.772	0.771	0.441
	Total effect	0.812	0.826	0.983	0.326
Zero-carbon action	Indirect effect	0.151	0.038	3.961	0.000
	Direct effect	0.513	0.098	5.243	0.000
	Total effect	0.664	0.102	6.532	0.000
Low-carbon action	Indirect effect	0.111	0.040	2.760	0.006
	Direct effect	0.541	0.089	6.094	0.000
	Total effect	0.653	0.093	7.003	0.000
High-carbon action	Indirect effect	−0.010	0.055	−0.190	0.853
	Direct effect	−0.187	0.140	−1.341	0.180
	Total effect	−0.197	0.150	−1.32	0.187
Gender	Indirect effect	0.117	0.132	0.890	0.374
	Direct effect	0.489	0.364	1.345	0.179
	Total effect	0.606	0.389	1.557	0.119
Divorced	Indirect effect	−0.148	0.620	−0.240	0.812
	Direct effect	−0.781	1.587	−0.492	0.622
	Total effect	−0.929	1.701	−0.546	0.585

Table 11. Bootstrap results for the high-income group.

Variables	Coefficient	Estimate	S.E.	Z	$p > z$
Zero-carbon action	Indirect effect	0.121	0.030	3.980	0.000
	Direct effect	0.436	0.074	5.920	0.000
	Total effect	0.556	0.077	7.213	0.000
Low-carbon action	Indirect effect	0.089	0.031	2.890	0.004
	Direct effect	0.413	0.073	5.639	0.000
	Total effect	0.502	0.077	6.490	0.000
High-carbon action	Indirect effect	0.001	0.038	0.027	0.978
	Direct effect	−0.227	0.100	−2.278	0.023
	Total effect	−0.226	0.107	−2.120	0.034
Gender	Indirect effect	0.102	0.101	1.020	0.310
	Direct effect	0.876	0.271	3.230	0.001
	Total effect	0.978	0.290	3.377	0.001

negative impact on residents' low-carbon awareness, but no mediating effect was detected. Previous studies reported that residents are more likely to choose low-carbon travel if they feel it is more convenient [9]. As the distance between the community and the subway station increases, the public traffic accessibility decreases, decreasing low-carbon awareness. The reason for this phenomenon may be that residents in areas with denser coverage of public transportation services are more likely to use more low-carbon travel modes, and low-carbon awareness in those regions is also higher [63]. The research results indicate that improving the external environment, such as transportation and community facilities, significantly enhances residents' low-carbon awareness, which is significant for constructing low-carbon cities in China's mega cities and the Guangdong Hong Kong Macao Greater Bay Area. In addition, different sociodemographic characteristics differently impact residents' low-carbon awareness. The female population has a higher low-carbon awareness, consistent with previous research [10]. Regarding education, groups with higher levels of education have higher low-carbon awareness, which may be due to receiving environmental education and promoting sustainable consumption habits, such as purchasing electric vehicles [20,64]. The higher the income group, the higher their low-carbon awareness. They have high requirements for environmental ecology, cherish their surroundings, and possess stronge environmental awareness [65].

However, this study also has some limitations and can be improved upon. On the one hand, due to the cross-sectional nature of our data, we could not analyze the factors that affect low-carbon awareness in various time periods, because travel behavior and low-carbon awareness may involve a long-term dynamic, including feelings and related behaviors that change with the evolution of low-carbon policies. On the other hand, due to data limitations, we have limitations in the classification of research subjects. Therefore, future research should further refine the research subjects and pay more attention to the differences in low-carbon awareness of different groups of people, while at the same time increasing the diversity of low-carbon awareness research data, employing more comprehensive and accurate models and methods to explore the impact mechanism of low-carbon awareness, and providing scientific reference for global low-carbon city decision making.

5.2. Conclusions

This study uses evidence from a survey of 1496 respondents in Guangzhou in 2022 to analyze the low-carbon awareness of Chinese urban residents. We explored the impact of residents' willingness to travel on their low-carbon awareness, mainly focusing on how residents' community satisfaction mediates this relationship. In addition, a comparison was made between the high-income and low-income groups in Guangzhou regarding low-carbon awareness levels and mediation mechanisms. We made the following findings through the stepwise regression of mediation modeling. First, travel behavior positively impacts residents' low-carbon awareness, and their community satisfaction mediates the impact of travel behavior on low-carbon awareness. Second, low-income individuals receive more positive feedback from community satisfaction. There is a mediating effect between the distance between low-income communities and commercial centers and low-carbon awareness, which is influenced by community satisfaction. These findings contribute to a deeper understanding of the relationship between travel behavior and low-carbon awareness. People with low-carbon travel behavior tend to have higher satisfaction with the community and more vital low-carbon awareness to maintain their community living environment. Public transportation has the advantages of convenience and economy, so low-income groups prefer low-carbon modes of transportation, so they often have more vital awareness of low carbon. In this sense, when studying low-carbon awareness, the heterogeneity of different income groups deserves more attention.

These findings provide new policy entry points for constructing low-carbon cities and transforming residents' lifestyles. Based on the above findings, we propose the following policy recommendations:

Firstly, the role of the community is of utmost importance in fostering awareness and understanding of low-carbon practices, and the intention of residents to choose low-carbon commuting modes is indirectly influenced by policy factors [66]. Thus, the primary focus of urban governments' community policies and urban governance should be on improving support services within the community, enhancing residents' satisfaction with their living environment, and promoting residents' awareness of low-carbon practices. Secondly, it is imperative for communities to actively engage in the organization of low-carbon education and publicity initiatives to enhance low-carbon awareness effectively. Residents are more likely to opt for low-carbon and public transportation when they know about the environmental impact associated with such modes of transportation [67]. Meanwhile, improving the public traffic accessibility of the community can encourage residents to travel low-carbon. We suggest that the municipal authorities add shared bicycles and electric shuttle buses at subway stations to solve the "last mile" travel problem [34]. Furthermore, community committees must prioritize ensuring equitable access to community public services and convenience facilities among diverse income groups. Promoting this equality can foster a sense of community satisfaction among residents, consequently bolstering their awareness and commitment to low-carbon practices.

Author Contributions: Conceptualization, Q.L. and M.D.; methodology, Y.Z. and Q.L.; software, Q.L.; validation, Y.Z.; formal analysis, Q.L. and M.D.; investigation, Q.L. and M.D.; resources, M.D. and Q.L.; data curation, Y.Z.; writing—original draft preparation, Q.L. and M.D.; supervision, R.W.; project administration, R.W.; funding acquisition, R.W. All authors have read and agreed to the published version of the manuscript.

Funding: This work was supported by the National Natural Science Foundation of China (Grant Nos. 41501184, 42001147), Guangzhou Science and Technology Program (Grant No. 202102020319) and Guangdong Province Natural Science Fund (Grant No. 2022A1515011728), Special Funds for the Cultivation of Guangdong College Students' Scientific and Technological Innovation. ('Climbing Program' Special Funds) (pdjh2022a0153).

Institutional Review Board Statement: Not applicable.

Informed Consent Statement: Not applicable.

Data Availability Statement: Data are contained within the article.

Conflicts of Interest: The authors declare no conflict of interest.

References

1. Hao, H.; Wang, H.; Yi, R. Hybrid modeling of China's vehicle ownership and projection through 2050. *Energy* **2011**, *36*, 1351–1361. [CrossRef]
2. Zhao, Y.B.; Gao, P.P.; Yang, W.D.; Ni, H.G. Vehicle exhaust: An overstated cause of haze in China. *Sci. Total Environ.* **2018**, *612*, 490–491. [CrossRef] [PubMed]
3. Wang, S.; Li, Q.; Fang, C.; Zhou, C. The relationship between economic growth, energy consumption, and CO_2 emissions: Empirical evidence from China. *Sci. Total Environ.* **2016**, *542*, 360–371. [CrossRef] [PubMed]
4. Wu, R.; Xie, Z. Identifying the impacts of income inequality on CO2 emissions: Empirical evidences from OECD countries and non-OECD countries. *J. Clean Prod.* **2020**, *277*, 123858. [CrossRef]
5. IEA. *CO2 Emissions from Fuel Combustion Highlights*; International Energy Agency: Paris, France, 2015.
6. Xi, J.P. In-depth knowledge and understanding of peak carbon and carbon neutrality. In *General Debate of the Seventy-Fifth Session of the United Nations General Assembly, Beijing*; UN: New York, NY, USA, 2020.
7. IEA (International Energy Agency). *Energy Technology Perspectives: Towards Sustainable Urban Energy Systems*; IEA: Paris, France, 2016.
8. UN (United Nations). *United Nations General Assembly Resolution Establishing World Bicycle Day*; UN-Habit: Nairobi, Kenya, 2022.
9. Nordfjærn, T.; Lind, H.B.; Şimşekoğlu, Ö.; Jørgensen, S.H.; Lund, I.O.; Rundmo, T. Habitual, safety and security factors related to mode use on two types of travels among urban Norwegians. *Saf. Sci.* **2015**, *76*, 151–159. [CrossRef]
10. Prillwitz, J.; Barr, S. Moving towards sustainability? Mobility styles, attitudes and individual travel behaviour. *J. Transp. Geogr.* **2011**, *19*, 1590–1600. [CrossRef]
11. Sun, C.; Zhang, Y.; Ma, W.; Wu, R.; Wang, S. The impacts of urban form on carbon emissions: A comprehensive review. *Land* **2022**, *11*, 1430. [CrossRef]

12. Carley, S.; Krause, R.M.; Lane, B W.; Graham, J.D. Intent to purchase a plug-in electric vehicle: A survey of early impressions in large US cites. *Transp. Res. Part D Transp. Environ.* **2013**, *18*, 39–45. [CrossRef]
13. Ajzen, I. The theory of planned behavior. *Organ. Behav. Hum. Dec.* **1991**, *50*, 179–211. [CrossRef]
14. Stern, P.C. Toward a coherent theory of environmentally significant behavior. *J. Soc. Issues* **2000**, *56*, 407–424. [CrossRef]
15. Guagnano, G.A.; Stern, P.C.; Dietz, T. Influences on attitude-behavior relationships: A natural experiment with curbside recycling. *Environ. Behav.* **1995**, *27*, 699–718. [CrossRef]
16. Salomon, G. *Distributed Cognitions: Psychological and Educational Considerations*; Cambridge University Press: Cambridge, UK, 1997.
17. Kahn, M.E.; Morris, E.A. Walking the walk: The association between community environmentalism and green travel behavior. *J. Am. Plan. Assoc.* **2009**, *75*, 389–405. [CrossRef]
18. Bamberg, S.; Schmidt, P. Incentives, morality, or habit? Predicting students' car use for university routes with the models of Ajzen, Schwartz, and Triandis. *Environ Behav.* **2003**, *35*, 264–285. [CrossRef]
19. Donald, I.J.; Cooper, S.R.; Conchie, S.M. An extended theory of planned behaviour model of the psychological factors affecting commuters' transport mode use. *J. Environ. Psychol.* **2014**, *40*, 39–48. [CrossRef]
20. Graham-Rowe, E.; Gardner, B.; Abraham, C.; Skippon, S.; Dittmar, H.; Hutchins, R.; Stannard, J. Mainstream consumers driving plug-in battery-electric and plug-in hybrid electric cars: A qualitative analysis of responses and evaluations. *Transp. Res. Part A Policy Pract.* **2012**, *46*, 140–153. [CrossRef]
21. Gärling, T.; Fujii, S.; Gärling, A.; Jakobsson, C. Moderating effects of social value orientation on determinants of proenvironmental behavior intention. *J. Environ. Psychol.* **2003**, *23*, 1–9. [CrossRef]
22. Jia, N.; Li, L.; Ling, S.; Ma, S.; Yao, W. Influence of attitudinal and low-carbon factors on behavioral intention of commuting mode choice–A cross-city study in China. *Transp. Res. Part A Policy Pract.* **2018**, *111*, 108–118. [CrossRef]
23. Lin, B.; Wang, X. Does low-carbon travel intention really lead to actual low-carbon travel? Evidence from urban residents in China. *Econ. Anal. Policy* **2021**, *72*, 743–756. [CrossRef]
24. Liu, T.; Wang, Y.; Li, H.; Qi, Y. China's low-carbon governance at community level: A case study in Min'an community, Beijing. *J. Clean. Prod.* **2021**, *311*, 127530. [CrossRef]
25. Maloney, M.P.; Ward, M.P. Ecology: Let's hear from the people: An objective scale for the measurement of ecological attitudes and knowledge. *Am. Psychol.* **1973**, *28*, 583. [CrossRef]
26. Abdul-Wahab, S.A. Level of environmental awareness towards depletion of the ozone layer among distributors and consumers in the solvent sector: A case study from Oman. *Clim. Change* **2010**, *103*, 503–517. [CrossRef]
27. Dunlap, R.E.; Van Liere, K.D.; Mertig, A.G.; Jones, R.E. New trends in measuring environmental attitudes: Measuring endorsement of the new ecological paradigm: A revised NEP scale. *J. Soc. Issues* **2000**, *56*, 425–442. [CrossRef]
28. Gagnon Thompson, S.C.; Barton, M.A. Ecocentric and anthropocentric attitudes toward the environment. *J. Environ. Psychol.* **1994**, *14*, 149–157. [CrossRef]
29. Van der Linden, S. Warm glow is associated with low-but not high-cost sustainable behaviour. *Nat. Sustain.* **2018**, *1*, 28–30. [CrossRef]
30. Gadenne, D.; Sharma, B.; Kerr, D.; Smith, T. The influence of consumers' environmental beliefs and attitudes on energy saving behaviours. *Energy Policy* **2011**, *39*, 7684–7694. [CrossRef]
31. Martinsson, J.; Lundqvist, L.J.; Sundström, A. Energy saving in Swedish households. The (relative) importance of environmental attitudes. *Energy Policy* **2011**, *39*, 5182–5191. [CrossRef]
32. Abrahamse, W.; Steg, L.; Vlek, C.; Rothengatter, T. The effect of tailored information, goal setting, and tailored feedback on household energy use, energy-related behaviors, and behavioral antecedents. *J. Environ. Psychol.* **2007**, *27*, 265–276. [CrossRef]
33. Kaplowitz, M.D.; Thorp, L.; Coleman, K.; Kwame Yeboah, F. Energy conservation attitudes, knowledge, and behaviors in science laboratories. *Energy Policy* **2012**, *50*, 581–591. [CrossRef]
34. Wright, H.A.; Ironside, J.E.; Gwynn-Jones, D. The current state of sustainability in bioscience laboratories: A statistical examination of a UK tertiary institute. *Int. J. Sust. High. Ed.* **2008**, *9*, 282–294. [CrossRef]
35. Hongyu, Z.; Yun, C. Study on Urban Design Strategy for Low-Carbon Trip. In Proceedings of the 2009 International Conference on Energy and Environment Technology, Guilin, China, 16–18 October 2009; Volume 1, pp. 373–377.
36. Barr, S.; Gilg, A.W.; Ford, N. The household energy gap: Examining the divide between habitual- and purchase-related conservation behaviours. *Energy Policy* **2005**, *33*, 1425–1444. [CrossRef]
37. Abrahamse, W.; Steg, L.; Gifford, R.; Vlek, C. Factors influencing car use for commuting and the intention to reduce it: A question of self-interest or morality? *Transp. Res. Part F Traffic Psychol. Behav.* **2009**, *12*, 317–324. [CrossRef]
38. Lopez-Mosquera, N.; Sanchez, M. Theory of planned behavior and the value-belief-norm theory explaining willingness to pay for a suburban park. *J. Environ. Manag.* **2012**, *113*, 251–262. [CrossRef] [PubMed]
39. Liu, D.; Du, H.; Southworth, F.; Ma, S. The influence of social-psychological factors on the intention to choose low-carbon travel modes in Tianjin, China. *Transp. Res. Part A Policy Pract.* **2017**, *105*, 42–53. [CrossRef]
40. Varela-Candamio, L.; Novo-Corti, I.; García-Álvarez, M.T. The importance of environmental education in the determinants of green behavior: A meta-analysis approach. *J. Clean. Prod.* **2018**, *170*, 1565–1578. [CrossRef]
41. Wang, T.; Shen, B.; Han Springer, C.; Hou, J. What prevents us from taking low carbon actions? A comprehensive review of influencing factors affecting low-carbon behaviors. *Energy Res. Soc. Sci.* **2021**, *71*, 101844. [CrossRef]

42. Ding, Z.; Wang, G.; Liu, Z.; Long, R. Research on differences in the factors influencing the energy-saving behavior of urban and rural residents in Chinaea case study of Jiangsu Province. *Energy Policy* **2017**, *100*, 252–259. [CrossRef]
43. Thogersen, J.; Gronhoj, A. Electricity saving in households—A social cognitive approach. *Energy Policy* **2010**, *38*, 7732–77 [CrossRef]
44. Best, H.; Kneip, T. The impact of attitudes and behavioral costs on environmental behavior: A natural experiment on household waste recycling. *Soc. Sci. Med.* **2011**, *40*, 917–930. [CrossRef]
45. Yongchun, Y.; Yan, S.; Jianxin, L.; Yan, T.; Wei, Z.; Weiwei, W. Urban spatial environmental cognition of both Tibetan and Han from a comparative perspective: A case study of Shigatse in Tibet. *Sci. Geogr. Sin.* **2019**, *39*, 334–341.
46. Jager, W. Stimulating the diffusion of photovoltaic systems: A behavioural perspective. *Energy Policy* **2006**, *34*, 1935–1943 [CrossRef]
47. Pickett-Baker, J.; Ozaki, R. Pro-environmental products: Marketing influence on consumer purchase decision. *J. Consum. Mark.* **2008**, *25*, 281–293. [CrossRef]
48. IPCC. *Climate Change 2014: Mitigation of Climate Change: Working Group III Contribution to the Fifth Assessment Report of the Intergovernmental Panel on Climate Change*; Cambridge University Press: Cambridge, UK, 2014.
49. Zhang, J.; Zhang, L.; Qin, Y.; Wang, X.; Zheng, Z. Influence of the built environment on urban residential low-carbon cognition in Zhengzhou, China. *J. Clean. Prod.* **2020**, *271*, 122429. [CrossRef]
50. Theodori, G.L. Examining the effects of community satisfaction and attachment on individual well-being. *Rural Sociol.* **2001**, *66* 618–628. [CrossRef]
51. Bureau of Statistics of Guangzhou. *Guangzhou Census Bulletin*; Guangzhou Statistics Press: Guangzhou, China, 2022; Available online: https://pan.baidu.com/s/1KBt1Wrp3_LvD6dbMHlnCEg?pwd=5smo (accessed on 12 June 2023).
52. Bureau of Statistics of Guangzhou. *Guangzhou Statistical Yearbook 2022*; Guangzhou Statistics Press: Guangzhou, China, 2022 Available online: http://tjj.gz.gov.cn/stats_newtjyw/zyxz/tjnjdzzz/content/post_8677056.html (accessed on 12 June 2023).
53. Baron, R.M.; Kenny, D.A. The moderator–mediator variable distinction in social psychological research: Conceptual, strategic and statistical considerations. *J. Pers. Soc. Psychol.* **1986**, *51*, 1173. [CrossRef] [PubMed]
54. Zhao, X.; Lynch, J.G., Jr.; Chen, Q. Reconsidering Baron and Kenny: Myths and truths about mediation analysis. *J. Consum. Res.* **2010**, *37*, 197–206. [CrossRef]
55. Krull, J.L.; MacKinnon, D.P. Multilevel modeling of individual and group level mediated effects. *Multivar. Behav. Res.* **2001**, *36* 249–277. [CrossRef]
56. Bai, Y.; Liu, Y. An exploration of residents' low-carbon awareness and behavior in Tianjin, China. *Energy Policy* **2013**, *61*, 1261–1270 [CrossRef]
57. Banister, D. The sustainable mobility paradigm. *Transp. Policy* **2008**, *15*, 73–80. [CrossRef]
58. Judd, C.M.; Kenny, D.A. Process analysis: Estimating mediation in treatment evaluations. *Eval. Rev.* **1981**, *5*, 602–619. [CrossRef]
59. MacKinnon, D.P.; Warsi, G.; Dwyer, J.H. A simulation study of mediated effect measures. *Multivar. Behav. Res.* **1995**, *30*, 41–62 [CrossRef]
60. Kenny, D.A.; Korchmaros, J.D.; Bolger, N. Lower level mediation in multilevel models. *Psychol. Methods* **2003**, *8*, 115–128 [CrossRef] [PubMed]
61. Kelly, T.C.; Mason, I.G.; Leiss, M.W.; Ganesh, S. University community responses to on-campus resource recycling. *Resour. Conserv. Recycl.* **2006**, *47*, 42–55. [CrossRef]
62. Shen, Y.; Chai, Y.W. Space-time flexibility of daily activities and gender differences: A case study of Beijing. *Acta Geog. Sin.* **2017**, *72*, 2214–2225.
63. Martin, E.W.; Shaheen, S.A. Evaluating public transit modal shift dynamics in response to bikesharing: A tale of two US cities. *J. Transp. Geogr.* **2014**, *41*, 315–324. [CrossRef]
64. Wang, P.; Liu, Q.; Qi, Y. Factors influencing sustainable consumption behaviors: A survey of the rural residents in China. *J. Clean. Prod.* **2014**, *63*, 152–165. [CrossRef]
65. Gu, P.; Ma, X. Investigation and analysis of a floating population's settlement intention and environmental concerns: A case study in the Shawan River Basin in Shenzhen, China. *Habitat Int.* **2013**, *39*, 170–178. [CrossRef]
66. Antimova, R.; Nawijn, J.; Peeters, P. The awareness/attitude-gap in sustainable tourism: A theoretical perspective. *Tour. Rev.* **2012**, *67*, 7–16. [CrossRef]
67. Axsen, J.; Mountain, D.C.; Jaccard, M. Combining stated and revealed choice research to simulate the neighbor effect: The case of hybrid-electric vehicles. *Resour. Energy Econ.* **2009**, *31*, 221–238. [CrossRef]

land

Article

Spatio-Temporal Dynamics of Carbon Emissions and Their Influencing Factors at the County Scale: A Case Study of Zhejiang Province, China

Xuanli Wang [1], Huifang Yu [1,2], Yiqun Wu [1,2,3,*], Congyue Zhou [3], Yonghua Li [3,4,5], Xingyu Lai [3] and Jiahao He [1,2,3]

1 College of Art and Archaeology, Hangzhou City University, Hangzhou 310015, China; 2231003017@stu.hzcu.edu.cn (X.W.); yuhf@hzcu.edu.cn (H.Y.); hejh@hzcu.edu.cn (J.H.)
2 Beautiful Hangzhou Environmental Planning and Architectural Design Research Center, Hangzhou City University, Hangzhou 310015, China
3 College of Civil Engineering and Architecture, Zhejiang University, Hangzhou 310058, China; 12012025@zju.edu.cn (C.Z.); lyh@zju.edu.cn (Y.L.); 22212126@zju.edu.cn (X.L.)
4 Center for Balance Architecture, Zhejiang University, Hangzhou 310058, China
5 Architectural Design and Research Institute of Zhejiang University Co., Ltd., Hangzhou 310058, China
* Correspondence: wuyq@hzcu.edu

Abstract: Significant carbon emissions, a key contributor to global climate warming, pose risks to ecosystems and human living conditions. It is crucial to monitor the spatial and temporal patterns of carbon emissions at the county level to reach the goals of carbon peak and neutrality. This study examines carbon emissions and economic and social problems data from 89 counties in Zhejiang Province. It employs analytical techniques such as LISA time path, spatio-temporal transition, and standard deviational ellipse to investigate the trends of carbon emissions from 2002 to 2022. Furthermore, it utilizes the GTWR model to evaluate the factors that influence these emissions on a county scale. The findings reveal the following: (1) The LISA time path analysis indicates a pronounced local spatial structure in the distribution of carbon emissions in Zhejiang Province from 2002 to 2022, characterized by increasing stability, notable path dependency, and some degree of spatial integration, albeit with a diminishing trend in overall integration. (2) The LISA spatio-temporal transition analysis indicates significant path dependency or lock-in effects in the county-level spatial clustering of carbon emissions. (3) Over the period 2002–2022, the centroid of carbon emissions in Zhejiang's counties mainly oscillated between 120°55′15″ E and 120°57′01″ E and between 29°55′52″ N and 29°59′11″ N, with a general northeastward shift forming a "V" pattern. This shift resulted in a stable "northeast–southwest" spatial distribution. (4) Factors such as population size, urbanization rate, and economic development level predominantly accelerate carbon emissions, whereas industrial structure tends to curb them. It is crucial to customize carbon mitigation plans to suit the circumstances of each county. This study provides insight into the spatial and temporal patterns of carbon emissions at the county level in Zhejiang Province. It offers crucial guidance for developing targeted and practical strategies to reduce carbon emissions.

Keywords: county level; carbon emissions; spatio-temporal dynamic; influencing factors; Zhejiang

Citation: Wang, X.; Yu, H.; Wu, Y.; Zhou, C.; Li, Y.; Lai, X.; He, J. Spatio-Temporal Dynamics of Carbon Emissions and Their Influencing Factors at the County Scale: A Case Study of Zhejiang Province, China. *Land* **2024**, *13*, 381. https://doi.org/10.3390/land13030381

Academic Editor: Nir Krakauer

Received: 6 February 2024
Revised: 10 March 2024
Accepted: 14 March 2024
Published: 17 March 2024

1. Introduction

Several crises, such as water scarcity, food insecurity, and public health emergencies, are being brought on by recent trends in global warming, which have led to a persistent degradation of the ecological environment and a threat to human survival [1–3]. The primary cause of global warming is the increase in carbon dioxide emissions. As a result, the world community has focused its efforts on reducing these emissions, encouraging low-carbon, sustainable growth and supporting ecological preservation [4–6]. Following a sharp increase in emissions, China overtook the US as the world's top carbon producer in

2007 [7]. By 2022, China's carbon emissions had surged to 11.48 billion tons, marking an annual growth rate of 3.7% since 2018 [8]. Predominantly, these emissions stem from fossil fuels, comprising 84.1% of China's total in 2020 [9]. As a result, China has actively looked for ways to reduce its carbon footprint, set several goals and policies [10,11], and, at the 75th UN General Assembly, declared its intention to achieve carbon neutrality by 2060 and a carbon peak by 2030, reiterating its commitment to a comprehensive green transition [12].

Counties, serving as the core administrative entities and spatial foundations for industrial and economic expansion in China [13], encompass 78% of the nation's landmass, are home to two-thirds of its population, and contribute approximately 51.8% to China's GDP [14]. The county-level administrative divisions here include counties (such as Deqing county, Cangnan county, and Panan county), county-level cities (Dongyang City, Cixi City and Yueqing City), and districts (Wucheng district, Yuhang district, and Haishu district) under city jurisdiction. Notably, they are responsible for over half of the nation's carbon emissions [15], underscoring the immense potential and urgent need for low-carbon initiatives at this level, which are crucial for China's carbon neutrality aspirations [16]. Achieving these reduction goals entails national macro-strategic planning and an emphasis on localized carbon emission strategies [17]. Hence, it is crucial to examine county-level carbon emissions thoroughly, accurately chart their spatio-temporal progression, and comprehend the factors that drive them. Acquiring this knowledge is essential for fostering regional cooperation in sustainable growth and improving the scientific accuracy, execution, and efficacy of energy conservation and emissions reduction strategies [15].

Researchers in the modern era have used a wide variety of analytical tools to focus more on the temporal and spatial patterns of carbon emissions. Duman [18] utilized the standard deviational ellipse (SDE). They observed a progressive change in the trajectory of urban carbon emissions in China, transitioning from a "northeast–southwest" to a "northwest–southeast" orientation. Concurrently, the standard deviation ellipse grew, indicating that air pollution and carbon dioxide emissions had expanded in space over time. Liu [19] utilized spatial autocorrelation (SA) to reveal notable geographical variation, self-correlation, and spillover effects associated with urbanization and carbon emissions. Zhang [20] utilized Kernel Density Estimation (KDE) for their study, effectively pinpointing spatial associations between carbon emission patterns and land-use-related emissions. In another investigation, Soares [21] utilized the Gini Index (GI) to analyze the world's 50 major economies and discovered a strong correlation between economic levels and carbon emissions, revealing that higher economic levels lead to greater carbon emissions. Williams [22] employed the Coefficient of Variation (CV) to analyze carbon emission patterns in the Southern Ocean and discovered that this region significantly contributes to the sequestration of heat and the absorption of anthropogenic carbon. These research works provide essential perspectives on the spatio-temporal patterns of carbon emissions. However, the main focus of these studies has been on the provincial and municipal levels, with significantly less attention paid to the county level. This discrepancy underscores the necessity for more detailed research at a finer scale. Delving into these aspects at a more granular level is crucial for understanding the subtle dynamics of spatial interactions and the specific impacts between neighboring areas. Such an approach can aid in the formulation of more effective and targeted decision-making and planning strategies.

Understanding the factors influencing carbon emissions is vital for advancing carbon reduction and environmental improvement. Scholars frequently employ methods like Spatial Econometric Regression Models (SEMs), Logarithmic Mean Divisia Index (LMDI), Geographically Weighted Regression (GWR), Multiscale Geographically Weighted Regression (MGWR) models, and STIRPAT models in their research. For example, Qu [23] analyzed the spatial response mechanism between carbon emission efficiency and ecosystem services using SEMs, finding that overall ecosystem services are influenced not only by local carbon emission efficiency but also by that of surrounding areas. Quan [24] and González [25], through their analysis with the LMDI model, discovered that in Spain, factors related to per capita output are predominant in the industrial sector, while factors related to carbonization effects play a critical role in the commercial sector. Khodakarami [26] used the GWR model to study how trees capture carbon, absorb CO_2, and produce oxygen across different areas and what affects these processes. He found that the carbon-absorbing services of urban green spaces are key to cutting greenhouse gas emissions and supporting sustainable environmental development. Li [27] utilized the MGWR model to investigate the determinants of urban carbon emissions in China. Their findings uncovered geographical variations in the effects of per capita GDP, secondary industry percentage, and population density on carbon emissions. Nosheen [28] used the STIRPAT model to analyze the factors affecting carbon emissions in some Asian countries and found that energy consumption and urbanization contribute to increasing carbon emissions. Although these studies provide valuable insights into the factors affecting carbon emissions, they often focus more on the overall positive and negative impacts during the entire phase, without delving into the temporal and spatial changes of these factors. Thus, there is a need for in-depth research from different temporal and spatial perspectives to clearly observe the changes in the time series of influencing factors and to reveal local factors causing regional disparities. By expanding the research scope to these aspects, it becomes possible to gain a more detailed understanding of the reasons affecting carbon emissions, thereby formulating more targeted and effective mitigation and adaptation strategies.

This article endeavors to clarify the spatial–temporal aspects of carbon emissions at the county scale in Zhejiang Province using ESTDA methods, including LISA time path and LISA spatio-temporal transition, as well as the standard deviational ellipse. The GTWR model is utilized to investigate the factors influencing carbon emissions at this level. This study is innovative in two key respects: Firstly, it adopts a micro-scale approach, analyzing the spatio-temporal attributes of carbon emissions at the county level in Zhejiang Province, marking a shift from static to dynamic research methodologies. Secondly, it leverages the GTWR model's capacity to account for the spatio-temporal heterogeneity of influencing factors. It enables a more nuanced understanding of these factors on carbon emissions, in contrast to traditional global regression methods. This approach is valuable for policymakers developing targeted carbon reduction strategies. The research objectives were as follows: (1) to collect and examine Zhejiang Province county-level carbon emission data in order to shed light on the spatio-temporal aspects of these emissions; (2) to create and employ a GTWR model for analyzing the factors that affect carbon emissions in different spatial and temporal dimensions within Zhejiang Province; and (3) to integrate the findings from both analyses to formulate viable strategies and recommendations for reducing carbon emissions at the county level.

The remainder of this work is organized in the following manner: Section 2 outlines the data sources and methodologies used in this research. Section 3 explores the spatial and temporal dynamics of carbon emissions, along with their determinants at the county level in Zhejiang Province. Section 4 consolidates the key results, discusses possible approaches and suggestions, recognizes the limitations of this study, and suggests future research avenues and areas of focus.

Based on this research gap, this study aimed to investigate the spatio-temporal patterns of carbon emissions and their influencing factors from a micro-perspective of county scale. Therefore, we proposed two research questions:

RQ1. How do the local development directions and global trajectories of the spatio-temporal pattern of carbon emissions at the county level in Zhejiang Province change?

RQ2. What factors drive the changes in the spatio-temporal pattern of carbon emissions at the county level in Zhejiang Province?

This paper contributes to the literature on carbon emissions in several ways: First, it delves into the county scale, moving beyond the provincial and city scales. Second, it examines the spatio-temporal pattern of carbon emissions at the county level in Zhejiang Province from both local and global dimensions. Third, it explores the factors affecting the spatio-temporal distribution of carbon emissions at the county level in Zhejiang Province and analyzes the spatio-temporal changes of significant influencing factors.

2. Materials and Methods

2.1. Research Framework

Our empirical research framework is illustrated in Figure 1. Firstly, we organized and processed data from the China Emissions Accounts and Datasets (CEAD), Zhejiang Statistical Yearbooks, and some municipal statistical yearbooks for the years 2002–2022, generating both total carbon emission data and data on influencing factors composed of multiple indicators. Secondly, based on the total carbon emissions of each county over the years, we comprehensively elucidated the spatio-temporal evolution of county-level carbon emissions in terms of spatial distribution, regional disparities, and temporal evolution using LISA time path, LISA spatio-temporal transitions, and standard deviation ellipse methods. The analysis of relative length in the LISA time path examined the dynamic nature of local spatial dependence and structure, while the analysis of curvature examined the direction of local spatial dependence, and the analysis of transfer direction assessed the competitive situation between neighboring counties. The LISA spatio-temporal transitions analysis evaluated the transfer of local spatial association types, thereby determining the stability of county-level spatial positions [15]. The analysis of centroid displacement, centroid distribution, and ellipse orientation in the standard deviation ellipse examined the overall spatial distribution and spatio-temporal trend [18]. The former two were analyzed from a local perspective, while the latter was analyzed from a global perspective. Finally, we conducted OLS and GWR analyses to examine and eliminate inappropriate indicators from the total carbon emission and influencing factor data, and then constructed a GTWR model to analyze the positive and negative impacts of key factors on county-level carbon emissions, which are of great significance for government policy formulation and intervention planning [26].

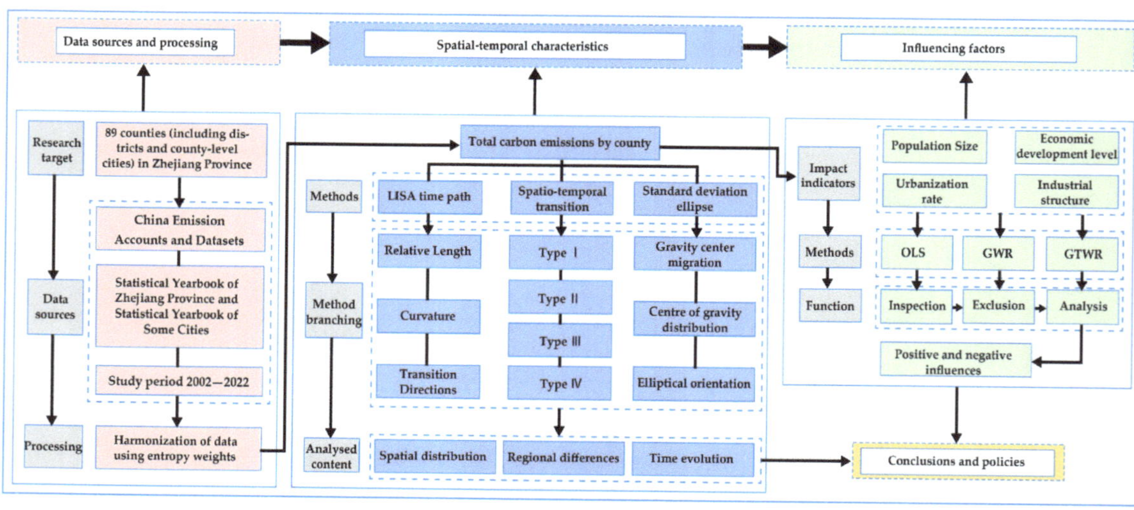

Figure 1. Schematic of research framework.

2.2. Research Area

Zhejiang Province is located on the southern bank of the Yangtze River Delta in China's southeastern coastline region. It extends from 27°02′ to 31°11′ N latitude and 118°01′ to 123°10′ E longitude. Comprising ten cities, Zhejiang Province added one county-level city in the 2021 administrative division adjustment, totaling 90 county-level administrative districts (Figure 2). Although Zhejiang's area accounts for less than 1.5% of the country's total, its per capita GDP in 2021 reached RMB 113,900, ranking fifth nationwide, with a gross domestic product amounting to RMB 6.5 trillion [29,30]. Nevertheless, the swift pace of development has brought about a multitude of environmental challenges, among which is the escalation of carbon emissions. As an "Ecological Civilization Demonstration Zone", the Zhejiang Provincial Government is committed to fostering a low-carbon economy, aiming to mitigate carbon emissions through strategic spatial planning [31]. Over the years, it has implemented policies such as "Implementation Opinions on Supporting Work for Carbon Peak and Carbon Neutrality" and "Zhejiang Province's 14th Five-Year Plan for Energy Conservation and Emission Reduction." The province's efforts in energy conservation and decreasing emissions have been particularly effective, making it a remarkable case for research with significant demonstrative and referential relevance.

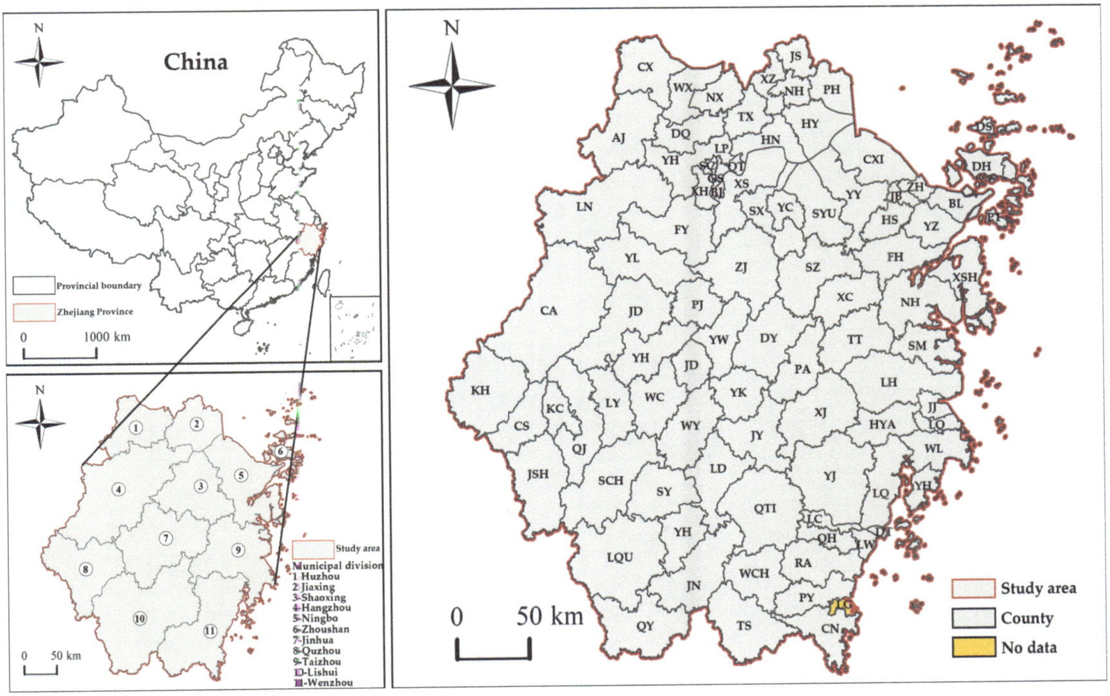

Figure 2. Study area and administrative division.

2.3 *Data Sources*

This study selected 90 districts and counties in Zhejiang Province as research subjects. To analyze the spatial and temporal patterns of carbon emissions in Zhejiang Province at the county level, this study employed carbon emission data from 2002 to 2022. The data were obtained from the China Carbon Accounting Database (CEAD), which may be accessed at https://www.ceads.net/data/county/ (accessed on 6 August 2023). Due to the absence of carbon emission data for Longgang City, this area was excluded from the data analysis, making the other 89 counties the focus of the study. This study examined the factors that influence carbon emissions in the counties of Zhejiang Province. It utilized social, economic, and industrial data sourced from the Zhejiang Statistical Yearbook (https://tjj.zj.gov.cn/col/col1525563/index.html accessed on 6 August 2023) as well as the statistical yearbooks of different cities and counties spanning from 2002 to 2022.

2.4. *Methods for Studying the Spatio-Temporal Dynamics of Carbon Emissions*

2.4.1. LISA Time Path

The LISA time path is a continuous representation of the Markov transition matrix. It dynamically displays spatial differential patterns over time by connecting changes in each period through the movement's length and angle in the Moran scatter plot. The LISA coordinates of each county unit are composed of Z and W_Z, derived from various

2.4.3. Standard Deviational Ellipse and Gravity Center Migration

The standard deviational ellipse (SDE), characterized by fundamental parameters like center, major axis, minor axis, and orientation angle, effectively unveils the general traits of geographic features' spatial distribution. It can quantitatively depict the spatial pattern distribution and the process of spatio-temporal evolution [36]. The equation for the standard deviational ellipse [37–39] is expressed as follows:

$$\overline{X}_w = \frac{\sum_{i=1}^{n} w_i x_i}{\sum_{i=1}^{n} w_i}, \overline{Y}_w = \frac{\sum_{i=1}^{n} w_i y_i}{\sum_{i=1}^{n} w_i} \tag{5}$$

In Equation (5), $(\overline{X}, \overline{Y})$ represents the centroid coordinates of carbon emissions in Zhejiang Province's county-level units, n is the number of county-level units in Zhejiang Province, and (x_i, y_i) represents the center coordinates of each county-level research unit, while w_i denotes the weight of the research unit.

$$\sigma_x = \sqrt{\frac{\sum_{i=1}^{n} \left(w_i \widetilde{x}_i \cos\alpha - w_i \widetilde{y}_i \sin\alpha \right)^2}{\sum_{i=1}^{n} w_i^2}} \tag{6}$$

$$\sigma_y = \sqrt{\frac{\sum_{i=1}^{n} \left(w_i \widetilde{x}_i \sin\alpha - w_i \widetilde{y}_i \cos\alpha \right)^2}{\sum_{i=1}^{n} w_i^2}} \tag{7}$$

In Equations (6) and (7), σ_x and σ_y reflect the degree of dispersion of carbon emissions in the major and minor directions, respectively.

2.5. Methods for Studying the Factors Influencing Carbon Emissions

2.5.1. Indicator Selection and Processing

Drawing from relevant research, several critical factors were recognized as significant contributors to carbon emissions. These include the size of the population, level of economic development, rate of urbanization, composition of industries, advancements in technology, investments in fixed assets, and intensity of energy usage [40–43]. Insufficient data accessibility has led to insufficient records for technological level and fixed asset investments at both the county and annual levels. Consequently, we chose to focus our investigation on variables such as population size, economic development level, urbanization rate, and industrial structure when exploring the spatio-temporal variations in carbon emission factors across Zhejiang Province's counties during the period spanning from 2002 to 2022, as detailed in Table 1. Specifically, population size was defined by the year-end resident population of each county within Zhejiang Province [44], economic development level was quantified using per capita GDP data for Zhejiang Province [45], urbanization rate was determined by the ratio of urban population to the total population at the year-end in each county [46], and industrial structure was expressed as the proportion of the secondary industry's value relative to each county's total GDP [47].

As the selected influencing factors had different units, it was necessary to normalize the raw data of each influencing factor to ensure data uniformity and eliminate the influence caused by unit heterogeneity [48]. Equation (8) was used to standardize positive indicators, while Equation (9) was used to normalize negative indicators.

$$Y_{ij} = \frac{x_{ij} - min(x_i)}{max(x_i) - min(x_i)} \tag{8}$$

$$Y_{ij} = \frac{max(x_i) - x_{ij}}{max(x_i) - min(x_i)} \tag{9}$$

In Formulas (8) and (9), Y_{ij} represents the jth indicator of the ith object among the four indicators. Y_{ij} stands for the coefficient of this indicator, with larger values indicating greater influence and vice versa.

Table 1. Definition of relevant indicators.

Indicators	Indicator Abbreviations	Connotation	Unit
Population size	POP	Resident population by county at the end of the year	Ten thousand
Economic development level	PGDP	GDP per capita	Ten thousand yuan
Urbanization rate	URB	Ratio of urban population to resident population	%
Industrial structure	INS	Ratio of secondary sector output to GDP	%

2.5.2. Spatio-Temporal Geographically Weighted Regression Model

The traditional linear regression approach does not effectively account for the spatial characteristics of regression parameters. To better assess the heterogeneity of elements impacting carbon emissions, the GWR model is employed, enabling the quantification of the spatial variability of these factors. However, the geographically weighted model lacks consideration of time factors and cannot accurately analyze cross-sectional data over a long time series. This issue is addressed in the GTWR model [49]. The GTWR model, building on the GWR framework, incorporates temporal effects, transforming cross-sectional data into panel data. This approach addresses issues of spatio-temporal non-stationarity and unestimated parameters, leading to more reliable outcomes [50]. The formula is as follows:

$$Y_i = \beta_0(u_i, v_i, t_i) + \sum_{k=1}^{p} \beta_k(u_i, v_i, t_i)X_{ik} + \varepsilon_i \tag{10}$$

In Equation (10) [51], Y_i represents the dependent variable for the ith sample area. X_{ik} denotes the observed value of the kth explanatory variable in the ith sample area. (u_i, v_i, t_i) represents the spatio-temporal coordinates of the ith sample area. $\beta_0(u_i, v_i, t_i)$ represents the spatio-temporal intercept for the ith sample area. $\beta_k(u_i, v_i, t_i)$ represents the regression coefficient of the kth explanatory variable in the ith sample area. $\beta > 0$ indicates that the explanatory variable is positively correlated with the dependent variable, while the opposite holds true for negative correlation. ε_i represents the random disturbance term.

3. Results

3.1. Analysis of LISA Time Paths

Based on the temporal variations in Moran's Index in ArcGIS, the study period was divided into four time intervals, 2002–2007, 2007–2012, 2012–2017, and 2017–2022, to analyze the geometric features of LISA time paths.

3.1.1. Analysis of Relative Length

This paper classifies the relative lengths of LISA time paths into six levels. To ensure comparability of data across the four time periods, the same division intervals and levels for relative lengths of time paths were applied to all four periods (Figure 3). From 2002 to 2007, counties with high relative lengths, ranging from 2.088 to 5.982, included Yu Hang, Ci Xi, and Yi Wu. From 2007 to 2012, additional counties, Tong Xiang and Hai Ning, were added to the list, along with Yu Hang and Ci Xi. From 2012 to 2017, although the number of counties remained the same, there were changes in the specific counties, namely Xi Hu, Xiao Shan, Jiang Bei, and Yi Wu. Similarly, from 2017 to 2022, the number of counties did not change, but they changed to Xi Hu, Xiao Shan, Ci Xi, and Yu Yao. This indicates a solid dynamic nature in these areas' local spatial structure of carbon emissions. The primary cause of this phenomenon may be ascribed to swift advancements in multiple areas, such as industry and services, propelled by the strategy of integrated growth in the Hangzhou metropolitan region. Among the four time periods, 144 counties, accounting for 53.93% of all counties, had low relative lengths falling between 0.246 and 0.911. The proportions for the first three periods experienced relatively minor changes, at 16.10%, 17.60%, and 16.48%, respectively, while the proportion for the 2017–2022 period showed a significant increase, rising to 21.35%. This suggests an overall more robust local spatial structure stability of carbon emissions in Zhejiang Province's counties from 2002 to 2022, with a trend of increasing stability. Spatially, it was observed that counties with low relative lengths in all four time periods were predominantly distributed in the western part of Wenzhou, Lishui in southern Zhejiang, Quzhou in Zhejiang's west, and the western regions of Taizhou and Hangzhou. This indicated that these areas exhibited the robust stability in the local spatial structure of carbon emissions and showed a trend of spreading from southern and western Zhejiang to northern and eastern regions. The primary cause of this occurrence could be attributed to the fact that these regions predominantly comprise hilly counties with relatively low levels of economic development, substantial forest resources, and greater capacity for carbon storage [17].

3.1.2. Analysis of Curvature

Likewise, the curvature of LISA time routes was categorized into six levels, and these same categorization intervals and levels were used for the four time periods (Figure 4). From 2002 to 2007, six counties had high curvature values ranging from 5.899 to 13.932, including Yuhang, Xiuzhou, Pinghu, Xinchang, and Xiangshan. For 2007–2012, it changed to Deqing, Cixi, Zhenhai, Beilun, Linhai, and Wencheng, totaling six counties. In 2012–2017, it was reduced to only one county, Lucheng. However, from 2017 to 2022, it increased to 19 counties, including Yuhang, Nanhu, Yuecheng, Jiangbei, and others. This indicates that these counties exhibited high fluctuations in both local spatial dependence direction and carbon emissions, and the counties with the most vital spatial dependence direction fluctuations were on the rise. In 2002–2007, there were 14 counties with relatively high curvature values ranging from 2.360 to 5.898. In the 2007–2012 and 2012–2017 periods, such counties remained relatively stable at 9 and 18, respectively. However, in the 2017–2022 period, there was a significant increase, reaching 44 counties. Spatially, the counties with relatively high curvature in the first three periods were mainly distributed in northern Zhejiang, western Zhejiang, and parts of eastern Zhejiang. In the 2017–2022 period, this trend expanded to central and southern Zhejiang. This indicates that these regions exhibited high fluctuations in both local spatial dependence direction and county-level carbon emissions, with an upward trend. The proportions of counties with low curvature values ranging from 1.075 to 2.359 in the four periods were 77.53%, 83.15%, 78.65%, and 29.21%, respectively. This suggests that carbon emissions and the direction of local spatial dependence in Zhejiang Province were relatively stable in the first three periods but experienced a sharp decline in the fourth.

3.1.3. Analysis of Transition Directions

Through the directional analysis of LISA time path transitions, this study revealed the phased characteristics of competitive and cooperative situations among neighboring counties in Zhejiang Province (Figure 5). From 2002 to 2007, 28 counties with synergistically high growth formed a strip-like pattern, primarily distributed in the northern Zhejiang counties like Yuhang and Deqing and southeastern Zhejiang counties like Yongjia and Wenling. This indicates strong spatial integration in the evolution of these areas' carbon emission spatial patterns. From 2007 to 2012, there was an increase in the number of counties experiencing extremely high development, reaching 29. This growth formed a block-like pattern, with a shift in distribution from the northern and southern regions to the western and eastern parts of Zhejiang. From 2012 to 2017, there were 27 counties with synergistically high growth, down by two from the previous phase, with significant changes in distribution from southeastern and western to northern and eastern Zhejiang.

The move can be ascribed to implementing the 13th Five-Year Plan, which aimed to promote integrated prosperity in north and east Zhejiang by accelerating economic growth and transforming industries. Counties with synergistically low growth increased to 36, forming a block-like pattern and shifting from northern to central, southern, and western Zhejiang. From 2017 to 2022, there were 28 counties with synergistically high growth, showing a fragmented pattern and shifting from northern and eastern to southern and western Zhejiang. In summary, when comparing the four stages, the count of counties experiencing synergistically significant growth remained relatively stable. In contrast, the count of counties with synergistically low growth exhibited more substantial variations. Overall, it appears that there is some regional coherence in the distribution of carbon emissions in Zhejiang Province. However, this tendency toward integration has been dwindling for some time now.

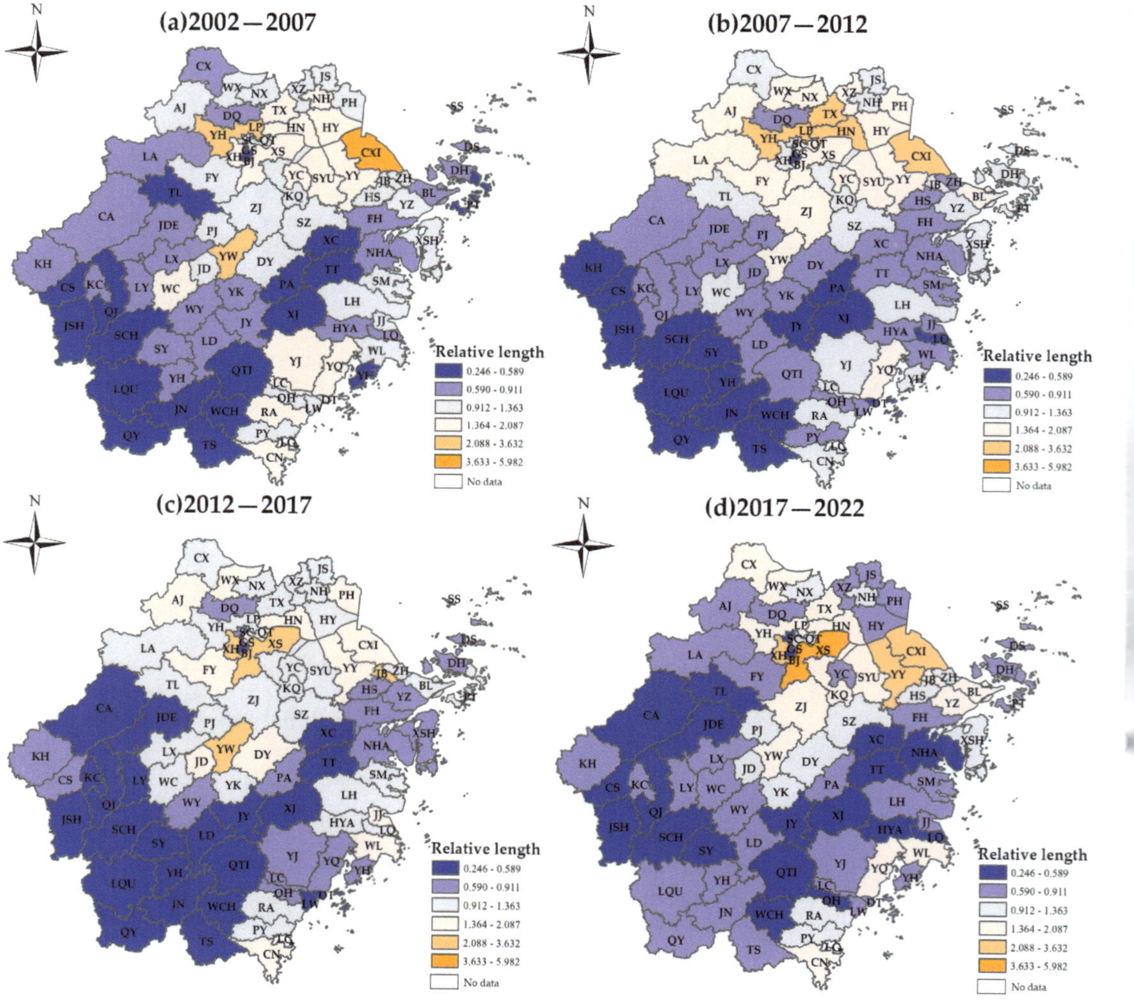

Figure 3. **(a–d)** The relative lengths of LISA time paths in counties of Zhejiang Province from 2002 to 2022.

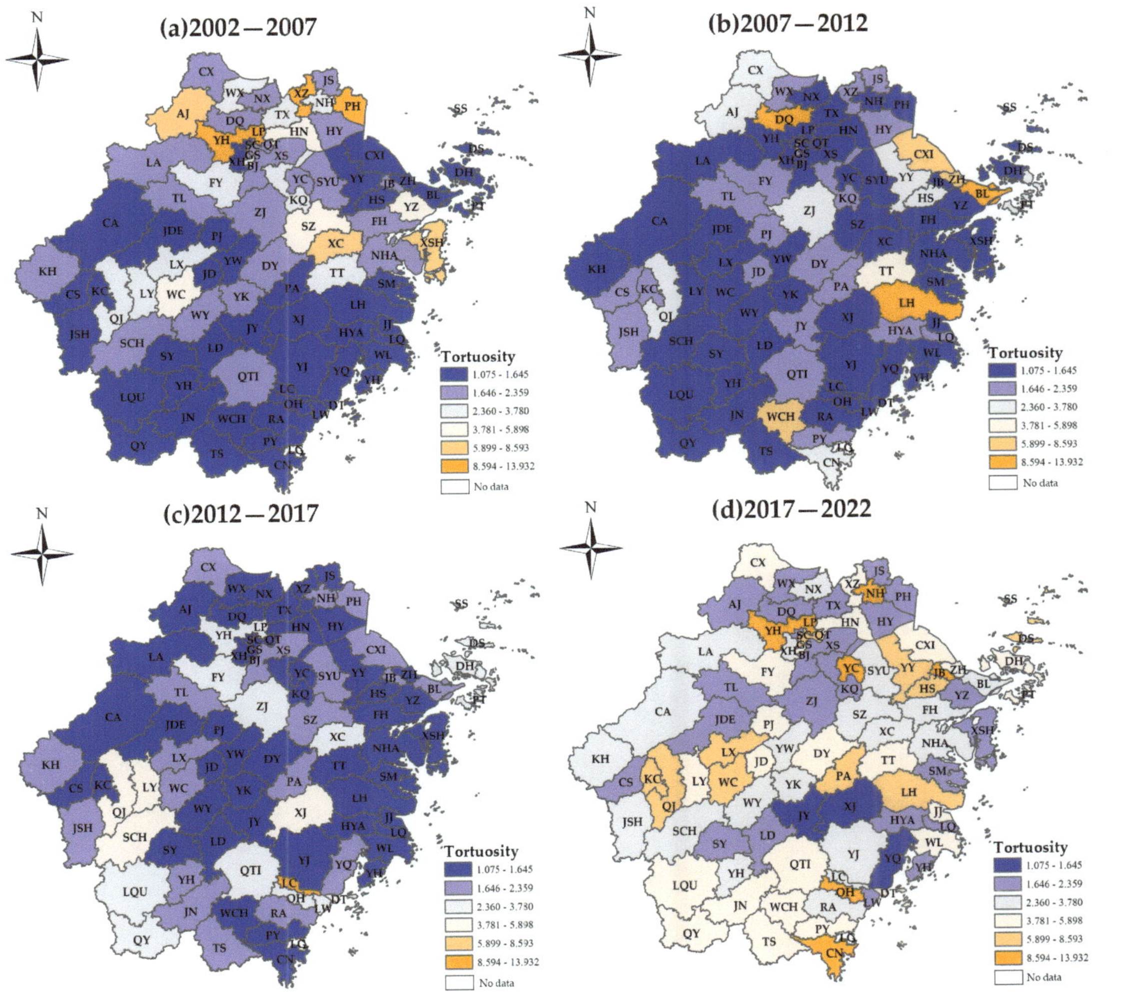

Figure 4. (**a–d**). The LISA time path curvature of counties in Zhejiang Province from 2002 to 2022.

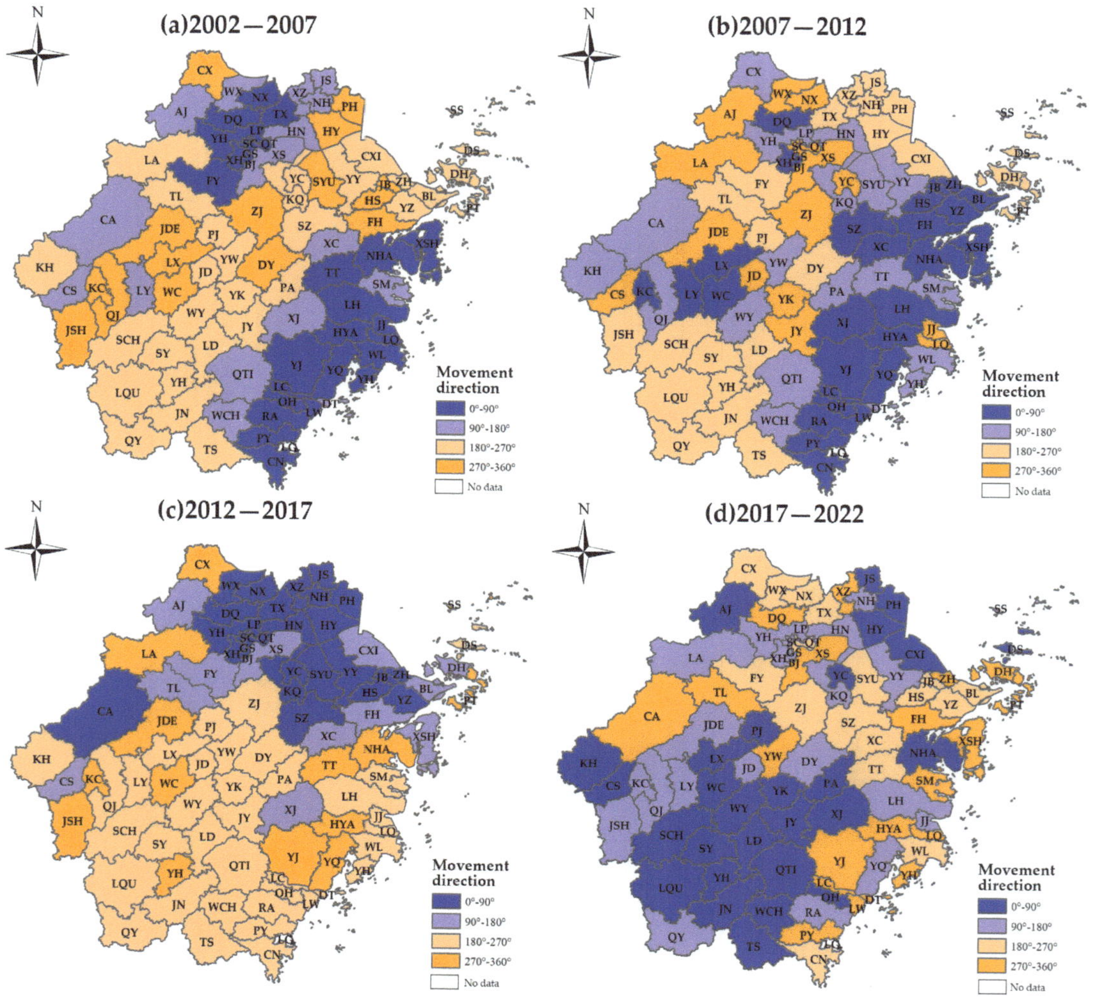

Figure 5. (a–d). The LISA spatio-temporal transition direction of counties in Zhejiang Province from 2002 to 2022.

3.2. *Analysis of Spatio-Temporal Transition*

Utilizing LISA space–time transitions allows for the investigation of the type transitions in spatial correlations of county-level carbon emissions in Zhejiang Province and the stability of the carbon emission space. As Table 2 indicates, the non-diagonal elements in the transition probability matrix were all smaller than the diagonal elements, suggesting transitions between types in all four categories of local spatial patterns. The most common transition during the four periods of 2002–2007, 2007–2012, 2012–2017, and 2017–2022 was of Type IV, with counties undergoing this type of transition accounting for 97.2%, 98.2%, 98.5%, and 97.6% of all counties, respectively. The stability in the Moran's scatterplot suggests a significant level of path dependency or lock-in features in the geographic aggregation of carbon emissions at the county level in Zhejiang Province. The proportions of transitions for Types I, II, and III were lower, indicating that the local county areas were less influenced by spillover effects from surrounding regions. Conversely, internal causes exerted a more significant influence on alterations in the geographical arrangement of carbon emissions. This shows that there is less reliance on neighboring regions and more emphasis on the role of internal dynamics in determining the geographic distribution of carbon emissions in Zhejiang Province at the county level.

Table 2. Spatio-temporal transitions and spatial autocorrelation values for each period from 2002 to 2022.

Period	t/t + 1	HH	LH	LL	HL	Types	Number	Proportions	St
2002–2007	HH	0.975	0.005	0.005	0.015	I	7	0.013	0.972
	LH	0.020	0.980	0.000	0.000	II	7	0.013	
	LL	0.000	0.005	0.989	0.005	III	1	0.002	
	HL	0.063	0.000	0.063	0.875	IV	520	0.972	
2007–2012	HH	0.973	0.016	0.000	0.011	I	4	0.007	0.985
	LH	0.000	0.991	0.009	0.000	II	4	0.007	
	LL	0.000	0.000	1.000	0.000	III	0	0.000	
	HL	0.025	0.000	0.025	0.950	IV	526	0.985	
2012–2017	HH	0.977	0.012	0.000	0.012	I	4	0.007	0.985
	LH	0.018	0.982	0.000	0.000	II	4	0.007	
	LL	0.000	0.000	1.000	0.000	III	0	0.000	
	HL	0.045	0.000	0.000	0.955	IV	526	0.985	
2017–2022	HH	0.971	0.018	0.006	0.006	I	7	0.013	0.976
	LH	0.017	0.983	0.000	0.000	II	4	0.007	
	LL	0.005	0.005	0.990	0.000	III	2	0.004	
	HL	0.047	0.000	0.047	0.907	IV	521	0.976	

3.3. *Standard Deviational Ellipse and Gravity Center Migration Analysis of Carbon Emissions*

The spatial distribution and centroid migration of county-level carbon emissions in Zhejiang Province can be further studied through the standard deviation ellipse method in ArcGIS. With the ellipse's primary axis modification, the spatial distribution produced a consistent 'northeast–southwest' pattern, as shown in Table 3 and Figure 6. Its reduction from 117.730 km to 117.487 km between 2002 and 2022 along the principal axis suggests a centripetal aggregation of carbon emissions in Zhejiang Province, which extended from the southwest to the northeast. One possible explanation is that the province of northern Zhejiang was one of the first in China to undergo fast industrialization after the country joined the WTO in 2001. The minor axis underwent a more significant change than the central axis, shrinking from 165.998 km to 161.923 km, demonstrating a pronounced centripetal distribution of county-level carbon emissions.

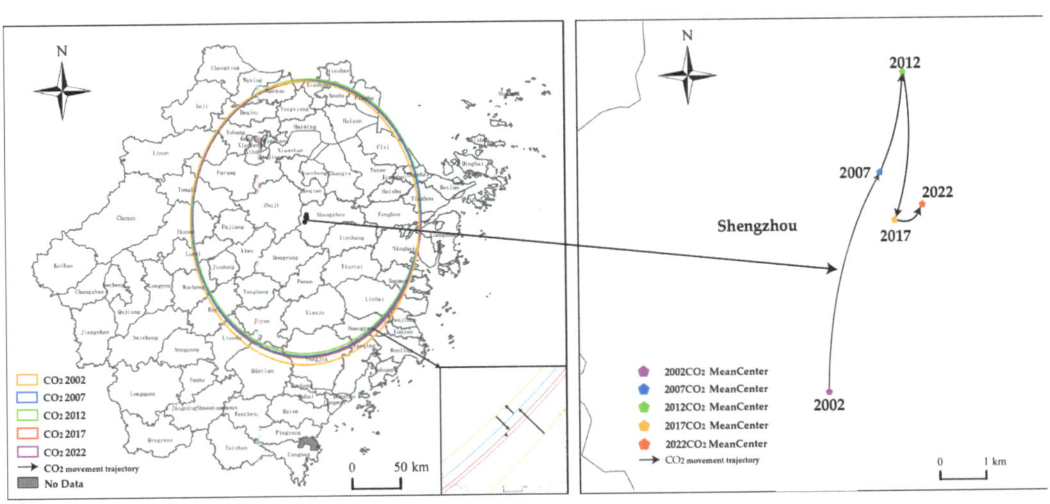

Figure 6. The distribution of standard deviational ellipse and gravity center shift trajectory.

Table 3. Spatio-temporal transition matrix and spatial cohesion value of each period.

Year	Coordinates of Gravity Center		Shift of Gravity Center		Long Axis/km	Short Axis/km	Angle°
	Longitudes	Latitudes	Orientations	Length/km			
2002	120°55′15″ E	29°55′52″ N			117.730	165.998	1.798
2007	120°56′18″ E	29°59′74″ N	northwestern	5.542	117.863	161.424	1.836
2012	120°56′65″ E	29°61′72″ N	northwestern	0.746	117.597	160.764	2.117
2017	120°56′47″ E	29°58′82″ N	southwestern	0.285	117.797	161.855	1.582
2022	120°57′01″ E	29°59′11″ N	northwestern	0.853	117.487	161.923	0.820

Regarding the centroid position, from 2002 to 2022, the centroid carbon emissions mainly varied between 120°55′15″ E and 120°57′01″ E and between 29°55′52″ N and 29°59′11″ N, with the centroid consistently located in Shengzhou City in northern Zhejiang. This suggests that the high carbon emission values have always been in the northern region during this period; hence, northern counties should assume more responsibility in formulating differentiated emission reduction policies. Analyzing the centroid's movement trajectory, the overall trend was northeastward, forming a 'V' shape. From 2002 to 2007, it moved 5.542 km northeastward; from 2007 to 2012, it shifted 0.746 km northeastward; from 2012 to 2017, it moved 0.285 km southwestward; and from 2017 to 2022, it turned 0.853 km northeastward. This suggests that the counties in Zhejiang's northeast are seeing a faster-than-average rise in carbon emissions and that the regions with the highest rates of overall carbon emissions increase continue expanding towards the northeast.

3.4. Analysis of Spatio-Temporal Heterogeneity of Carbon Emission Influencing Factors

3.4.1. Construction of GTWR Model and Analysis of Regression Model Results

This article utilized panel data from 2002 to 2022, specifically focusing on county-level locations in Zhejiang Province. The indicator variable was the carbon emissions of each district and county, while the explanatory variables included population size, economic development level, urbanization rate, and industrial structure. To determine the parameters of the factors affecting carbon emissions in various districts and counties throughout time, the GTWR model was built at the county level. Due to the issue of multicollinearity among the independent variables affecting regression analysis, this study first conducted an OLS analysis of the four explanatory variables. Generally, a VIF more significant than ten and a tolerance less than 0.1 indicate the potential presence of multicollinearity. As seen in Table 4, the VIF values for each explanatory variable were all below 10, confirming the absence of significant multicollinearity, thereby justifying the progression to GWR model regression analysis. Subsequently, the data were input into the GWR model for testing. It was found that these four explanatory variables could be included in the GWR model, thus proving the absence of both explicit and implicit multicollinearity among these variables. This validation allowed for the progression of GTWR model regression analysis.

Table 4. Multicollinearity test.

Variable	Population Size	Level of Economic Development	Urbanization Rate	Industrial Structure
VIF	1.155	1.411	1.230	1.211

Utilizing the GTWR model, this study selected the goodness-of-fit R^2, adjusted R^2, and AICc as indicators to evaluate the model's confidence. An analysis of the parameter estimation results was conducted. Higher values of R^2, a measure of fit that ranges from 0.0 to 1.0, indicate that the model has greater explanatory ability; the AICc value is a measure of model performance, with smaller values suggesting better model fit to the observed data. This study also provided regression results from OLS and GWR models for comparison to validate the applicability and accuracy of the GTWR model. According to Table 5, the GTWR model shows R^2 and adjusted R^2 values of 0.8846 and 0.8844, respectively, representing significant improvements compared to OLS and GWR; the AICc value is

−4784.4224, indicating a substantial reduction compared to OLS and GWR. These findings show that the GTWR model was better able to account for the spatial and temporal variety of the many factors impacting carbon emissions in Zhejiang Province's county-level locations.

Table 5. Comparison of model evaluation metrics.

Model Parameters	OLS	GWR	GTWR
R^2	0.7487	0.8556	0.8846
Adjusted R^2	0.7368	0.8203	0.8844
AICc	−157.3643	−180.3784	−4784.4224

3.4.2. Spatio-Temporal Evolution of Various Influences

Using the GTWR model, this study derived the contribution coefficients of various influencing factors on carbon emissions across different counties during the research period, and constructed a diagram illustrating the temporal and spatial variations of these factors (Figures 7 and 8). The results showed the following:

(1) The population size contributes the most to carbon emissions in the county-level regions of Zhejiang Province, with a positive coefficient. Temporally (Figure 7a), the regression coefficient of population size showed an upward trend, rising gently between 2002 and 2008 and more sharply from 2008 to 2022. This phenomenon could be attributed to rapid economic development after 2008, an improvement in living standards, and the opening of the two-child policy in 2016, leading to a sharp increase in population once again and intensified consumption of resources and energy. Spatially (Figure 8a), the areas most affected by population size were mainly located in the northern counties and cities of Zhejiang Province, including Hangzhou, Huzhou, and Jiaxing. Bui Minh [52] also proposed in his research that the degree of population aggregation affects regional carbon emissions. Counties less impacted by population size included two categories: one was counties with smaller population sizes and relatively backward economic development, which have a smaller impact on carbon emissions, such as Kaihua, Qingyuan, and Qujiang counties; the other included counties with larger population sizes but higher degrees of aggregation, where population aggregation has effectively improved energy utilization efficiency and reduced carbon emissions, such as Ruian, Yongjia, and Cangnan counties. Rehman's [53] study also mentioned that high population density can develop renewable energy to reduce carbon emissions.

(2) The level of economic development has a positive overall impact on carbon emissions at the county level in Zhejiang Province. Temporally (Figure 7b), the coefficient of overall economic development level gradually decreased, significantly from 2002 to 2017, and then more gently from 2017 to 2022. This decrease may be attributed to the "Eleventh Five-Year Plan" and the "Thirteenth Five-Year Plan" introduced after 2017, which promoted further optimization of the economic structure and the development of a low-carbon economy. Spatially (Figure 8b), counties in the central part of Zhejiang Province, such as Yiwu, Zhuji, and Dongyang, have seen significant positive impacts of economic growth factors on carbon emissions. The main reason might be the continuous expansion of construction land, development of industry and manufacturing, and increasing energy demands, thus leading to rapid growth in carbon emissions. Gershon's [54] study also found that economic growth significantly increases energy consumption, being a major factor promoting carbon emissions. Counties where economic growth hurt carbon emissions were mainly concentrated in the eastern areas of Taizhou, Wenzhou, and the eastern counties of Jiaxing. The reasons might be twofold: firstly, these countries might focus more on optimizing energy structure and encouraging technological innovation while growing economically, thereby improving energy efficiency and reducing carbon emissions, similar to Balta-Ozkan's [55] findings in the UK; secondly, by developing the tourism economy, this economic model can reduce environmental degradation and carbon emissions, a finding confirmed by Raihan's [56]

study on the positive role of developing a low-carbon tourism economy in reducing carbon emissions.

(3) Urbanization rates are positively correlated with overall carbon emissions in the county-level regions of Zhejiang Province. From a temporal perspective (Figure 7c), there was an upward trend from 2002 to 2012, and a stable state from 2012 to 2022, overall showing an upward trend. The reason might be that after 2002, with the rapid advancement of urbanization, the surge in urban population led to increased energy use in transportation, industry, and urban land, resulting in continuous increases in carbon emissions. Wojewodzki's [57] study also mentioned that rapid urban development intensified land use and resource consumption, strengthening carbon emissions. Spatially (Figure 8c), in 62% of the counties, urbanization rates have promoted carbon emissions, while the negative impact of urbanization development on carbon emissions was mainly distributed in the northern part of Hangzhou, the western part of Jiaxing, and Huzhou. The reason might be that as the level of urbanization continuously improves, the quality of urbanization also shifts from inefficient and extensive to ecologically sustainable, leading to high operational efficiency of public facilities, industrial aggregation, improved production efficiency, and energy utilization, thereby resulting in lower carbon emissions. This aligns with Awan's [58] study which found that cities with rapid urbanization, developed transportation facilities, and high efficiency emit less carbon.

(4) The industrial structure has suppressed the overall county-level carbon emissions in Zhejiang Province, indicating that the optimization of industrial structure can reduce county-level carbon emissions. Temporally (Figure 7d), the suppressive effect weakened from 2002 to 2013, but gradually strengthened from 2013 to 2022. Qi's [17] study shows that the secondary industry occupies a high proportion of the total carbon emissions, but the tertiary industry has gradually become the main "contributor" to the increase in carbon emissions, with carbon emissions from the tertiary industry sectors such as transportation, logistics, and residential life sharply rising. Spatially (Figure 8d), the areas most negatively affected by the industrial structure were mainly concentrated in the eastern part of Shaoxing City, the northeastern part of Taizhou City, and the counties of Ningbo City. This may be because these counties have a smaller proportion of secondary industries and a larger proportion of tertiary industries, while focusing on upgrading the internal structure of industries and sustainable industrial development, which has reduced carbon emissions. Mehmood's [59] study also found that green industrial transformation and upgrading are related to carbon emission reduction. However, the areas positively affected by the industrial structure were mainly concentrated in the eastern part of Jinhua, the central part of Hangzhou, and the eastern part of Huzhou, indicating that in these counties, industries with high carbon emissions still dominated, increasing carbon emissions.

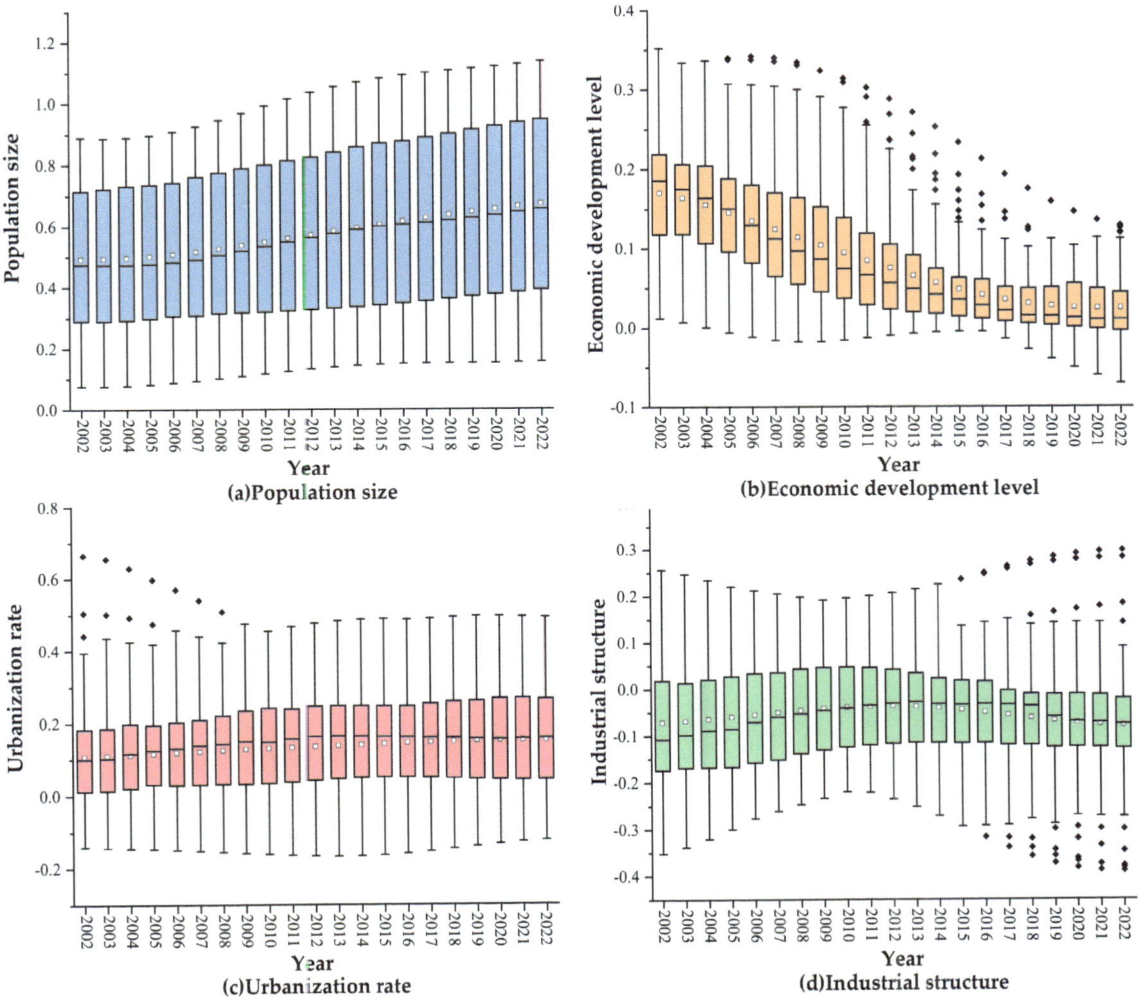

Figure 7. (**a**–**d**). Time series trend of GTWR regression coefficients.

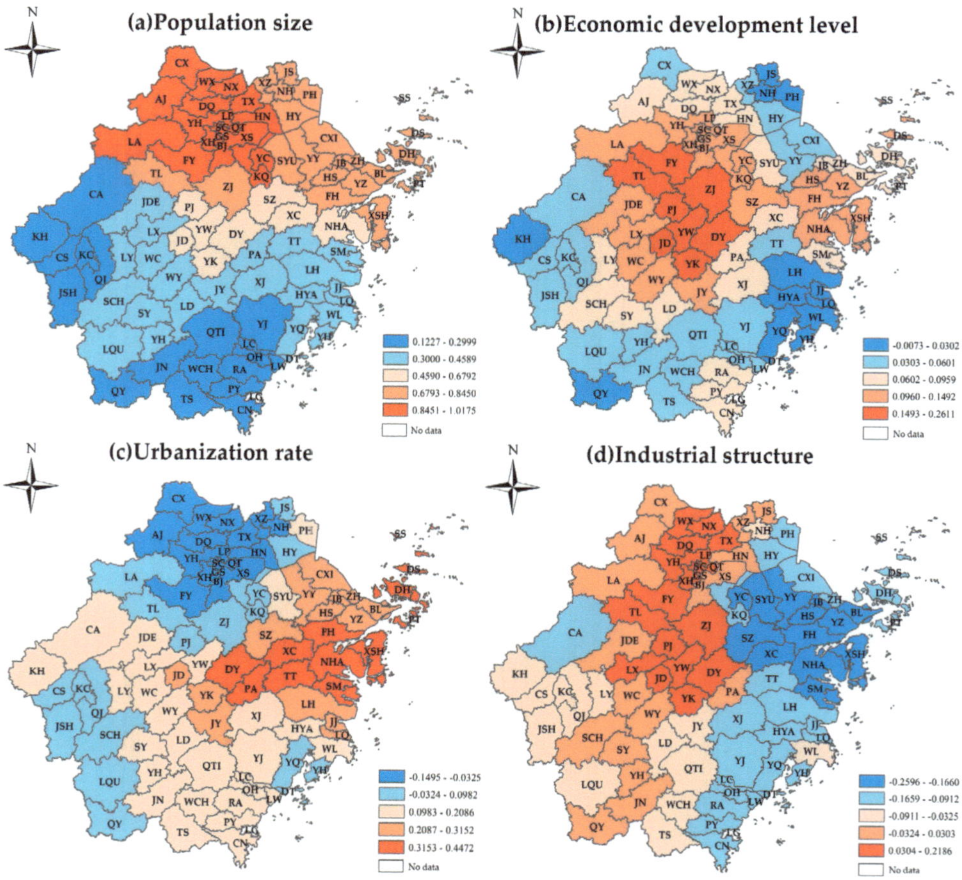

Figure 8. (**a–d**). Spatial distribution pattern of coefficients of factors influencing carbon emissions in the county areas of Zhejiang Province.

4. Conclusions

This study examined the spatial and temporal patterns of carbon emissions in the county areas of Zhejiang Province, considering global climate warming and China's goals of achieving carbon peak and carbon neutrality. Utilizing the GTWR model, it examined the factors influencing carbon emissions in these county areas. The conclusions are as follows:

(1) LISA time path analysis results indicated that, overall, the county-level carbon emissions in Zhejiang Province from 2002 to 2022 exhibited strong local spatial structure stability, with an upward trend in this stability. Spatially, the most robust regional spatial structural stability of carbon emissions was found in western Wenzhou and Lishui in southern Zhejiang, Quzhou in Zhejiang's west, and western Taizhou and western Hangzhou, showing a trend of spreading from south and west Zhejiang to northern and eastern Zhejiang. The spatial pattern of carbon emissions in the county areas of Zhejiang Province possesses specific spatial integrative characteristics. However, this integrative tendency is generally weakening, with non-synergistic growth predominantly exhibiting a fragmented pattern.

(2) The spatio-temporal transition results from the Moran's scatterplot indicated that the most common transition was Type IV, with solid stability in the Moran's scatterplot. This suggests a relatively high spatial accumulation of carbon emissions in the county areas of Zhejiang Province, a high degree of path dependence in carbon emissions spatial aggregation, and relative stability in the spatial structure of carbon emissions. This reflects

that spillover effects from surrounding regions influence local county areas less, and internal factors significantly impact changes in their carbon emission spatial structure.

(3) A consistent "northeast–southwest" pattern emerged in the spatial distribution as the axis length varied. From 2002 to 2022, the centroid of county-level carbon emissions in Zhejiang Province varied within the range of 120.551° to 120.570° E and 29.555° to 29.591° N, moving generally northeastward and overall forming a 'V' shape. This indicated that the rate of increase in carbon emissions in the northeastern counties of Zhejiang was higher than the average, with the areas of high growth in overall carbon emissions still developing towards the northeast direction.

(4) The elements that influence carbon emissions vary significantly in terms of space and time. These factors, ranked in descending order of their degree of influence are population size, urbanization rate, industrial structure, and economic development level. Among these, population size, urbanization rate, and economic development level mainly promote carbon emissions, while industrial structure acts to suppress emissions. While Zhu [29] argued that the growth of the secondary industry has significantly worsened carbon emissions, our research shows that the secondary sector reduces carbon emissions. This finding aligns with the studies of Qi [17] and Yang [60].

5. Discussions

5.1. Implications

Exploring the spatio-temporal dynamics and determinants of carbon emissions in Zhejiang Province's counties through the lens of spatio-temporal interactions helps in the detailed analysis of these emissions from a micro-perspective. This approach also offers a basis for developing tailored emission reduction policies. Drawing from empirical findings and relevant literature, we present the following insights:

(1) The results of the standard deviation ellipse indicated that the northeastern counties of Zhejiang have higher carbon emissions, while the LISA time path results showed that the spatial structure stability of carbon emissions in the southwestern counties of Zhejiang was the strongest, with counties such as Yinzhou and Xiaoshan having the lowest stability. Therefore, differentiated carbon governance measures need to be adopted for different regional countries. For the northeastern counties of Zhejiang, more attention should be paid to reducing the total carbon emissions, which can be achieved by enhancing technological innovation [60], promoting the development of low-carbon industries, and accelerating the modernization, integration, and transformation of traditional high-energy-consuming enterprises [17]. For the southwestern counties of Zhejiang such as Wencheng and Kaihua, more emphasis should be placed on the stability of carbon emissions through long-term, progressive (rather than short-term, volatile) policy guidance and control. At the same time, it is crucial to promote the transformation of the ecological economy in the long term, by optimizing the energy structure, such as increasing the proportion of clean energy and reducing reliance on fossil fuels, to ensure a stable decline in carbon emissions across counties [61].

(2) Spatio-temporal transitions indicated that the spatial agglomeration stability of carbon emissions in Zhejiang's counties was strong, with weak influences between adjacent counties, and effective inter-regional linkages had not yet formed. Therefore, it is necessary to strengthen collaborative development between counties, breaking the highly stable spatial structure and path dependence of carbon emissions. Different counties have varying development modes and actual situations, which can complement each other's advantages for joint development. This can be achieved by improving resource utilization efficiency, achieving regional coordinated development, and sharing ecological advantages [62]. Establishing a cross-regional cooperation mechanism can promote the sharing of experiences, technologies, and resources between different regions, strengthening collaborative efforts in carbon reduction. High-carbon emitting regions can collaborate with low-carbon emitting regions to reduce carbon emissions through resource transfer and technical support.

Promoting cross-regional energy interaction and the development of the electricity networ can enhance energy use efficiency and reduce carbon emissions [63].

(3) The factor of population size has a positive effect on carbon emissions, so attentio should be paid to densely populated counties, transforming large population size an concentration into an advantage to improve per capita energy use efficiency. In additio increased fiscal support, talent cultivation, and innovation and progress in low-carbo industries and products should be emphasized [64]. Similarly, the factor of economic de velopment increases carbon emissions, so the quality of economic development should b emphasized over speed. Zhejiang should strategically implement the digital economy an promote industrial digital transformation. By optimizing production processes and improv ing energy use efficiency, energy consumption can be optimized, thus effectively controllin and reducing carbon emissions while promoting local economic development [65]. Th urbanization rate factor exacerbates carbon dioxide emissions; hence, high-quality nev urbanization development focusing on sustainable and eco-friendly urbanization growt models should be pursued, contrary to traditional demolition and construction urbaniza tion patterns [66]. Meanwhile, the industrial structure factor inhibits carbon emissions, s the transformation of industrial structure should continue, vigorously developing the low carbon and environmentally friendly tertiary industry, and accelerating the transformatio of energy-intensive industries [31].

5.2. Limitations and Future Research Direction

This study comprehensively discusses the influencing factors of carbon emission in the county areas of Zhejiang Province; however, due to the limited number of factor considered, more than this depth of research might be required. Future research wil explore more relevant influencing factors on carbon emissions at the county level, aiming tc provide differentiated strategies and suggestions for various counties. This study provide theoretical and empirical bases for formulating carbon emission reduction policies in Zhe jiang Province. Nonetheless, the features of county regions vary considerably throughou Chinese provinces and cities. Consequently, other countries' unique characteristics should inform the careful formulation of policies aimed at reducing carbon emissions. Future research directions should be more comprehensive than the study of carbon emissions ir the county areas of Zhejiang Province. Still, they should extend to more provincial region to explore county-level carbon emissions.

Author Contributions: Conceptualization, X.W. and Y.W.; Methodology, X.W. and Y.W.; Software X.W.; Validation, X.W. and Y.W.; Formal analysis, H.Y.; Resources, H.Y.; Data curation, C.Z. and X.L. Writing—original draft preparation, X.W.; Writing—review and editing, X.W. and Y.W.; Visualization Y.L.; Supervision, Y.W.; Project administration, J.H.; Funding acquisition, X.W. and Y.W. All authors have read and agreed to the published version of the manuscript.

Funding: This research was supported by the Soft Science Research Plan Project of Zhejiang (No 2024C35003), the Philosophy and Social Science Planning Project of Zhejiang (No. 24NDQN042YB), the Scientific Research and Nurturing Foundation of Hangzhou City University (No. J-202308), the Foundation of Hangzhou Federation of Humanities & Social Sciences (No. 2023HZSL-ZC011), the National Natural Science Foundation of China (No. 51878593), the Center for Balance Architecture, Zhejiang University (No. KH-20212946), the Artificial Intelligence Key Technologies R&D Program of Hangzhou (No. 2022AIZD0057), and the Strategic Research and Consulting Project of the Chinese Academy of Engineering (No. 2022-XZ-51).

Data Availability Statement: The data employed for this study were obtained from the Statistical Yearbook of Zhejiang and the China Emission Accounts and Dataset (CEADs). Access to these datasets is available at the following URLs: http://data.cnki.net and https://www.ceads.net (accessed on 6 August 2023).

Acknowledgments: The authors gratefully acknowledge the support of the funding.

Conflicts of Interest: All authors declare that the research was conducted in the absence of any commercial or financial relationships that could be construed as a potential conflict of interest.

Abbreviations

Attachment: To improve recognition in images, the names of counties are abbreviated. Below are the corresponding abbreviations for each county.

Counties	**Abbreviations**
Shangcheng	SC
Gongshu	GS
Xihu	XH
Binjiang	BJ
Xiaoshan	XS
Yuhang	YHA
Fuyang	FY
Linan	LA
Tonglu	TL
Chunan	CA
Jiande	JDE
Haishu	HS
Jiangbei	JB
Zhenhai	ZH
Beilun	BL
Yinzhou	YZ
Fenghua	FH
Yuyao	YY
Cixi	CXI
Xiangshan	XSH
Ninghai	NHA
Lucheng	LC
Taishun	TS
Ruian	RA
Yueqing	YQ
Nanhu	NH
Xiuzhou	XZ
Jiashan	JS
Haiyan	HY
Haining	HN
Pinghu	PH
Tongxiang	TX
Wuxing	WX
Nanxun	NX
Deqing	DQ
Changxing	CX
Anji	AJ
Yuecheng	YC
Kecheng	KQ
Shangyu	SYU
Zhuji	ZJ
Shengzhou	SZ
Xinchang	XC
Wucheng	WC
Yongkang	YK
Kecheng	KC
Qujiang	QJ
Jiangshan	JSH
Changshan	CS
Kaihua	KH
Longyou	LY
Dinghai	DH
Putuo	PT

Daishan	DS
Shengsi	SS
Jiaojiang	JJ
Huangyan	HYA
Luqiao	LQ
Wenling	WL
Linhai	LH
Yuhuan	YHU
Sanmen	SM
Tiantai	TT
Xianju	XJ
Liandu	LD
Longquan	LQU
Longwan	LW
Ouhai	OH
Dongtou	DT
Yongjia	YJ
Pingyang	PY
Cangnan	CN
Wencheng	WCH
Jindong	JD
Wuyi	WY
Pujiang	PJ
Panan	PA
Lanxi	LX
Yiwu	YW
Dongyang	DY
Qingtian	QTI
Yunhe	YH
Qingyuan	QY
Jinyun	JY
Suichang	SCH
Songyang	SY
Jingning	JN
Longgang	LG

References

1. Calvia, M. Worldwide Fluctuations in Carbon Emissions: Characterization and Synchronization. *Clean. Prod. Lett.* **2024**, *6*, 100054. [CrossRef]
2. Oyewo, B.; Tauringana, V.; Tawiah, V.; Aju, O. Impact of Country Governance Mechanisms on Carbon Emissions Performance of Multinational Entities. *J. Environ. Manag.* **2024**, *352*, 120000. [CrossRef]
3. Perdichizzi, S.; Buchetti, B.; Cicchiello, A.F.; Dal Maso, L. Carbon Emission and Firms' Value: Evidence from Europe. *Energy Econ.* **2024**, *131*, 107324. [CrossRef]
4. Perera, K.; Kuruppuarachchi, D.; Kumarasinghe, S.; Suleman, M.T. The Impact of Carbon Disclosure and Carbon Emissions Intensity on Firms' Idiosyncratic Volatility. *Energy Econ.* **2023**, *128*, 107053. [CrossRef]
5. Khan, A.A.; Mei, B.; Khan, S.U.; Ali, M.A.S.; Luo, J. Agri-Food Evolution and Carbon Emissions in Chinese Residential Consumption: A Life Cycle Analysis of Urban-Rural Disparities and Socioeconomic Influences. *Environ. Impact Assess. Rev.* **2024**, *105*, 107387. [CrossRef]
6. Kotsompolis, G.; Konstantakis, K.N.; Xidonas, P.; Michaelides, P.G.; Thomakos, D.D. Climate Change Economics and the Determinants of Carbon Emissions' Futures Returns: A Regime-Driven ARDL Model. *Financ. Res. Lett.* **2023**, *58*, 104485. [CrossRef]
7. Zhou, K.; Yang, J.; Yang, T.; Ding, T. Spatial and Temporal Evolution Characteristics and Spillover Effects of China's Regional Carbon Emissions. *J. Environ. Manag.* **2023**, *325*, 116423. [CrossRef] [PubMed]
8. Liu, H.; Song, Y. Financial Development and Carbon Emissions in China since the Recent World Financial Crisis: Evidence from a Spatial-Temporal Analysis and a Spatial Durbin Model. *Sci. Total Environ.* **2020**, *715*, 136771. [CrossRef] [PubMed]
9. Zhou, C.; Chen, X. Forecasting China's Energy Consumption and Carbon Emission Based on Multiple Decomposition Strategy. *Energy Strategy Rev.* **2023**, *49*, 101160. [CrossRef]

10. Dong, F.; Yu, B.; Hadachin, T.; Dai, Y.; Wang, Y.; Zhang, S.; Long, R. Drivers of Carbon Emission Intensity Change in China. *Resour. Conserv. Recycl.* **2018**, *129*, 187–201. [CrossRef]
11. Pang, J.; Li, N.; Mu, H.; Zhang, M.; Zhao, H. Study on the Spatial Interaction between Carbon Emission Intensity and Shadow Economy in China. *Sci. Total Environ.* **2022**, *813*, 152616. [CrossRef] [PubMed]
12. Zheng, J.; Mi, Z.; Coffman, D.; Milcheva, S.; Shan, Y.; Guan, D.; Wang, S. Regional Development and Carbon Emissions in China. *Energy Econ.* **2019**, *81*, 25–36. [CrossRef]
13. He, Q.; Wei, F.; Deng, X.; Kong, F.; Li, C.; Yan, Z.; Qi, Y. Spatiotemporal Pattern of Carbon Productivity and Carbon Offset Potential in Chinese Counties. *Sci. Total Environ.* **2022**, *846*, 157153. [CrossRef]
14. NBS. *China Statistical Yearbook (County-Level)*; China Statistics Press: Beijing, China, 2018.
15. Liu, X.; Jin, X.; Luo, X.; Zhou, Y. Quantifying the Spatiotemporal Dynamics and Impact Factors of China's County-Level Carbon Emissions Using ESTDA and Spatial Econometric Models. *J. Clean. Prod.* **2023**, *410*, 137203. [CrossRef]
16. Zhao, J.; Chen, Y.; Ji, G.; Wang, Z. Residential Carbon Dioxide Emissions at the Urban Scale for County-Level Cities in China: A Comparative Study of Nighttime Light Data. *J. Clean. Prod.* **2018**, *180*, 198–209. [CrossRef]
17. Qi, H.; Shen, X.; Long, F.; Liu, M.; Gao, X. Spatial–Temporal Characteristics and Influencing Factors of County-Level Carbon Emissions in Zhejiang Province, China. *Environ. Sci. Pollut. Res.* **2022**, *30*, 10136–10148. [CrossRef] [PubMed]
18. Duman, Z.; Mao, X.; Cai, B.; Zhang, Q.; Chen, Y.; Gao, Y.; Guo, Z. Exploring the Spatiotemporal Pattern Evolution of Carbon Emissions and Air Pollution in Chinese Cities. *J. Environ. Manag.* **2023**, *345*, 118870. [CrossRef]
19. Liu, X.; Xu, H.; Zhang, M. The Effects of Urban Expansion on Carbon Emissions: Based on the Spatial Interaction and Transmission Mechanism. *J. Clean. Prod.* **2024**, *434*, 140019. [CrossRef]
20. Zhang, Y.; Lin, W.; Ren, E.; Yu, Y. Evaluation of Spatial Distribution of Carbon Emissions from Land Use and Environmental Parameters: A Case Study in the Yangtze River Delta Demonstration Zone. *Ecol. Indic.* **2024**, *158*, 111496. [CrossRef]
21. Soares, T.C.; Fernandes, E.A.; Toyoshima, S.H. The CO_2 Emission Gini Index and the Environmental Efficiency: An Analysis for 60 Leading World Economies. *EconomiA* **2018**, *19*, 266–277. [CrossRef]
22. Williams, R.G.; Ceppi, P.; Roussenov, V.; Katavouta, A.; Meijers, A.J.S. The Role of the Southern Ocean in the Global Climate Response to Carbon Emissions. *Phil. Trans. R. Soc. A.* **2023**, *381*, 20220062. [CrossRef] [PubMed]
23. Qu, H.; You, C.; Feng, C.-C.; Guo, L. Spatial Responses of Ecosystem Services to Carbon Emission Efficiency Based on Spatial Panel Models in the Yangtze River Economic Belt. *J. Clean. Prod.* **2024**, *434*, 139953. [CrossRef]
24. Quan, C.; Cheng, X.; Yu, S.; Ye, X. Analysis on the Influencing Factors of Carbon Emission in China's Logistics Industry Based on LMDI Method. *Sci. Total Environ.* **2020**, *734*, 138473. [CrossRef] [PubMed]
25. González, P.F.; Presno, M.J.; Landajo, M. Tracking the Change in Spanish Greenhouse Gas Emissions through an LMDI Decomposition Model: A Global and Sectoral Approach. *J. Environ. Sci.* **2024**, *139*, 114–122. [CrossRef] [PubMed]
26. Khodakarami, L.; Pourmanafi, S.; Soffianian, A.R.; Lotfi, A. Modeling Spatial Distribution of Carbon Sequestration, CO_2 Absorption, and O_2 Production in an Urban Area: Integrating Ground-Based Data, Remote Sensing Technique, and GWR Model. *Earth Space Sci.* **2022**, *9*, e2022EA002261. [CrossRef]
27. Li, W.; Dong, F.; Ji, Z. Research on Coordination Level and Influencing Factors Spatial Heterogeneity of China's Urban CO_2 Emissions. *Sustain. Cities Soc.* **2021**, *75*, 103323. [CrossRef]
28. Nosheen, M.; Abbasi, M.A.; Iqbal, J. Analyzing Extended STIRPAT Model of Urbanization and CO_2 Emissions in Asian Countries. *Environ. Sci. Pollut. Res.* **2020**, *27*, 45911–45924. [CrossRef]
29. Zhu, E.; Qi, Q.; Chen, L.; Wu, X. The Spatial-Temporal Patterns and Multiple Driving Mechanisms of Carbon Emissions in the Process of Urbanization: A Case Study in Zhejiang, China. *J. Clean. Prod.* **2022**, *358*, 131954. [CrossRef]
30. NBS. *China Statistical Yearbook (County-Level)*; China Statistics Press: Beijing, China, 2022.
31. Xia, C.; Li, Y.; Ye, Y.; Shi, Z.; Liu, J. Decomposed Driving Factors of Carbon Emissions and Scenario Analyses of Low-Carbon Transformation in 2020 and 2030 for Zhejiang Province. *Energies* **2017**, *10*, 1747. [CrossRef]
32. Rey, S.J. Spatial Empirics for Economic Growth and Convergence. *Geogr. Anal.* **2001**, *33*, 195–214. [CrossRef]
33. Gui, D.; He, H.; Liu, C.; Han, S. Spatio-Temporal Dynamic Evolution of Carbon Emissions from Land Use Change in Guangdong Province, China, 2000–2020. *Ecol. Indic.* **2023**, *156*, 111131. [CrossRef]
34. Jiang, H.; Yin, J.; Qiu, Y.; Zhang, B.; Ding, Y.; Xia, R. Industrial Carbon Emission Efficiency of Cities in the Pearl River Basin: Spatiotemporal Dynamics and Driving Forces. *Land* **2022**, *11*, 1129. [CrossRef]
35. Ye, X.; Rey, S. A Framework for Exploratory Space-Time Analysis of Economic Data. *Ann. Reg. Sci.* **2013**, *50*, 315–339. [CrossRef]
36. Razzaq, A.; Sharif, A.; Yang, X.; Dogan, E. Influence Mechanism of Electricity Price Distortion on Industrial Green Transformation: A Spatial Analysis of Chinese Regions. *Energy Econ.* **2024**, *130*, 107308. [CrossRef]
37. Güneyli, H.; Saleh Ahmed, S.M. Detecting Abnormal Seismic Activity Areas of Anatolian Plate and Deformation Directions Using Python Geospatial Libraries. *Heliyon* **2023**, *9*, e14394. [CrossRef] [PubMed]
38. Lefever, D.W. Measuring Geographic Concentration by Means of the Standard Deviational Ellipse. *Am. J. Sociol.* **1926**, *32*, 88–94. [CrossRef]
39. Talkhabi, H.; Ghalehteimouri, K.J.; Mehranjani, M.S.; Zanganeh, A.; Karami, T. Spatial and Temporal Population Change in the Tehran Metropolitan Region and Its Consequences on Urban Decline and Sprawl. *Ecol. Inform.* **2022**, *70*, 101731. [CrossRef]
40. Qin, Q.; Yan, H.; Li, B.; Lv, W.; Wasif Zafar, M. A Novel Temporal-Spatial Decomposition on Drivers of China's Carbon Emissions. *Gondwana Res.* **2022**, *109*, 274–284. [CrossRef]

41. Tang, D.; Ma, T.; Li, Z.; Tang, J.; Bethel, B. Trend Prediction and Decomposed Driving Factors of Carbon Emissions in Jiangsu Province during 2015–2020. *Sustainability* **2016**, *8*, 1018. [CrossRef]
42. Wang, S.; Ma, Y. Influencing Factors and Regional Discrepancies of the Efficiency of Carbon Dioxide Emissions in Jiangsu, China. *Ecol. Indic.* **2018**, *90*, 460–468. [CrossRef]
43. Zhang, F.; Deng, X.; Phillips, F.; Fang, C.; Wang, C. Impacts of Industrial Structure and Technical Progress on Carbon Emission Intensity: Evidence from 281 Cities in China. *Technol. Forecast. Soc. Chang.* **2020**, *154*, 119949. [CrossRef]
44. Huang, Y.; Matsumoto, K. Drivers of the Change in Carbon Dioxide Emissions under the Progress of Urbanization in 30 Provinces in China: A Decomposition Analysis. *J. Clean. Prod.* **2021**, *322*, 129000. [CrossRef]
45. Zhang, H.; Zhang, X.; Yuan, J. Driving Forces of Carbon Emissions in China: A Provincial Analysis. *Environ. Sci. Pollut. Res.* **2021**, *28*, 21455–21470. [CrossRef] [PubMed]
46. Liu, Y.; Jiang, Y.; Liu, H.; Li, B.; Yuan, J. Driving Factors of Carbon Emissions in China's Municipalities: A LMDI Approach. *Environ. Sci. Pollut. Res.* **2022**, *29*, 21789–21802. [CrossRef]
47. Shum, W.Y.; Ma, N.; Lin, X.; Han, T. The Major Driving Factors of Carbon Emissions in China and Their Relative Importance: An Application of the LASSO Model. *Front. Energy Res.* **2021**, *9*, 726127. [CrossRef]
48. Sitorus, F.; Brito-Parada, P.R. A Multiple Criteria Decision Making Method to Weight the Sustainability Criteria of Renewable Energy Technologies under Uncertainty. *Renew. Sustain. Energy Rev.* **2020**, *127*, 109891. [CrossRef]
49. Hohl, A.; Lotfata, A. Modeling Spatiotemporal Associations of Obesity Prevalence with Biking, Housing Cost and Green Spaces in Chicago, IL, USA, 2015–2017. *J. Transp. Health* **2022**, *26*, 101412. [CrossRef]
50. Ariken, M.; Zhang, F.; Chan, N.W.; Kung, H. Coupling Coordination Analysis and Spatio-Temporal Heterogeneity between Urbanization and Eco-Environment along the Silk Road Economic Belt in China. *Ecol. Indic.* **2021**, *121*, 107014. [CrossRef]
51. Mohammadnazar, A.; Mahdinia, I.; Ahmad, N.; Khattak, A.J.; Liu, J. Understanding How Relationships between Crash Frequency and Correlates Vary for Multilane Rural Highways: Estimating Geographically and Temporally Weighted Regression Models. *Accid. Anal. Prev.* **2021**, *157*, 106146. [CrossRef]
52. Bui Minh, T.; Nguyen Ngoc, T.; Bui Van, H. Relationship between Carbon Emissions, Economic Growth, Renewable Energy Consumption, Foreign Direct Investment, and Urban Population in Vietnam. *Heliyon* **2023**, *9*, e17544. [CrossRef]
53. Rehman, E.; Rehman, S. Modeling the Nexus between Carbon Emissions, Urbanization, Population Growth, Energy Consumption and Economic Development in Asia: Evidence from Grey Relational Analysis. *Energy Rep.* **2022**, *8*, 5430–5442. [CrossRef]
54. Gershon, O.; Asafo, J.K.; Nyarko-Asomani, A.; Koranteng, E.F. Investigating the Nexus of Energy Consumption, Economic Growth and Carbon Emissions in Selected African Countries. *Energy Strategy Rev.* **2024**, *51*, 101269. [CrossRef]
55. Balta-Ozkan, N.; Watson, T.; Mocca, E. Spatially Uneven Development and Low Carbon Transitions: Insights from Urban and Regional Planning. *Energy Policy* **2015**, *85*, 500–510. [CrossRef]
56. Wojewodzki, M.; Wei, Y.; Cheong, T.S.; Shi, X. Urbanisation, Agriculture and Convergence of Carbon Emissions Nexus: Global Distribution Dynamics Analysis. *J. Clean. Prod.* **2023**, *385*, 135697. [CrossRef]
57. Raihan, A.; Ibrahim, S.; Muhtasim, D.A. Dynamic Impacts of Economic Growth, Energy Use, Tourism, and Agricultural Productivity on Carbon Dioxide Emissions in Egypt. *World Dev. Sustain.* **2023**, *2*, 100059. [CrossRef]
58. Awan, A.; Alnour, M.; Jahanger, A.; Onwe, J.C. Do Technological Innovation and Urbanization Mitigate Carbon Dioxide Emissions from the Transport Sector? *Technol. Soc.* **2022**, *71*, 102128. [CrossRef]
59. Mehmood, S.; Zaman, K.; Khan, S.; Ali, Z.; Khan, H.U.R. The Role of Green Industrial Transformation in Mitigating Carbon Emissions: Exploring the Channels of Technological Innovation and Environmental Regulation. *Energy Built Environ.* **2024**, *5*, 464–479. [CrossRef]
60. Yang, L.; Xia, H.; Zhang, X.; Yuan, S. What Matters for Carbon Emissions in Regional Sectors? A China Study of Extended STIRPAT Model. *J. Clean. Prod.* **2018**, *180*, 595–602. [CrossRef]
61. Tao, A. Research on the Realization Path of Carbon Emission Reduction in Zhejiang Province. *Energy Rep.* **2022**, *8*, 501–506. [CrossRef]
62. Zeng, S.; Wang, M. Theoretical and Empirical Analyses on the Factors Affecting Carbon Emissions: Case of Zhejiang Province, China. *Environ. Dev. Sustain.* **2023**, *25*, 2522–2549. [CrossRef]
63. Di, K.; Chen, W.; Zhang, X.; Shi, Q.; Cai, Q.; Li, D.; Liu, C.; Di, Z. Regional Unevenness and Synergy of Carbon Emission Reduction in China's Green Low-Carbon Circular Economy. *J. Clean. Prod.* **2023**, *420*, 138436. [CrossRef]
64. Ke, S.; Zhang, Z.; Wang, Y. China's Forest Carbon Sinks and Mitigation Potential from Carbon Sequestration Trading Perspective. *Ecol. Indic.* **2023**, *148*, 110054. [CrossRef]
65. Li, Z.; Lu, Z.; Xu, L.; Shi, Y.; Ma, Q.; Wu, Y.; Cao, Y.; Sheng, B. Examining the Decoupling of Economic Growth with Land Expansion and Carbon Emissions in Zhejiang Province, China. *Land* **2023**, *12*, 1618. [CrossRef]
66. Chen, W.; Wang, G.; Xu, N.; Ji, M.; Zeng, J. Promoting or Inhibiting? New-Type Urbanization and Urban Carbon Emissions Efficiency in China. *Cities* **2023**, *140*, 104429. [CrossRef]

MDPI AG
Grosspeteranlage 5
4052 Basel
Switzerland
Tel.: +41 61 683 77 34

Land Editorial Office
E-mail: land@mdpi.com
www.mdpi.com/journal/land

www.ingramcontent.com/pod-product-compliance
Lightning Source LLC
Chambersburg PA
CBHW041402210926
51850CB00013B/3156